Modeling Language, Cognition and Action

PROGRESS IN NEURAL PROCESSING*

Series Advisor

Alan Murray *(University of Edinburgh)*

Vol. 4: Analogue Imprecision in MLP Training
by Peter J. Edwards & Alan F. Murray

Vol. 5: Applications of Neural Networks in Environment, Energy, and Health
Eds. Paul E. Keller, Sherif Hashem, Lars J. Kangas & Richard T. Kouzes

Vol. 6: Neural Modeling of Brain and Cognitive Disorders
Eds. James A. Reggia, Eytan Ruppin & Rita Sloan Berndt

Vol. 7: Decision Technologies for Financial Engineering
Eds. Andreas S. Weigend, Yaser Abu-Mostafa & A.-Paul N. Refenes

Vol. 8: Neural Networks: Best Practice in Europe
Eds. Bert Kappen & Stan Gielen

Vol. 9: RAM-Based Neural Networks
Ed. James Austin

Vol. 10: Neuromorphic Systems: Engineering Silicon from Neurobiology
Eds. Leslie S. Smith & Alister Hamilton

Vol. 11: Radial Basis Function Neural Networks with Sequential Learning
Eds. N. Sundararajan, P. Saratchandran & Y.-W. Lu

Vol. 12: Disorder Versus Order in Brain Function: Essays in Theoretical Neurobiology
Eds. P. Århem, C. Blomberg & H. Liljenström

Vol. 13: Business Applications of Neural Networks: The State-of-the-Art of Real-World Applications
Eds. Paulo J. G. Lisboa, Bill Edisbury & Alfredo Vellido

Vol. 14: Connectionist Models of Cognition and Perception
Eds. John A. Bullinaria & Will Lowe

Vol. 15: Connectionist Models of Cognition and Perception II
Eds. Howard Bowman & Christophe Labiouse

*For the complete list of titles in this series, please write to the Publisher.

Progress in Neural Processing • 16

Proceedings of the Ninth Neural Computation and Psychology Workshop

Modeling Language, Cognition and Action

University of Plymouth, UK 8 – 10 September 2004

NCPW9

PLYMOUTH 2004

Editors

Angelo Cangelosi

Guido Bugmann

Roman Borisyuk

University of Plymouth, UK

World Scientific

NEW JERSEY • LONDON • SINGAPORE • BEIJING • SHANGHAI • HONG KONG • TAIPEI • CHENNAI

Published by

World Scientific Publishing Co. Pte. Ltd.
5 Toh Tuck Link, Singapore 596224
USA office: 27 Warren Street, Suite 401-402, Hackensack, NJ 07601
UK office: 57 Shelton Street, Covent Garden, London WC2H 9HE

British Library Cataloguing-in-Publication Data
A catalogue record for this book is available from the British Library.

MODELING LANGUAGE, COGNITION AND ACTION
Proceedings of the Ninth Neural Computation and Psychology Workshop

ISBN 981-256-324-5

Printed in Singapore by Mainland Press

Preface

This volume collects together peer reviewed versions of papers presented at the Ninth Neural Computation and Psychology Workshop (NCPW9), held at the University of Plymouth (England) in September 8-10, 2004. The NCPW series continues to be a highly focused and lively conference that brings together researchers from such diverse disciplines as psychology, artificial intelligence, cognitive science, computer science, neuroscience, and philosophy. The selection of high quality papers, the workshop's limited size, and its informal setting, were specifically chosen to encourage interaction among the researchers present.

The conference invited submissions on neural computation models of all cognitive and psychological processes. However, the special theme of this year's workshop was "Modeling of Language, Cognition and Action". The selection of this special topic aimed to extend the appeal of NCPW from the usual connectionist psychology community to experts in neuroscience, robotics and cognitive systems design.

The structure of this book broadly follows the workshop program structure. However, some papers have been repositioned in order to increase the coherence of publication. Long papers were presented orally, whilst short papers were presented as posters. The book is organized into eleven sections. These cover the breadth of research in neural computation and psychology, although a great number of papers have focused on language modeling due to this year's special theme. The sections are organized as follows: Language, perception and action; Categorization and language; Language processing; Cooperation and communication; Motor representations and cognition; Attention and perception; Cognitive processing and control; Face processing and classification; Developmental processes; Neural network methodology; Future challenges on thinking and reasoning. The introductory chapter (page 3) provides a brief overview of individual papers.

The NCPW9 event and the preparation of this volume have been possible through the invaluable help of many people. First, we would like to thank all the participants, who made this workshop a highly productive scientific meeting and a very sociable event. In particular, we would like to thank our seven keynote speakers, who represented some of the top international experts in the research fields of connectionist modeling, robotics, cognitive psychology and neuroscience. The list of keynote speakers included: Nick Chater (Warwick University, UK), who presented a new, challenging approach to modeling language; Bob French (University of Liege, Belgium) who gave an update on the methodologies for dealing with catastrophic forgetting in neural network models; Art Glenberg (University of Wisconsin – Madison, USA) who gave a critical overview of current connectionist studies and identified future modeling directions for the inclusion of embodiment factors in cognitive systems; Daniel Wolpert (University College London, UK) who presented neuroscientific evidence of a probabilistic framework for sensorimotor control in which prediction plays a key role; and Jonathan Evans (Plymouth University, UK) who launched new challenges to neural computation for the modeling of reasoning and thinking abilities. Their participation has significantly contributed to the very high scientific quality of the workshop.

We also wish to thank all those that helped in the organization of the conference. First of all, our thanks go to John Bullinaria, who helped us in the organization of the workshop. His presence in the organizing committee ensured that this continued the good traditions of the NCPW series as a high quality scientific meeting in a friendly and relaxed atmosphere. We would also like to thank Davi Bugmann, for her efficiency and kindness in managing the conference registration and help desk. We also received invaluable support from various people at the University of Plymouth, in particular from Denise Horne and the Short Course Unit team, Nicola Griffin, Annette Silsbury and their colleagues at the catering and accommodation services, the University's technical support team and Sue Kendall at the School of Computing, Communications and Electronics. A special thank you also goes to Chelsea Chin, at the World Scientific Headquarters in Singapore, for her continuous support in the preparation of this post-conference proceedings volume.

We are very grateful to the funding agencies that sponsored NCPW9. The special focus on Language, Cognition and Action, and the interdisciplinarity of the workshop, attracted the interest of the UK research funding institutions. NCPW9 was sponsored by grants from the BBSRC (Biotechnology and Biological Research Sciences Council), the EPSRC (Engineering and Physical Sciences Research Councils) and The British Academy. We wish to acknowledge their significant financial sponsorship as this supported the award of 12 conference travel grants, the funding of international keynote speakers, and the reduction of the conference fee which included the proceedings and a series of social events.

Finally, we are grateful to our own dear spouses and families for coping with our extra time out of home during the conference organization, its smooth running, and during the production of this volume.

Grazie mille! Merci Beaucoup! Спасибо!

Angelo Cangelosi, Guido Bugmann & Roman Borisyuk
January 2005

List or Reviewers

Roman Borisyuk (University of Plymouth, UK)
Howard Bowman (University of Kent, UK)
Joanna Bryson (University of Bath, UK)
Guido Bugmann (University of Plymouth, UK)
John Bullinaria (University of Birmingham, UK)
Angelo Cangelosi (University of Plymouth, UK)
Axel Cleeremans (Université Libre de Bruxelles, Belgium)
Kenny R. Coventry (University of Plymouth, UK)
Eddy J. Davelaar (Birkbeck, University of London, UK)
Art Glenberg (University of Wisconsin - Madison, USA)
Peter Hancock (University of Stirling, UK)
Timo Honkela (Helsinki University of Technology, Finland)
Yakov B. Kazanovich (Russian Academy of Sciences, Pushchino, Russia)
Eleni Koutsomitopoulou (LexisNexis UK – Georgetown University, USA)
Joe Levy (University of Surrey Roehampton, UK)
Christophe Labiouse (University of Liege, Belgium)
Will Lowe (University of Bath, UK)
Davide Marocco (ISTC-CNR Rome, Italy)
Gianluca Massera (University of Plymouth, UK)
Marco Mirolli (ISTC-CNR Rome – University of Siena, Italy)
Stefano Nolfi (ISTC-CNR Rome, Italy)
Liz Stuart (University of Plymouth, UK)
Robert Ward (University of Bangor, UK)
Brad Wyble (University of Kent, UK)
Rod Adams (University of Hertfordshire, UK)

Contents

Language Processing

Cooperation and Communication

Motor Representations and Cognition

Attention and Perception

Cognitive Processing and Control

Introduction

MODELING LANGUAGE, COGNITION AND ACTION: FROM CONNECTIONIST SIMULATIONS TO EMBODIED NEURAL COGNITIVE SYSTEMS

ANGELO CANGELOSI

Adaptive Behaviour and Cognition Research Group
School of Computing, Communications and Electronics, University of Plymouth,
Drake Circus, Plymouth PL4 8AA, UK
acangelos@plymouth.ac.uk

This chapter introduces the research topic of the modeling language cognition and action in neural computation and psychology. It highlights the shift from classical connectionist simulations of isolated language and cognitive processes towards the modeling of autonomous cognitive systems in which linguistic, cognitive and sensorimotor abilities interact and co-exist. In the second part, a brief overview of the sections and individual chapters of the book is given.

1. Integrating Language, Cognition and Action in Neural Computation Models

The view that language is closely integrated with other cognitive and sensorimotor abilities of the individual has gained significant momentum in recent years (e.g. Rizzolatti & Arbib, 1998; Greenfield, 1991). This theory is supported by a variety of data from neuroscience (Cappa & Perani, 2003; Pulvermuller, 2003), psychology (Glenberg & Kaschak, 2002; Reilly in press; Barsalou, 1999; Coventry & Garrod, 2004) and artificial intelligence (Steels, 1999; Cangelosi & Parisi, 2004; Lauria et al., 2002; Roy, 2003).

Classical connectionist models of language (Christiansen et al., 1999) have generally studied language in isolation from other cognitive abilities and from the sensorimotor interactions of the organism with the environment. Prototypical examples of connectionist linguistic models are those studying morphology (English past tense model – e.g. Plunkett & Marchman 1991), or phonology (reading models – e.g. Seidenberg & McClelland1989) or syntax (e.g. word prediction model – Elman, 1990). These various linguistic abilities are often studied in isolation. Seldom, if ever, have there been models where language abilities strictly interact with other cognitive abilities. Even in connectionist models that deal with both phonetic and semantic factors (e.g. deep dyslexia model – Plaut & Shallice, 1993), a feature-based representation of semantic knowledge is used, that cannot easily incorporate known psychological phenomena such as those on language embodiment (Glenberg, this volume).

More recent approaches to modeling language and cognition tend to include the simultaneous simulation of linguistic, cognitive and motor abilities. This approach focuses on the development of autonomous cognitive systems (e.g. a simulated agent or a robot) capable of performing a range of tasks including action, categorization and cognitive processing, linguistic tasks etc. Cognitive systems models often involve hybrid architecture with some connectionist component (Steels, in press). For example, great efforts have been devoted lately to the development of innovative approaches to the modeling of communication and language in autonomous cognitive systems. These also are interactive systems because allow interaction between humans and intelligent devices (including robots and virtual agents), or between agents themselves. These pioneering approaches (Steels, 1999; Steels & Kaplan, 2000; Cangelosi & Parisi, 2004; Marocco et al., 2003; Wermter et al., 2003; Lauria et al., 2002) are based strongly on the latest cognitive and neuroscience advancements on the grounding of language in cognition and action. This approach is referred at as embodied and situated cognition (Varela et al., 1991; Clark, 1997; Pfeifer & Sheier, 1999) and it includes both empirical investigation into embodied cognition (e.g. Glenberg & Kaschak, 2002) and computational studies for the development of autonomous cognitive systems (Steels, 2003).

This new integrative modeling approach to language, cognition and action can provide significant benefits to the understanding the relation between language and action, and cognition in general. The development and study of cognitive systems, in particular those also including neural computation techniques (e.g. through the use of neural controllers for the agent's own behavioral and cognitive system) can permit: (i) the investigation of the neural mechanisms that form the basis of the evolution and development of language, (ii) the understanding of the interaction between the neural control of language and that of other behavioral and cognitive abilities, (iii) the fact that language is directly related to the organism's sensorimotor and cognitive abilities, (iv) the relationship between language structure (syntax) and cognitive and sensorimotor representations.

2. Neural Cognitive Systems Modeling

A deeper understanding of the relationships between language and other sensorimotor and cognitive abilities is of central importance not only in psychology and cognitive science, but also in the fields of computer science and engineering. This is particularly important for the development of new and effective methods and applications in artificial intelligence, robotics, and interactive systems design. Such a focus on biologically- and psychologically-inspired design of interactive cognitive systems has also been the focus of recent international R&D projects, such as the UK Foresight Cognitive Systems Project and the EU Framework VI Cognitive Systems Programme.

In consideration of these scientific and technological developments, the main aim of the 9th Neural Computation and Psychology Workshop (NCPW9) was to bring together researchers from the various fields of cognitive and engineering sciences to stimulate more integrated research approaches. Previous NCPW workshops have mainly concentrated on the connectionist modeling of psychological processes. The NCPW9 held in Plymouth in September 2004 introduced a new focus on the chain of processes linking language, cognition and action, and the resulting cross-systems design constraints, with regards to both natural and artificial cognitive systems. The workshop, still keeping its focus on fundamental issues and theoretical modeling (in contrast to other events with a more applied flavor), aimed at (i) overcoming the current limitations of classical connectionist approaches to language modeling, (ii) enhancing our understanding of the relationship between language, cognition and action and (iii) providing novel computing and IT systems and applications for the design of communicating and interactive devices and autonomous artificial cognitive systems.

3. Overview of the Book

The 2005 Neural Computation and Psychology Workshop attracted numerous submissions in all areas of the neural computation models of cognitive and psychological processes. However, due to the special theme on "Modeling Language, Cognition and Action", and the focus on the development of cognitive systems, a good proportion of papers were dedicated to studies on the integration of language and other cognitive and sensorimotor capabilities.

The book is organized into eleven sections. The first five sections mainly collect papers on the special theme: Language, perception and action; Categorization and language; Language processing; Cooperation and communication, Motor representations and cognition. The other sections cover the wide spectrum of connectionist modeling: Attention and perception; Cognitive processing and control; Face processing and classification; Developmental processes; Neural network methodology; Future challenges on thinking and reasoning.

Language, Perception and Action

The first section on Language, Perception and Action collects most of the papers dedicated to the grounding of language in perception and action. The first two chapters are by the keynote speakers Art Glenberg and Stefan Wermter. In his chapter, Glenberg first reviews the limits of feature-based connectionist models of language (e.g. the classical reading models based that only use phonetic, graphemic and explicit semantic features) for modeling the well known effects of embodiment on language processing. He then suggests that the new modeling approaches based on embodied and situated systems (such as the grounded neural models and the cognitive robotics models presented in this volume) can help moving towards a better integration of neural and embodiment processes. In the next chapter, Stefan Wermter and collaborators present recent work on a mirror-neuron robot for the learning and integration of the multimodal inputs of language, vision and motor programs. Their approach employs diverse learning methodologies (e.g. reinforcement learning, self-organization and associative learning) to perform cognitive robotic operations such as language grounding in actions, object recognition, localization and docking. The other chapters propose additional models of language grounding. Cangelosi, Coventry and collaborators present a psychologically-plausible model for the grounding of spatial language in perception. This grounded neural approach permits the integration of both geometric and extra-geometric factors (such as object knowledge and function) for the processing of spatial language prepositions. Chang, Feldman and Narayanan present the neural theory of language and give a review of their structured connectionist approach to modeling language, cognition and action. They describe several biologically plausible models that explore the role of embodied representations and simulative inference in language understanding and acquisition. Klein and Doya introduce a multi-agent system to study compositional languages. They use neural architectures which are capable of learning to use language in accordance with desires and to understand the intentions expressed by utterances of other people. They also discuss the relationship between the agents' neural architecture and the real brain, by pointing out functional imaging and other experiments in which subjects had to carry out tasks related to the capabilities of the simulation model. Finally, Knoblauch et al. describe the implementation of a cell assembly-based model of several visual, language, planning, and motor areas to enable a robot to understand and react to simple spoken commands.

Categorization and Language

This section is dedicated to connectionist and cognitive agent models of the effects of language on the acquisition of categorical representations. Gary Lupyan presents a model of object categorization and labeling. His simulation demonstrates that labels can augment perceptual information of entities whose perceptual features alone are insufficient for reliable classification. Mirolli and Parisi present a cognitive agent model, based on evolutionary neural networks, which integrates language, categorization and sensorimotor abilities. The simulation shows that language causes the internal representations of different inputs that must be responded to with the same action to be more similar, whilst the internal representations of inputs that must be responded to with different actions vary more.

Language Processing

This additional section on language collects a variety of papers on connectionist modeling of various linguistic abilities. In the first paper by Hayes and Howell, a connectionist evaluation of two schemes for measuring the phonological complexity of words (the Index of Phonetic Complexity and Analysis of Phonological Structure) is presented. The networks used the continuous stream of spontaneous speech produced by people who stutter. While the model trained using the first codes did not learn to discriminate between fluent and dysfluent items, the second method, based on the analysis of phonological structure, showed evidence of learning. In the next chapter by Stefan Frank, a sentence comprehension model is presented. It consists of a neural network that transforms sentences into representations of the events they describe. An investigation of these representations reveals that they can encode propositional information without implementing propositional structure. Honkela et al. study how independent component analysis can be used to create automatically syntactic and semantic features based on analyzing words in contexts. Lagus and colleagues also use independent component analysis and study the properties of morphemes by analyzing their use in a large Finnish text corpus. Analyses reveal that main syntactic categories are observed on a coarse level, whilst the potential thematic roles of the morphemes emerge at a more detailed level. Koutsomitopoulou studies the case of homonymy in a Cognitive Linguistic Adaptive Resonant Network (CLAR-NET), where conceptual linguistic associations are analyzed as dynamic neuronal activation resonant patterns. Grüning analyzes how simple recurrent networks (SRN) represent and process stack-like and queue-like structures and finds an explanation of why queue-like structures are harder to process even in an analogue computing or connectionist setting. Finally, Giolito proposes some philosophy of language considerations on the function of proper names. She uses evidence from neural network models of language to demonstrate the active role of names when they act as a point of attention around which properties can be clustered.

Cooperation and Communication

This part includes two papers on the emergence of cooperation and communication abilities in simulated cognitive agents. Marocco and Nolfi use a population of embodied cognitive agents (Khepera robots) that have to evolve communication capabilities in order to solve a collective navigation problem. The analysis of the evolved behaviors shows the emergence of tightly co-adapted communicative and non-communicative behaviors. It also shows the exploitation of properties resulting from the dynamical interactions between motor and signaling behaviors produced by interacting robots. In the paper by Sugimotu et al., simulated agents use on a modular control scheme based on the reinforcement learning MOSAIC architecture. Agents are trained to estimate discrete intentions of others by observing continuous trajectories in the cart-pole swing up task. The authors discuss how these estimation abilities can be used for imitative and collaborative tasks.

Motor Representation and Cognition

This section includes papers on various models of motor and spatial representations and their relationship to other cognitive processes. Kühn and Cruse present a neural network model used for generation of action and as a basis for mental representations. The new learning algorithm proposed by the authors leads to a self-organized weight distribution for stable attractor states. The internal models represented within the network can then be used as the basis to produce sequences of behavior. Panayi and Roy present a preliminary model for the development of spatial cognition as action. Neural network modeling techniques, video analysis and simulated animation were applied to the recognition of dynamic gesture of people with severe speech and motor impairment. Massera and colleagues propose a new robotic model for object manipulation. Results demonstrate that evolutionary robotics techniques might scale up to situations involving robots with several degrees of freedom and problems that require the ability to produce sequential behavior. Finally, Ponticorvo and Miglino present another evolutionary robotic model to test the hypothesis that the observed stages in the development of spatial orientation abilities may reflect changes in the frequency of exposition to different classes of stimuli. By manipulating the frequency with which the robots come into contact with different classes of spatial information they show that, at specific frequencies, the sequence observed in the robots corresponds to the pattern observed in human beings.

Attention and Perception

The connectionist modeling of attentional and perceptual mechanisms is another research area that attracted numerous paper submissions. In the first paper, Davelaar proposes a neurocognitive theory of semantic priming based on the attentional modulation of spread of activation. He suggests that, at the neural level of description, the behavioral definitions of automatic and controlled processing, facilitation and inhibition are not suitable. At this level of description, "permitted" semantic activation can capture some of the internal semantic dynamics. In the next chapter, Bowman and Wyble present a computational model of the attentional blink constructed using a dual-stage model. Predictions from the model, followed by ad-hoc experimental investigations, show various phenomena such as the fact that enhancing the distinctiveness of T2 can impair T1 performance and also provoke order inversions of T1 and T2. Lanyon and Denham propose a model of spatial and object-based attention for active visual search. This model applies the biased competition hypothesis in active vision. It replicates behavior observed during the active visual search for orientation and color conjunction targets, where the scan path is guided to target colored locations in preference to locations containing the target orientation or blank areas. Di Ferdinando and colleagues study hemispheric asymmetries in the neglect syndrome. They use a series of neural networks simulations which implement different theories of neglect and show that asymmetries are best explained by a right hemispheric dominance for spatial representations, expressed in terms of a higher number of neurons involved in spatial tasks. Kazanovich and Borisyuk propose an oscillatory neural model of attention that can be used to track a moving target among a set of distractors. They implement a scheme of attention focus formation and switching through the interplay of synchronizing and desynchronizing interactions in a neural network with a central element. Ward and Ward use an evolutionary agent model to study selective attention. The analysis of the simulation shows the agent selectively responds to targets through reactive inhibition of non-target items. This suggests that reactive inhibition may be a fundamental process in selective control of action. Finally, Connolly and Reilly propose a model of repetition blindness which draws on the dichotomous division of the visual system into two subsystems which process identity and location information.

They report how these networks are capable of learning translation invariant letter representations and learning to locate a stimulus in the input array using the SOM learning algorithm.

Cognitive Processing and Control

This section includes papers on a variety of models of cognitive processing tasks and architectures. Wyble and colleagues present a connectionist model that provides a theoretical explanation for the existence of slow and fast emotional Stroop effects, and depicts them as independent but interacting phenomena. Results suggests that slow emotional Stroop effects are caused by activation of the affective portion of the Anterior Cingulate Cortex, which inhibits the cognitive division, reducing top-down cognitive control on the subsequent trial. Joanna Bryson's paper focuses on the important issue of modularity from primate errors during task learning. She presents a modular model of primate learning of the transitive inference task, and proposes an extension to this model which would explain task-learning results in other domains. Andrew Coward uses the Recommendation Architecture cognitive model to give a qualitative explanation for experimental observations of episodic, semantic and procedural memory based on the different ways in which information derived from the senses is recorded and accessed. Banfield and Christodoulou implemented a two-player model, simulating a top-level conscious processor for effortful reasoning, and an intuitive processor for heuristic problem solving. The results support the theory that self-control can be explained in terms of learning to co-operate with oneself, where behavior variability is a factor. Kalidindi et al. present an abstract simulation of the neural mechanisms that could underlie decision-making in reinforcement learning tasks. This is used to investigate the validity of the 'somatic marker' theory, which proposes that body states act as a valence associated with potential choices based on prior outcomes; and thus aid decision-making. Their model refers to clinical data on subjects with ventromedial prefrontal cortex (VMF) lesions and their performance on the Iowa 'Gambling Task'.

Face Processing and Classification

This section includes two papers on pain recognition and on gender classification through analysis of faces. Hancock and colleagues propose a method of automatic recognition of facial expressions of pain for medical use. A simple computer model consisting of a Gabor filtering front end and a linear discriminator network is used. This is able to emulate human performance with high accuracy. Buchala et al. study gender classification based on both a global and feature based representations of face images. They use dimensionality reduction techniques and a support vector machine classifier to demonstrate that this method performs better than either global or feature based representations alone.

Developmental Processes

This short section includes two papers which deal directly with models of cognitive development. The first paper by Mermillod and colleagues presents a distributed model of visual categorization to study the role of head preference in infants. They start from the empirical evidence that 3- to 4-month-old infants mainly use head information to categorize basic-level exemplars like cats and dogs. The simulation shows that the infants may well use mainly head information to create their category representations and that a model based on purely bottom-up processing is not able to explain the preference for head-related information reported in 3- to 4-month-old infants. They propose that an attentional focus on the head information might be necessary to simulate available behavioral data. In the second paper, Musca studies causal inferences in children using backwards blocking tasks. In the literature it had been demonstrated that children aged 3½ exhibit less backwards blocking effect than those aged 4½ because they develop a mechanism for Bayesian structure learning. The author proposes an alternative explanation based on memory limitation, with retroactive interference stronger in younger children, but adult-like in older children. This claim is supported by simulations with a memory self-refreshing neural networks architecture.

Neural Network Methodology

This section includes papers on a variety of studies on neural network methodological issues, such as embryological modeling and cell assembly simulations. In the first chapter, Bowers and Bullinaria propose a method for embryological modeling of evolutionary neural networks. They present some preliminary work on the use of computational embryogeny to model the evolution of neural architecture at a less coarse level of description. They suggest that such an approach will lead to much more reliable and biologically realistic simulations of brain evolution and modularity. In the next paper, Seipone and Bullinaria present a preliminary study for the use of simulated evolution to generate neural networks that suffer significantly less from catastrophic forgetting. Huyck and Ghalib use a cell assembly simulator based on neurons that are fatiguing leaky integrators with Hebbian learning rules. This paper aims to resolve the problem of incorrectly recruited neurons in cell assemblies. The found that modifying the fatigue mechanism reduces the likelihood that an incorrectly recruited neuron fires. In the next chapter, Gale and colleagues study visual similarity within different categories of objects, with a view to using such measures to predict impaired recognition performance. Results show that features extracted by calculating the standard deviation of wavelet coefficients provides the closest fit to the human rating data of all the methods we applied here. Finally, Morse proposes the Enactive Distributed Associationism (EDA) method, consisting of a simple and general purpose approach to cognitive modeling using an artificial life framework. The author discusses the fact that this approach provides a high level of interpretability and an understanding of the structures developed in relation to their formal psychological properties.

Future Challenges: Thinking and Reasoning

This final section includes the paper of one of the keynote speakers, Jonathan Evans. He was invited to launch a challenge in neural computation for an area of research which has received minimal interest from connectionist researchers. Evans discusses the fact that the domination of the deduction paradigm in psychology of thinking has resulted in the use of computational models that implement syntactic or semantic forms of mental logic. He highlights the fact that, more recently, theorists have given much greater emphasis to pragmatic influences on reasoning. This has resulted in the development of a dual-system theory of human reasoning that reflects the operation of two distinct cognitive systems. This new theoretical approach opens new possibilities for the computational modeling of human reasoning. For example, he suggests that the interaction of heuristic and analytic processes, in particular cognitive tasks, should be explored with the aid of computational models.

References

Barsalou L. (1999) Perceptual symbol systems. *Behavioral and Brain Sciences*, 22, 577-609.

Borisyuk R., Kazanovich Y. (in press) Oscillatory Model of Attention-Guided Object Selection and Novelty Detection. *Neural Networks*

Cangelosi A., Parisi D. (2004). The processing of verbs and nouns in neural networks: Insights from Synthetic Brain Imaging. *Brain & Language*, 89(2), 401-408

Cappa S.F., Perani, D. (2003). The neural correlates of noun and verb processing. *Journal of Neurolinguistics*, 16(2-3), 183-189

Christiansen M.H., Chater N., Seidenberg M.S. (Eds.) (1999). Connectionist models of human language processing: Progress and prospects (Special Issue). *Cognitive Science*, 23(4), 415-634.

Clark A. (1997). *Being There: Putting Brain Body and World Together Again.* Cambridge, MA: MIT Press.

Coventry K.R., Garrod S.C. (2004). *Saying, Seeing and Acting: The Psychological Semantics of Spatial Prepositions*. Psychology Press, Hove.

Elman J.L. (1990). Finding structure in time. *Cognitive Science*, 14, 179-211.

Glenberg A., Kaschak M. (2002). Grounding languagc in action. *Psychonomic Bulletin & Review*, 9 (3), 558-565.

Greenfield P. (1991). Language, tool and brain: The ontogeny and phylogeny of hierarchically organized sequential behavior. *Behavioral and Brain Sciences*, 14, 531-595.

Lauria S., Bugmann G., Kyriacou T. & Klein E. 2002 Mobile robot programming using natural language. *Robotics and Autonomous Systems*, 38(3-4), 171-181.

Marocco D., Cangelosi A., Nolfi S. (2003). The emergence of communication is evolutionary robots. *Philosophical Transactions of the Royal Society London – A*, 361, 2397-2421

Pfeifer R., Scheier C. (1999). *Understanding Intelligence*. Cambridge, MA: MIT Press.

Plaut D.C., Shallice T. (1993). Deep dyslexia: A case study of connectionist neuropsychology. *Cognitive Neuropsychology*, 10, 377-500.

Plunkett K., Marchman V. (1991). U-shaped learning and frequency effects in a multi-layered perceptron: implication for child language acquisition. *Cognition*, 38, 1-60.

Pulvermuller F. (2003). *The neuroscience of language. On brain circuits of words and serial order.* Cambridge: Cambridge University Press.

Reilly R.G. (in press). The relationship between object manipulation and language development in Broca's area: A connectionist simulation of Greenfield's hypothesis. *Behavioral and Brain Sciences.*

Rizzolatti G., Arbib M. (1998). Language within our grasp. *Trends in Neuroscience*, 21, 188-194.

Roy D. (2003). Grounded Spoken Language Acquisition: Experiments in Word Learning. *IEEE Transactions on Multimedia.*

Seidenberg M.S., McClelland J.L. (1989). A distributed, developmental model of word recognition and naming. *Psychological Review*, 96, 523-568.

Steels L. (1999). *The Talking Heads Experiment (Volume 1: Words and Meanings)*. Laboratorium: Antwerpen.

Steels L. (2003). Evolving grounded communication for robots. *Trends in Cognitive Science*. 7(7), 308-312.

Steels L. (in press). The emergence of linguistic structure: From minimal to grammatical communication systems. *Connection Science*, special issue on the Emergence of Language: Neural and Adaptive Agent Models.

Steels L., Kaplan F. (2002). AIBO's first words: The social learning of language and meaning. *Evolution of Communication*, 4(1).

Varela F., Thompson E., Rosch E. (1991). *The Embodied Mind*. Cambridge, MA: MIT Press

Wermter S., Elshaw M., Farrand S., (2003). A modular approach to self-organization of robot control based on language instruction, *Connection Science*, 15(2-3), 73-94.

Language, Perception and Action

LESSONS FROM THE EMBODIMENT OF LANGUAGE: WHY SIMULATING HUMAN LANGUAGE COMPREHENSION IS HARD

ARTHUR M. GLENBERG

Department of Psychology, University of Wisconsin-Madison,
1202 West Johnson Street, Madison, WI 53706, USA
glenberg@wisc.edu

Research from psycholinguistic investigations of the embodiment of language converges on two conclusions. First, models of real language use require flexible semantic features in contrast to a fixed set of features. Second, successful models cannot depend exclusively on frequency-based constraints. Instead, successful models of real language use will need to be sensitive to constraints imposed by bodily processes of perception, action, and emotion.

1. Embodied language comprehension

1.1. *Overview*

I begin with an introduction to a particular embodied account of language comprehension, the Indexical Hypothesis (IH, Glenberg & Robertson, 1999, 2000). Next, I review research demonstrating the contribution of perception and action systems to language comprehension not just in the periphery, but in the central activities that allow for combinatoric semantics. This research is distilled into two arguments concluding that simulations of language comprehension will fail if they depend on a fixed set of semantic features or frequency-based constraints (as the only constraints on processing). Instead, successful models of comprehension will need to be grounded in bodily processes of perception, action, and emotion.

1.2. *The Indexical Hypothesis of embodied language comprehension*

Analyses of symbol grounding (e.g., Harnad, 1990; Searle, 1980) lead to the conclusion that symbols (such as variables in an equation or words in an utterance) can only carry meaning when they are grounded outside of the formal system (e.g., outside of mathematics or language). Consequently, I developed the hypothesis that human meaning is grounded in action (Glenberg, 1997). That is, the meaning of a situation to a particular person consists of the set of goal-directed actions available to that person in that situation. But how are the words and syntax of language to contact this type of grounded meaning? According to the IH, the same neural processes that contribute to perception, emotion, and action play an integral role in translating utterances into meanings. Namely, language is understood when we are able simulate sentences using the same neural systems as those used in perception, action, emotion and perhaps other bodily states. Thus, the symbols of language are grounded by relating them to bodily processes that ultimately lead to action. The IH proposes that three overlapping processes accomplish this translation from symbols to embodied meaning.

Consider how we might understand a sentence such as "Art stood on the chair to screw in a light bulb." According to the IH, the first process is to index (or map) words and phrases to objects in the environment or to analogical perceptual symbols. Thus, if you heard the sentence in context, you might glance at a particular (e.g., kitchen) chair to index the phrase "the chair." Or if you read the sentence out of context, you might think of an example of a chair. According the Perceptual Symbol hypothesis (Barsalou,1999), such thoughts include traces of the perceptual activity generated when experiencing the object (see also Roy's "image schema" described in this volume). The second process is deriving affordances (cf. Gibson, 1979) from the object or perceptual symbol. Affordances are interactions between a situation and an agent with a particular morphology. Thus, an adult can sit in a kitchen chair (the chair affords sitting), but the adult could also stand on the chair or even heft it to ward off an attacking dog. In contrast, for a child (but not an adult) the chair affords hiding in a game of hide-and-seek. Derivation of affordances (from a scene or from a perceptual symbol) requires use of the neural systems used in perception.

The third process is to mesh (or integrate) the affordances into a smooth and doable (that is, coherent) set of actions. Mesh is guided by syntax (e.g, Art is standing on the chair, rather than the chair standing on Art) as described more fully in Kaschak and Glenberg (2000). Furthermore, the mesh process is constrained by the body which determines which affordances can be integrated and how. Thus, the affordances of a light bulb (e.g., that it can be held in one hand) can be integrated with the affordances of a chair (e.g., that it affords standing-on without using both hands), resulting in successful comprehension of the sentence about Art. In contrast, if "Art" had been indexed to a pet fish, or if "the chair" had been indexed to a doll-house chair, the resulting affordances could not be meshed, and the sentence would be difficult or impossible to understand. Note that the process of mesh solves the symbol grounding problem by completing the translation of symbols (words and syntax) into goal-directed (potential) actions. What experimental data support these claims?

2. Lesson 1: language, perception, and action

In this section, I review empirical work that supports the basic claim of the IH that language comprehension requires the same neural systems as used in perception and action. Section 3 will make the same point in regard to comprehension of language about novel situations and comprehension of linguistic innovations. Section 4 will do the same for language about abstract domains.

2.1 *Language and action*

Borghi, Glenberg, and Kaschak (in press) demonstrated an Action-Sentence Compatibility Effect (ACE) in the comprehension of simple sentences and words. Participants read a sentence such as, "There is a car in front of you." Then, by pressing a central button in a three-button panel mounted vertically, the participant exposed a word such as "roof" or "wheel" or "road." The task was to judge if the word named a part of the object mentioned in the sentence (the car). In one condition, participants answered Yes by moving to the upper button on the response panel, and in another condition, they answered Yes by moving to the lower button.

The IH proposes that words such as "roof" are indexed to perceptual symbols from which affordances are derived. Many affordances of a car's roof require reaching up, whereas affordances of a car's wheel require reaching down. Consequently, participants should verify that a roof is part of a car more quickly when the Yes response requires moving upwards than when the Yes response requires moving downwards; just the opposite should be found for verifying that a wheel is part of a car. Borghi et al. (in press) reported just this interaction.

The results are consistent with the IH, but they do not strongly imply that derivation of affordances is necessary for comprehension and compositional semantics. That is, the ACE interaction may reflect interference of a previously learned response (e.g., reach up when interacting with a roof) with the response required in the experiment (e.g., move down to respond Yes), rather than something intrinsic to language comprehension. An experiment reported by Glenberg and Kaschak (2002) goes a step further. In that experiment, participants judged whether sentences were sensible (e.g., "Courtney gave you the notebook" and "You gave Courtney the notebook") or nonsense (e.g., "You drank the house to Joe"). The same response panel was used in this experiment, except it was oriented horizontally and held in the lap. Thus, in one condition the Yes (sensible) response required moving from the middle button in a direction away from the body, and in the other condition the Yes response required moving toward the body. Glenberg and Kaschak (2002) also observed an ACE interaction: Participants were faster to judge the sensibility of sentences implying motion toward the body ("Courtney gave you the notebook") when responding Yes by moving toward the body, and when the sentence implied movement away from the body, participants were faster to respond Yes by literally moving away from their bodies. Note that both versions of the sentence used exactly the same words, so the interference cannot be due to a previously learned response to particular words. Instead, the results arise from compositional principles, and apparently, from the simulation of the sentence using perception and action systems.

2.2 *Language and perception*

There are several impressive demonstrations of the interaction of perceptual systems and language comprehension (e.g., Coventry & Garrod, 2004; Pecher, Zeelenberg, & Barsalou, 2003), but one that is particularly striking is reported in Kaschak, Madden, Therriault, Yaxley, Aveyard, Blanchard, and Zwaan, (in press). Participants listened to sentences describing motion either toward or away from the reader (e.g., "The car approached you") or up or down (e.g., "The rocket blasted off"), and judged the sensibility of the sentence. While listening to the sentence, the participant was watching a visual display that simulated motion toward or away from the participant (using a rotating spiral) or up and down (using horizontal bars that appeared to move up or down). The time to judge the sensibility of the sentence describing motion along one dimension was affected by the visually perceived motion in the same dimension. It is hard to resist the conclusion that language comprehension required access to the same perceptual systems used to detect and analyze the motion.

2.3 *Language, perception, action and syntax*

Another particularly compelling demonstration of the embodiment of language comprehension was reported by Chambers, Tanenhaus, and Magneson (2004). Participants listened to sentences describing an action, and then they took the relevant action. While listening and acting, the participant's eye movements were monitored. Consider the following situation. The participant was facing four items, a whistle on a plate, an empty box, an empty folder, and a whistle (with a lanyard) on a folder. The participant listened and acted on two types of sentences. (1) "Put the whistle that's on the folder in the box," and (2) "Put the whistle on the folder in the box." Sentence (2) uses a reduced relative clause (the "that's on" is eliminated). Sentence (2) is perfectly acceptable in some situations, namely when there are **two** whistles in the environment that need to be differentiated. In this case, a relative clause ("that's on the folder" or the reduced "on the folder") is expected because it differentiates between the two whistles. If there had been only **one** whistle, however, then Sentence (2) is temporarily ambiguous. That is, while listening to the beginning of the sentence, "Put the whistle on the folder..." one could reasonably interpret "on the folder" as the place to put the **one** whistle. In fact, when there were **two** whistles, participants spent little time looking at the empty folder, presumably because the relative clause (full or reduced) was expected.

In another condition of the experiment, the participants were given a hook to hold, and they were required to use the hook to carry out the instructions. In this condition, there is **functionally only one** whistle because only the whistle with the lanyard affords picking up and moving with the hook. In this condition, when people heard the Sentence (2), they spent substantially more time looking at the empty folder than when they heard Sentence (1). The interpretation of this effect is quite striking: Holding the hook changes the affordances of the situation so that functionally there is only one whistle. This change in affordances changes the preferred parsing of the reduced relative clause so that "...on the folder" is interpreted as a goal location (hence the time spent looking at the empty folder) rather than a relative clause differentiating the two whistles. Furthermore, Chambers et al (2004) demonstrated that the effects of changing the affordances were observable as soon as 150 msec. after the word "whistle" was uttered. In other words, the affordances affected the immediate, on-line parsing of the sentence.

2.4 *Lessons for modeling language comprehension*

These results suggest both positive and negative conclusions. On the positive side, they support the claims that language comprehension involves perception and action systems and that something akin to affordances are derived while comprehending. Perhaps more important for modeling, are the theoretical ideas that are ruled out.

1. Language comprehension cannot be modular. Derivation of meaning is affected by concurrent perception (Kaschak et al., in press), and on-line syntactic parsing is affected by interactions between bodily capabilities and the environment (e.g., how many whistles can be picked up using a hook; Chambers, et al. 2004).
2. Language comprehension requires perception and action systems not just in the periphery, but as a central component to compositional semantics.
3. What counts as a semantic feature changes with bodily capabilities (e.g., holding a hook) and goals. Consequently, no fixed set of features will successfully model language comprehension. (See French, 1997, for a similar argument.)

3.0 Lesson 2: Novel situations and novel language

In Section 3, I will review work that strongly implies that language comprehension cannot be modeled with a fixed set of semantic features nor with systems that rely solely on frequency-based constraints. Both of these claims are supported by work on the comprehension of language about novel situations and linguistic innovations. In these cases, the novelty ensures that previously learned features and constraints cannot be the only important mechanisms driving comprehension.

3.1 Comprehension of language about novel situations.

Glenberg and Robertson (2000) asked participants to judge the sensibility of language describing novel situations (that is, situations in which the participant was unlikely to have experience and in which tools were used in novel ways). The description was in one of three forms, such as:

Bill needed to paint the top of his barn wall so he stood on his…

> Afforded-and-associated: ladder.
> afforded: tractor.
> non-afforded: hammer.

In the afforded-and-associated condition, the tool was one that is commonly used in the situation. In the afforded condition, the tool was one that has the requisite affordances for accomplishing the goal implicit in the sentence (e.g., one can climb on a tractor to reach the top of a wall). In the non-afforded condition, the tool does not have the affordances needed to accomplish the goal. The tools mentioned in the afforded and the non-afforded conditions were equally (poorly) associated with the context as demonstrated by LSA (Landuaer & Dumais, 1997).

The participants' judgments were obvious, namely both the afforded-and-associated and afforded descriptions were judged as sensible, whereas the non-afforded description was judged as nonsense. One explanation of this finding is that people consider affordances when judging the sensibility of novel descriptions. In fact, given that none of the participants had ever climbed on a tractor to paint a wall, and none had tried to stand on a hammer to paint a wall, it is hard to think of an account of these judgments that does not involve some sort of simulation using bodily capabilities.

These results speak strongly against accounts of language comprehension based solely on fixed features and frequency constraints. First, the afforded and non-afforded endings cannot be discriminated on the basis of standard, fixed "semantic" features such as frequency, grammatical class, animacy, semantic category (e.g., tools) and so on. Second, it is unlikely that any list of features would contain things such as "can stand on it to paint the top of a wall" or "cannot stand on it to paint the top of a wall." Third, given that the situations are novel, it is unlikely that any participant had actually attempted these activities, so there is very little chance that the sensibility judgments were based on particular previous episodic experiences. Finally, Robertson and Glenberg (2000), demonstrated that the time to judge the afforded condition as sensible was statistically comparable to the time needed to judge the afforded-and-associated condition as sensible. People did not engage in any complex reasoning to arrive at the conclusion that a tractor affords raising the body to a significant degree. Instead, it appears that people judged the sensibility of both the ladder and tractor sentences the same way: by deriving and meshing affordances.

3.2 Comprehension of linguistic innovations

One might argue that there is no strong guarantee that close analogs of the novel situations used by Glenberg and Robertson (2000) were never experienced by the participants, thus the judgments might have been based on analogy rather than the derivation of affordances. Glenberg and Robertson (2000, Experiment

3) and Kaschak and Glenberg (2000) addressed this argument by using linguistic innovations, that is, language invented for the experiment. In English, it is possible to change nouns into verbs (called denominal verbs), such as "bicycle" becoming "to bicycle" and "bottle" becoming "to bottle." In fact, one can innovate denominal verbs, and the innovative uses can be understood in context. For example, Clark and Clark (1979) discuss sentences such as "The newsboy porched the newspaper."

Glenberg and Robertson (2000) and Kaschak and Glenberg (2000) used innovative denominal verbs in contexts similar to the following: "A man with a broken foot was sitting on a park bench peeling his hard-boiled egg. A soccer ball rolled up to him from a nearby game being played by some young girls." Following this context, participants judged the sensibility of sentences containing innovative denominal verbs, such as "He crutched the girls the ball" or "He egg-shelled the girls the ball." Why is the first sentence so much easier to understand than the second? Note that it cannot be related to frequency because neither "to crutch" nor "to egg-shell" is a verb in standard English. Thus, low (or zero) frequency does not rule out understanding, but neither does it guarantee it. Apparently, in understanding these sorts of innovations, people consider affordances, and a crutch has the affordances needed to transfer a soccer ball from one location to another, whereas an egg-shell does not.

3.3 Lessons for modeling language comprehension

Two lessons can be derived from this work on novel situations and novel language. First (and again), no fixed set of semantic features can capture the phenomena; instead the features needed for understanding (e.g., that a tractor affords raising the body, or that a crutch affords transfer of a soccer ball) arise from an interaction between types of bodies, human goals, and components of the situation. Second, the combinatoric semantics (e.g., how we arrive at a meaning for an innovation such as "to crutch") arises from the same sort of body/goal/situation interaction: A sensible combination cannot result from combining elements using only frequency constraints.

4. Lesson 3: Language about abstract ideas is concretely embodied

Several research programs are directed toward developing the idea that all language (in fact, all cognition) is concretely embodied. Thus, Barsalou (1999) discussed how operations on perceptual symbols can result in abstract ideas such as "truth" and "negation." Lakoff (1987) developed the notion that logical ideas such as "p or not-p" are based on concrete, embodied schemata, and Lakoff and Nunez (2000) attempt to account for several areas of mathematics using embodied schemata. Roth (1999) has demonstrated how understanding of physics by high-school students may evolve from literal bodily action taking place during discussions of a laboratory exercise. In a similar vein, Ochs, Gonzales, and Jacoby (1996) analyze the use of embodied language during discussions of theoretical, high-energy physics among a PI and post-doctoral students. In the remainder of Section 4, I will briefly discuss data from my own laboratory demonstrating the embodiment of language about emotional states.

4.1 *Embodiment of language implying emotional states.*

Work reviewed in previous sections has demonstrated that language comprehension calls on the same neural systems as used in perception and action. Might the same be true for language about emotion? That is, does understanding language about an emotional situation require simulation of the emotion using the same neural and bodily systems that contribute to emotional reactivity? Glenberg, Havas, Becker, & Rinck (in press) tested this idea using a procedure developed by Strack, Martin, and Stepper (1988). In this procedure, a participant holds a pen in the mouth (pointing out from the body) using only the teeth, thus forcing a smile, or using only the lips, thus forcing a frown. Strack et al. (1988) demonstrated that this bodily manipulation reliably elevates (when smiling) or depresses (when frowning) mood. In our experiments, participants held the pen in their mouths while judging sentences along several dimensions (e.g., ease of understanding). The sentences described either pleasant situations (e.g., "You and your lover embrace after a long separation") or unpleasant situations (e.g., "The police car rapidly pulls up behind you, siren blaring"). If understanding of language about emotional situations requires the simulation of emotion using the same neural systems involved in emotional processing, then we should observe an ACE-type effect, that is, judgments should be faster when the sentence content and pen manipulation are congruent than when incongruent. In fact, this interaction was observed in two experiments.

Bower (1981; Bower & Forgas, 2000) has suggested that these sorts of effects of bodily processes on conceptual processing reflect a spreading activation mechanism. That is, the smile produced by holding the pen in the teeth activates a cognitive node corresponding to a happy state, and this node spreads activation to associatead nodes representing words such as "lover" and "embrace." Because these word nodes are activated, participants can read and judge sentences containing these words faster than sentences containing non-activated nodes corresponding to words such as "police" and "blaring." In contrast to this spreading activation account, another experiment (reported by Havas, Glenberg, & Rinck, in preparation), demonstrates that the ACE-type effect cannot be due to a simple spread of activation. In that experiment, we demonstrated that the pen manipulation does not prime the lexical decision of individual words (e.g., "lover") taken from the sentences. Apparently, the ACE-type effect reflects a simulation of the sentence content that calls upon systems involved in emotional processing, not a simple spread of activation to particular words.

4.2 *Lessons for modeling language comprehension*

Understanding language about emotions may require a bodily simulation of components of the emotional state. Once again, it appears that language comprehension is intimately connected to bodily states and activities; it is not simply the manipulation of symbols.

5. Two, quasi-formal arguments

5.1 *An argument against exclusive use of fixed semantic features*

1. The "features" relevant to language comprehension depend on human abilities (e.g., climbing on a tractor; crutching a soccer ball; moving a whistle with a hand or hook) and current goals.
2. Human abilities and goals vary from person to person, time to time, and situation to situation; arguably, the combinations are infinite.
Conclusion: No fixed list of semantic features can account for language comprehension. Instead, the comprehension system must be sensitive to the human interpretations (i.e., actions and goals) of situations. A similar argument is made by French (1997).

5.2 *An argument against exclusive use of frequency-based constraints*

1. Language comprehension requires bodily states and combinations of bodily states (e.g., to comprehend sentences about transfer, visual motion, emotions).

2. These combinations are governed by biological constraints (e.g., on what movements can be combined into coherent action).
3. Frequency constraints do not correspond to biological constraints (e.g., we know that standing on a hammer will not allow us to paint the top of a barn, but standing on a tractor might, even if we have never done either).
Conclusion: Models that depend exclusively on frequency constraints cannot accurately simulate how language is comprehended.

6. Challenges for models given the realities of embodiment

If the quasi-formal arguments in Section 5 are correct, they demonstrate that a) most current connectionist models of reading for meaning (e.g, Harm & Seidenberg, 2004) and meaning (e.g, Rogers, Lambopn Ralph, & Garrod, 2004) are inadequate, and b) most current non-connectionist formal models of language and meaning such as LSA (Landauer & Dumais, 1997), HAL (Burgess & Lund, 1997), and Construction-Integration theory (Kintsch, 1988) are inadequate. These models use fixed feature sets or frequency-based constraints (or both) exclusively.

6.1 *Why are so many types of theories inadequate?*

An answer to this question would require thorough historical, sociological, and scientific analyses well beyond the scope of this chapter. Instead, I will simply assert two ideas. The first is that modern cognitive psychology has, for the most part, ignored the symbol grounding problem (e.g., Harnad, 1990; Searle, 1980). This stance led to the belief that fixed and limited feature sets could be used to model meaning. Furthermore, this belief led to simple experiments (e.g., lexical decision, word and picture naming, category judgments) that could be adequately (and elegantly) modeled with fixed features and frequency-based constraints. Unfortunately, because those models are based on fixed feature sets and frequency-based constraints, they cannot scale up to model language as used in most communicative situations.

6.2 *What is the solution?*

Undoubtedly, there are many ways of dealing with the current state of affairs. Here I offer an empirical and a theoretical guess. Empirically, cognitive psychologists need to focus on tasks that use language in more realistic contexts, such as the tasks used by Chambers et al. (2004) in which participants literally acted as directed by utterances. Once psychologists have built a rich, numerical data base, it will become grist for the theoretician/modeler.

My theoretical guess is that accounting for such language use will force the construction of embodied simulations. In such a model, conceptual information (e.g., what a word means) will be encoded in terms of perceptual, action, and other body systems. Thus, what X means will be in terms of what can be done with X to accomplish goals and how action can change the perceptual information (cf. O'Regan, & Noë, 2001) arising from X. Several papers in this volume (e.g., Chang, 2004; Roy, 2004) attest to the power of this approach.

Acknowledgments

This chapter is based upon work supported by the National Science Foundation under Grant No. BCS-0315434 and INT-0233175 and the Department of Education under Grant No. R305H030266 to Arthur Glenberg. Any opinions, findings, and conclusions or recommendations expressed in this material are those of the author and do not necessarily reflect the views of the funding agencies. The ideas presented here benefited from discussion with Gerry Altmann, Lawrence Barsalou, Anna Borghi, Mike Kaschak, Maryellen MacDonald, Mark Seidenberg, Manuel de Vega, and Rolf Zwaan.

References

Barsalou, L. W. (1999). Perceptual symbols systems. *Behavioral & Brain Sciences*, 22, 577-660.

Borghi, A. M., Glenberg, A. M., & Kaschak, M. P. (in press). Putting words in perspective. *Memory & Cognition*.

Bower, G. H. (1981). Mood and memory. *American Psychologist*, 36, 129-148.

Bower, G. H., & Forgas, J.P. (2000). Affect, memory, and social cognition. In J. Forgas, K. Eich, G. H. Bower, & P. M. Niedenthal (Eds.) *Cognition and emotion*. Oxford: Oxford University Press (pp 87-168).

Burgess, C. & Lund, K. (1997). Modelling parsing constraints with high-dimensional context space. *Language & Cognitive Processes*, 12, 177-210.

Clark, E. V., & Clark, H. H. (1979). When nouns surface as verbs. *Language*, 55, 767-811.

Chang, N. (this volume)

Coventry, K. R. & Garrod, S. C. (2004). *Saying, seeing and acting: The psychological semantics of spatial prepositions*. London: Taylor & Francis, Psychology Press.

Chambers, C. G., Tanenhaus, M. K., & Magnuson, J. S. (2004). Actions and affordances in syntactic ambiguity resolution. *Journal of Memory and Language*, 30, 687-696.

French, R. M. (1997). When coffee cups are like old elephants or why representation modules don't make sense. In A. Riegler & M. Peschl (Eds.) *Understanding representation in the Cognitive Sciences* (93-100). New York: Plenum Press.

Gibson, J. J. (1979). *The ecological approach to visual perception*. New York: Houghton Mifflin.

Glenberg, A. M. (1997). What memory is for. *Behavioral & Brain Sciences*, 20, 1-55.

Glenberg, A. M., Havas, D., Becker, R., & Rinck, M. (in press). Grounding Language in Bodily States: The Case for Emotion. R. Zwaan and D. Pecher (Eds.) *The grounding of cognition: The role of perception and action in memory, language, and thinking*. Cambridge: Cambridge University Press.

Glenberg A.M., Kaschak M.P. (2002). Grounding language in action. *Psychonomic Bulletin & Review*, 9, 558-565

Glenberg, A. M., & Robertson, D. A. (1999). Indexical understanding of instructions. *Discourse Processes*, 28, 1-26.

Glenberg, A. M. & Robertson, D. A. (2000). Symbol grounding and meaning: A comparison of high-dimensional and embodied theories of meaning. *Journal of Memory & Language*, 43, 379-401.

Harm, M. W., & Seidenberg, M. S. (2004). Computing the Meanings of Words in Reading: Cooperative Division of Labor Between Visual and Phonological Processes. *Psychological Review*, 111, 662-720.

Harnad, S. (1990). The symbol grounding problem. *Physica D*, 42, pp 335-346.

Havas, D., Glenberg, A. M., & Rinck, M. (in preparation). Emotional systems in language comprehension.

Kaschak, M. P. & Glenberg, A. M. (2000). Constructing meaning: The role of affordances and grammatical constructions in sentence comprehension. *Journal of Memory and Language*, 43, 508-529.

Kaschak, M. P., Madden, C. J., Therriault, D. J., Yaxley, R. H., Aveyard, M., Blanchard, A., & Zwaan, R. A. (in press). Perception of motion affects language processing. *Cognition*.

Kintsch, W. (1988). The role of knowledge in discourse comprehension: A construction-integration model. *Psychological Review*, **95**, 163-182. (4)

Lakoff, G. (1987). *Women, fire, and dangerous things: What categories reveal about the mind.* Chicago: University of Chicago Press.

Lakoff, G., & Núñez, R. E. (2000). *Where mathematics comes from: How the emobodied mind brings mathematics into being*. New York: Basic Books.

Landauer, T. K. & Dumais, S. T. (1997). A solution to Plato's problem: the Latent Semantic Analysis theory of acquisition, induction and representation of knowledge. *Psychological Review*, 104, 211- 240.

O'Regan, J. K., & Noë , A, & (2001). A sensorimotor account of vision and visual cognition. *Behavioral and Brain Sciences*, 24, 939-1031.

Ochs, E., Gonzales, P., & Jacoby, S. (1996). "When I come down I'm in the domain state":grammar and graphic represwentation in the interpretive activity of physicists. In E. A. Schegloff and S. A. Thompson (Eds.), Interaction and grammar. Cambridge: Cambridge University Press. (pp. 328-369)

Pecher, D., Zeelenberg, R., & Barsalou, L.W. (2003). Verifying properties from different modalities for concepts produces switching costs. *Psychological Science*, 14, 119-124.

Robertson, D. A., & Glenberg, A. M. (2000). Grounding symbols and computing meaning: A supplement to Glenberg & Robertson (2000). http://psych.wisc.edu/glenberg/supplement_jml_gandr.html.

Rogers, T.T., Lambon Ralph, M. A., & Garrod, P. Structure and deterioration of semantic memory: A neuropsychological and computational investigation. *Psychological Review*, 111, 205-235.

Roth, W.-M. (1999). Discourse and agency in school science laboratories. Discourse Processes, 28, 27-60.

Roy, D. (this volume)

Searle, J. R. (1980). Minds, brains and programs. *Behavioral & Brain Sciences*, 3, 417-424.

Strack, F., Martin, L. L., & Stepper, S. (1988). Inhibiting and facilitating condition of facial expressions: A non-obtrusive test of the facial feedback hypothesis. *Journal of Personality and Social Psychology*, 54, 768-777.

ASSOCIATIVE NEURAL MODELS FOR BIOMIMETIC MULTI-MODAL LEARNING IN A MIRROR NEURON-BASED ROBOT

STEFAN WERMTER, CORNELIUS WEBER, MARK ELSHAW
Hybrid Intelligent Systems, School of Computing and Technology,
University of Sunderland,
Sunderland, SR6 0DD, UK
[Stefan.Wermter, Cornelius.Weber, Mark.Elshaw]@Sunderland.ac.uk
www.his.sunderland.ac.uk

By using neurocognitive evidence on mirror neuron system concepts the MirrorBot project has developed neural models for intelligent robot behaviour. These models employ diverse learning approaches such as reinforcement learning, self-organisation and associative learning to perform cognitive robotic operations such as language grounding in actions, object recognition, localisation and docking. In this paper we describe architectures based on an associative self-organising framework which were designed to combine multimodal inputs of language, vision and motor programs to produce complex robot behaviours.

1. Introduction

In this paper we will describe some research performed as part of the "biomimetic multimodal learning in a mirror neuron-based robot" (MirrorBot) project for neuroscience-based models for an intelligent robot. Theories and experiments in neuroscience have indicated that a neuroscience-oriented approach for multimodal processing is promising for new computational techniques to associate vision, language and motor control (Pulvermuller 1999, Rizzolatti and Arbib 1998, Wermter et al. 2001). In particular the neuroscience motivation for our models comes from the neurocognitive evidence on action verb processing in the brain and the mirror neuron system found in primates and humans.

Neurocognitive evidence on word processing shows that cortical assemblies have been identified in the cortex that activate in response to the performance of motor tasks at a semantic level (Pulvermuller 1999, Rizzolatti and Arbib 1998, Hauk and Pulvermüller 2004). The meaning of words is critical for determining the cortical populations or cell assemblies that form a functional unit that are activated to implement the cognitive task (Pulvermüller 1999, Wermter and Elshaw 2003). For instance, perception words are represented by perisylvian assemblies and posterior cortex, and nouns related to animals activate the inferior temporal or occipital cortices (Pulvermüller 1999).

Neurocognitive evidence on action verb processing sees a division of representation in the brain based on whether the action is performed by a specific body part (Hauk et al. 2004, Pulvermüller 1999, Pulvermüller 2001). Various EEG experiments on processing of action verbs were carried out to test this hypothesis by looking at words related to the action performed by the leg, arm and face/head. The differences based on different body parts for action verbs processing can be seen from the average response times for lexical decisions. They are faster for face-associated words compared to arm-associated words, and the arm-associated words are faster than leg-associated words. Leg-words generated greater activation in the central brain region around the vertex, while face-words activated inferior-frontal areas, thereby suggesting that the relevant body part representations are differentially activated when words for different actions are being comprehended. This is schematically depicted in Figure 1, with the light circled cell assemblies shared by all body types and the darker assemblies being specific to the particular body type.

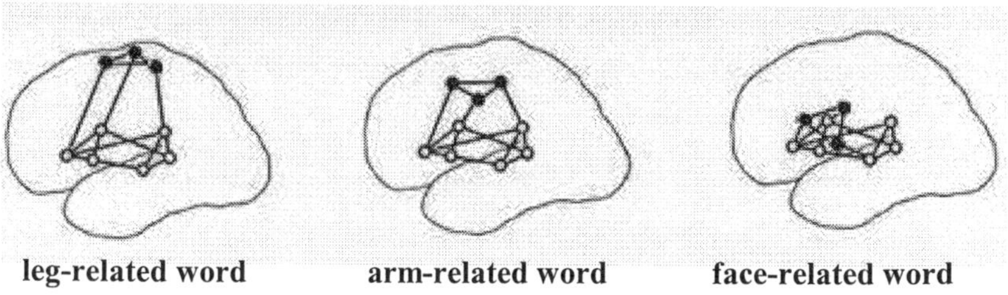

Figure 1. The cell assemblies found to be associated with the processing of action words based on body parts (Pulvermüller 2003).

Focusing on individual neurons, Rizzolatti and Arbib (1998) found two types of neurons located in the F5 motor cortex region of primates, the classical motor neurons which only respond to the performance of the action and the mirror neurons which respond not only when performing an action but also when seeing or hearing the action performed (Kohler et al. 2002, Rizzolatti and Arbib 1998).

This ability to understand actions allows the observer to learn through imitation, to predict the actions and to act accordingly (Gallese 1998). The concepts from mirror neurons suggest that own actions, observed actions and language are very much interrelated since the same mirror neurons fire when any of these three modalities is an input.

In response to this neuroscience evidence we have developed and will describe various neural models for biomimetic multimodal learning in a mirror neuron-based robot. These models perform activities such as object localization, grounding of language in actions and robot docking. These models are based on diverse learning approaches that are proposed for distinct regions of the brain and we will describe our overall associator approach. Finally, we will consider a self-organising architecture to combine concepts from these neural models to produce a multimodal behaviour in a robot (Figure 2). This architecture takes as inputs language, vision and actions. This architecture is able to associate these so that it can produce or recognize the appropriate action. The architecture either takes a language instruction and produces the behaviour or receives the visual input and action at the particular time-step and produces the language representation (Wermter et al. 2004).

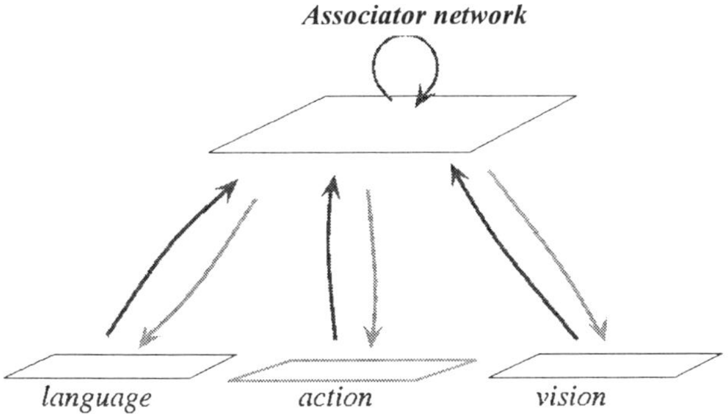

Figure 2. The feature extractor and associator architecture for multimodal integration.

2. Associator model for object localization

Our model for associative object localisation in Figure 3 consists of a "what" pathway on the left, and a "where" pathway on the right (Weber and Wermter 2003). The "what" pathway consists of an input area and a hidden area. The hidden area of the "what" pathway consists of two layers. The lower layer receives bottom-up connections W^{bu} from the input. These have been trained according to a sparse coding Helmholtz machine and represent feature detectors similar to Gabor functions, but also with colour sensitive elements. The depicted top-down weights W^{td} were needed to train W^{bu}, but are not used further on. The upper layer receives a one-to-one copy of the output of the lower layer cells (denoted by the 3 thin arrows in Figure 3). After it receives this initial input, it functions as an attractor network which solely updates its activations based on its previous activations. Each cell receives its input from all other neurons via recurrent weights V^{22}. In addition, input arrives from the laterally connected area of the "where" pathway via weights V^{23}.

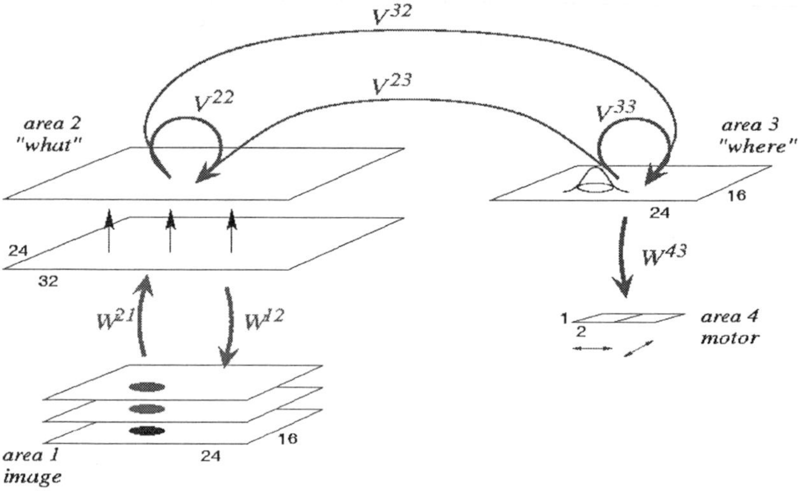

Figure 3. Associator model architecture. Left, the pathway of the lower visual system: below, the three red-, green- and blue-layers of the image, then the feature detecting cells and on top, the attractor cells of the "what" area. On the right, the "where" area displays the location of the object of interest. Small numbers denote simulated area sizes. Thick arrows denote trained weights. The two motor units in this simple setup control the pan- and tilt-position of the robot camera. Their weights W^{43} have been trained in an error driven fashion so that the camera focuses on the object.

The "where" pathway on the right of Figure 3 consists of just one area. The "where" neurons are fully connected via recurrent weights V^{33} and in addition receive input from the highest "what" layer via V^{32}. In the following, we will refer to all connections V^{22}, V^{33}, V^{23} and V^{32} collectively as V^{lat}, because they are lateral weights and receive the same treatment, during training as well as during activation update. The activation update of the "where" and highest level "what" neurons is governed by the following equation:

$$u_i(t+1) = \mathrm{f}\left(\sum_l \mathrm{v}_{il}^{\mathrm{lat}} u_l(t)\right), \quad \text{where } \mathrm{f}(x) = e^{\beta x} / (e^{\beta x} + n) \qquad (1)$$

Parameters in the neuronal transfer function f are the slope β = 2 and the sparseness factor n = 8 which leads to little average neuronal activity, if there is no input. The lateral weights are trained from the bottom-up input. Their purpose is to memorise the incoming activities $\boldsymbol{u}_i(t=0)$ as activation patterns which they maintain. Learning maximises the log-likelihood to generate the incoming data distribution by the internal activations $\boldsymbol{u}_i(t)$ if Eq. (1) is applied repeatedly:

$$\Delta v_{il}^{lat} \approx \sum_t (\mathrm{u}_i(t=0) - u_i(t))\,\mathrm{u}_i(t-1). \qquad (2)$$

The contribution to learning can be seen in the difference term where $\boldsymbol{u}_i(t=0)$ is the data while $\boldsymbol{u}_i(t)$ is produced through recurrent application of Eq. (1) which corresponds to a continuous-valued attractor basin of the attractor network.

First, the weight matrices $\boldsymbol{W}^{td}$ and $\boldsymbol{W}^{bu}$ were trained on small patches randomly cut out from natural images. Lateral weights V^{lat} were then trained with $\boldsymbol{W}^{td}$ and $\boldsymbol{W}^{bu}$ fixed. For this, within each data point (an image patch), an artificially generated orange fruit was placed at a randomly chosen position. The "where" area received a Gaussian hill of activity on the location which corresponds to the one in the input where the orange is presented.
The representation of the image with an orange obtained through $\boldsymbol{W}^{bu}$ on the lower layer was copied to the upper layer cells. This together with the Gaussian hill on the "where" area was used as target training vector. Relaxations were done for four time steps, which had been discovered in empirical tests.

Figures 4 and 5 show the relaxation of the network activities after initialisation with sample stimuli. In all cases, the "where" area neuron's activations were initialised to zero at time $t = 0$. The relaxation procedure therefore completes a pattern which spans both, the "what" and the "where" area, but which is incomplete at time $t = 0$, as can be seen in Figure 4.

All weights in the model have been trained on the basis of real images and are therefore irregular. Localisation quality may vary at slightly different object locations within the image. The third frame in Figure 5, for example, leads to a blurred "where" representation. If information from the second frame would be taken into account, this may be cleaned up. However, for simplicity and consistency with the training procedure, the algorithm processes only one frame at a time.

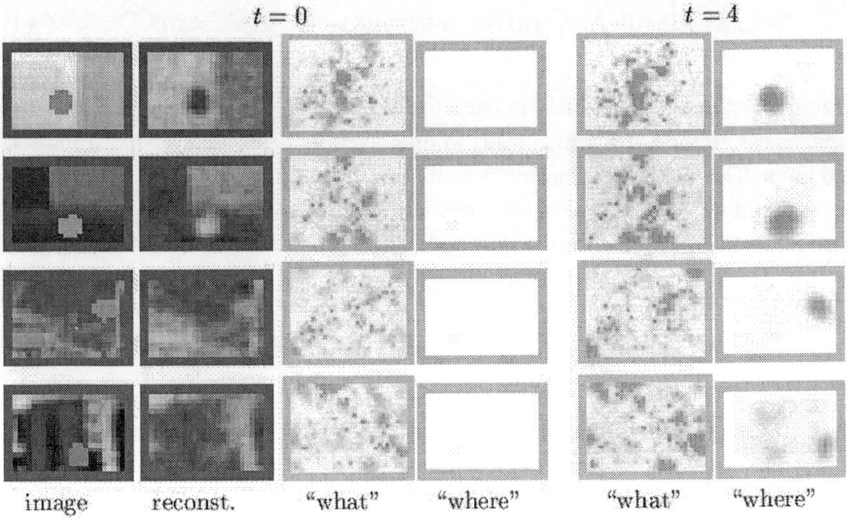

Figure 4. Each row shows the network response to a color image which contains an artificially generated orange fruit. From left to right: the image, the reconstruction of the image via W^{12} (cf. Figure 3), the representation on the "what" area (used for the image reconstruction), the initial zero activities on the "where" area at time $t = 0$. Then the activations on the "what" and "where" areas at $t = 4$ time steps which is the relaxation time used for training. The estimated position of the orange on the "where" area is correct in the upper three rows.

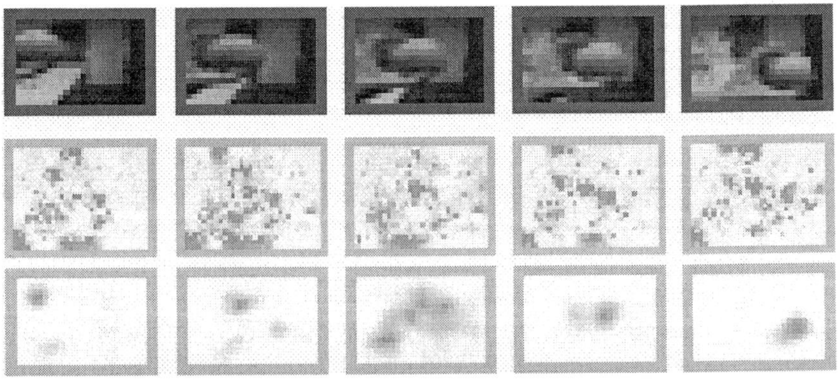

Figure 5. Localisation for real images taken from the robot camera. In each column, the top row shows an image, the middle row shows the response on the "what" area and lower row the response of the "where" area at iteration time $t = 4$ to the presentation of the image in the upper row.

3. Associator architecture for robot docking

After the successful completion of the localization associator network it was expanded as a learning approach to perform the docking behaviour (Weber et al. 2004b). This extended model uses neural vision and reinforcement learning as a solution for robotic docking, which moves the PeopleBot robot toward a table so that it can grasp an object. As the robot has a short non-extendable gripper and wide "shoulders" it must approach the table at a perpendicular angle so that the gripper can reach over it.

Our robot selects the action which leads to the largest expected reward. This makes the action selection network the core part. The model includes four motor neurons that are "on" depending on if the robot is to move forward, backward, left or right, respectively. The peripheral vision module is trained before the action selection network so that it can supply the necessary visually obtained perception as input.

Overall, we have three training phases: stages one and two perform the training for the object localization outlined above, stage three trains the weights W^c and W^m from the conceptual space to the critic of the motor outputs, respectively (Figure 6). During training, the motor units are guided by the firing rate of one "value function" unit which assigns a fitness value to any state. Together, these four neurons are trained by reinforcement learning, in which a scalar reinforcement signal is given only at the end of each training action sequence. The value of the signal is positive, if the robot docks at the object in parallel to the table, or negative, if the robot's shoulders bump into the table at an angle or if the object is lost out of sight.

The input to the action selection network is the robot's visual perceptual state, defined by its relative position to the target, an orange fruit at the border of a table as shown in Figure 7. The localization model described in the previous section is used to provide the location of the object for docking. Additional input to the action selection network is the robot rotation angle φ, supplied by the robot's internal odometry.

Each trial thus constitutes a robot's experience about either reaching the goal or loosing the target or hitting the table. It consists of a sequence of actions and several learning steps. The more trials have been done, the better the robot performs.

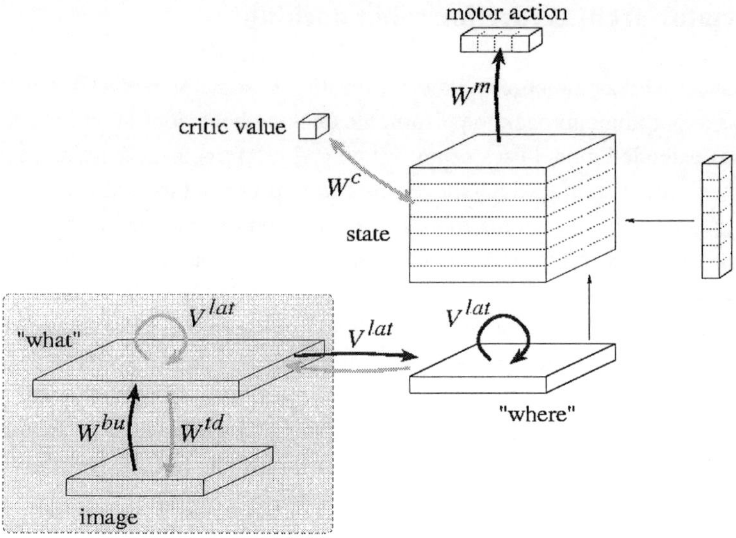

Figure 6 Associator network for action selection. Thick arrows denote trained weights W of which only the dark ones are used during performance, while those depicted bright are involved in training. The "what"-"where" network is the same as in Figure 3, but the object location is used here together with the robot angle as input to an action-critic network that learns the strategic docking manoeuvre by adapting $\boldsymbol{W}^m$ and $\boldsymbol{W}^c$. The shaded part of the network uses the same learning principle as the associator architecture for multimodal integration, described in the next chapter.

The state description (Figure 6) consists of a Gaussian covering several state space units simultaneously. A critic weight w_j^c is thus updated close to weight $w_{j'}^c$ if j and j' are neighbours in the state space; analogously motor weights w_{ij}^m. This topological relation exploits the fact that similar states imply similar optimal actions and speeds up learning. A successful example of simulated docking performance is depicted in Figure 7b which displays a successful movement.

The experiments described in this section of the paper have shown the suitability of associator models for learning robot docking behaviour. Such models offer great flexibility compared to approaches for robot behaviour that rely on preprogramming. They also indicate that such models could be linked by an associative architecture to achieve multimodal input fusion to achieve complex behaviours based on a goal given as a language instruction.

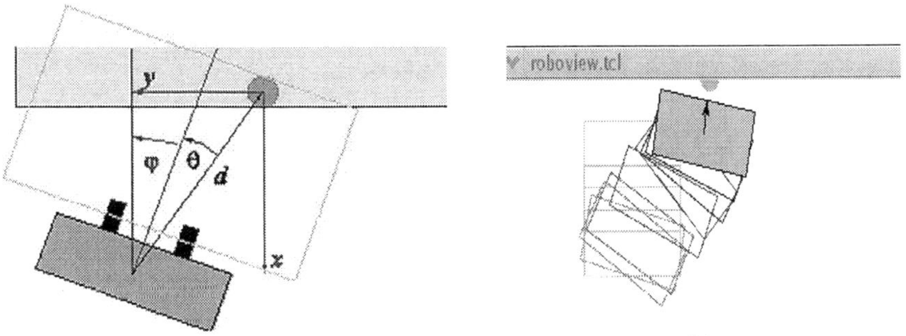

(a) (b)

Figure 7 a). The figure depicts the table and the target on it. The robot is shown with short black grippers and its field of vision is outlined by the dotted line. Real world coordinates (x, y, φ) specify the position and rotation angle of the robot and the perceived position of the target within the robot's visual field is then defined by angle θ and distance d. 7 b). The simulated trained robot during the docking to the orange object on the edge of a table.

4. Associator architecture for multimodal integration

An overall self-organising associator architecture for multimodal integration of vision, action and language was designed that contains many of the features of the models described above (Elshaw et al. 2004, Weber et al. 2004b). Furthermore, the approach takes inspiration from the mirror neuron system by a student robot recognising and producing the behaviours of a teacher. In our approach a student robot learns from a teacher robot how to perform three separate behaviours 'pick', 'lift' and 'go' based on multimodal inputs. These behaviours were selected as they are typical ones that a robot has to perform and are complex in that the same visual input would require different actions based on the language instruction.

First, a robot simulator was produced with a teacher robot performing 'go', 'pick' and 'lift' actions continuously in an environment (Figure 8). The student robot observed the teacher robot performing the behaviours and was trained by receiving multimodal inputs. These multimodal inputs were the higher-level visual inputs of the x and y coordinates and the rotation angle φ of the teacher robot, the motor directives of the robot and the language instruction.

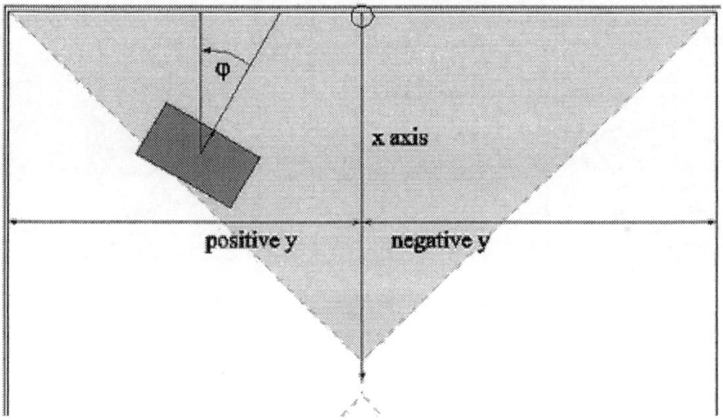

Figure 8. The simulated environment containing the robot. The x, y and φ coordinates of the robot are taken with respect to the nearest wall. Dashed lines indicate the borders of areas "belonging" to a wall. The area belonging to the top wall is depicted in light grey.

The simulated teacher robot performs the three behaviours in reoccurring loops and in the following order: the behaviour represented by the word 'go' involves moving around the environment until it reaches a wall and then turns away from the wall at a set angle. It switches to the docking behaviour outlined above represented by the word 'pick' if it comes close to the target at the top of the arena. The final behaviour follows the behaviour represented by the word 'lift' involving moving backward to leave the table and then turning around to face toward the middle of the arena. When receiving the multimodal inputs, the student robot was required to learn these behaviours so that it could recognise them in the future or perform them based on a language instruction.

The imitation model (Figure 9) used an associator network based on the Helmholtz machine approach (Hinton et al. 1995). The Helmholtz machine generates representations of data using unsupervised learning. Bottom-up weights W^{bu} generate a hidden representation $\vec{r}$ of some input data $\vec{z}$. Conversely, top-down weights W^{bu} $\tilde{z}$ reconstruct an approximation of the data $\vec{z}$ from the hidden representation.

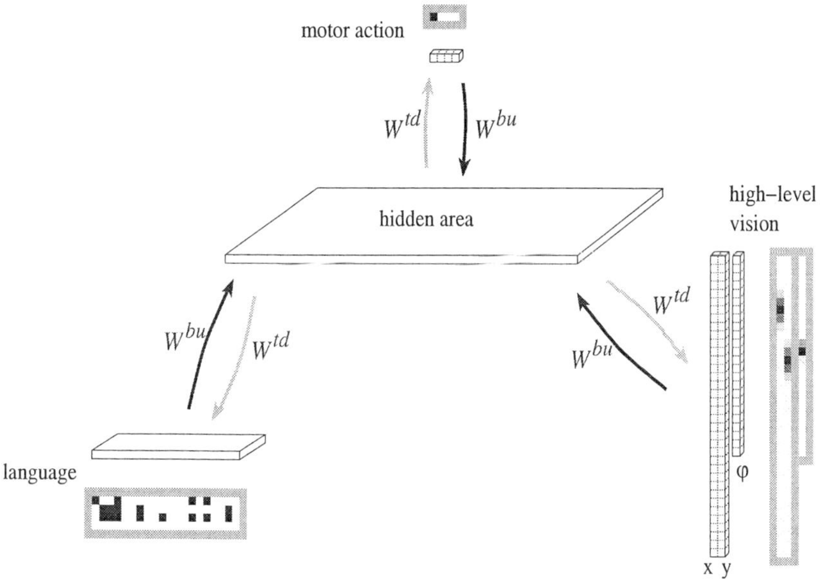

Figure 9. The associator model for robot imitation learning. A stimulus example is shown near each of the three input areas.

Both sets of weights are trained by the unsupervised wake-sleep algorithm which uses the local delta rule. In the wake phase, a full data point $\vec{z}$ is presented which consists of the full motor, language and higher-level visual components. The linear hidden representation $\vec{r} = W^{bu}\vec{z}$ is obtained from which a competitive version $\vec{r}^c$ is obtained by taking the winning unit of $\vec{r}$ (given by the strongest active unit) and assigning activation values under a Gaussian envelope to the units around the winner. Thus, $\vec{r}^c$ is effectively a smoothed localist code. The reconstruction of the data is obtained by $\tilde{\vec{z}} = w^{td}\vec{r}^c$ and the top-down weights from units j to units i are modified according to

$$\Delta w_{ij}^{td} = \eta r_j^c (z_i - \tilde{z}_i) \tag{3}$$

with an empirically determined learning rate $\eta = 0.001$. The learning rate was increased 5-fold whenever the active motor unit of the teacher changed.

In the sleep phase, a random hidden code $\vec{r}^s$ is produced by assigning activation values under a Gaussian envelope centred on a random position on the hidden layer. Its linear input representation $\vec{r}^{s} = W^{td}\vec{r}^{s}$ is obtained, and then the reconstructed hidden representation $\tilde{r}^{s} = W^{bu}\vec{z}^{s}$ is obtained from which a competitive version $\tilde{r}^{c}$ is obtained by assigning activation values under a Gaussian envelope centred around the winner. All weights W^{td} and W^{bu} were rectified to be non-negative at every learning step and the bottom-up weights W^{bu} of each hidden unit were normalised to unit length.

The hidden layer of the associator network in Figure 9 that acted as the student robot's cortex had 16 by 48 units. In the wake phases of training it received multimodal inputs $\vec{z}$ based on observing the actions of the teacher robot performing the three behaviours.

These inputs included first the higher-level vision which represents the x and y coordinates and rotation angle φ of the teacher robot. The x, y and φ coordinates in the environment were represented by two arrays of 36 units and one array of 24 units, respectively. For a close distance of the robot to the nearest wall, the x position was a Gaussian of activation centred near the first unit while for a robot position near the middle of the arena the Gaussian was centred near the last unit of the first column of 36 units. The next column of 36 units represented the y coordinates so that a Gaussian centred near the middle unit represented the robot to be in the centre of the environment along the y axis. Rotation angles φ from -180° to 180° were represented along 24 units with the Gaussian centred on the centre unit if $\varphi = 0°$.

The bottom-up weights from units i to units j are modified according to

$$\Delta w_{ij}^{bu} = \in (w_{ji}^{bu} - z_i^{s})\tilde{r}_j^{c} \tag{4}$$

with an empirically determined learning rate $\varepsilon = 0.01$.

For the language region representations of phonemes were presented. This approach used a phoneme representation consisting of 20 phonetic features, which produced a different binary activation pattern in the language input region

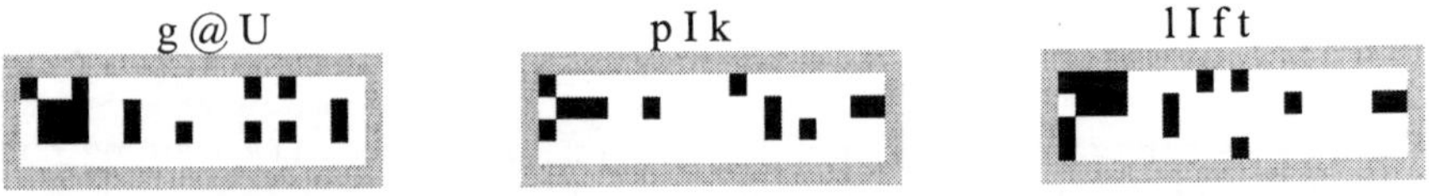

Figure 10. The phonemes and the corresponding 4 x 20-dimensional vectors representing 'go', 'pick' and 'lift'.

for each phoneme. A region of 4 rows by 20 columns was used to represent the words with the first row representing the first phoneme and the second row the second phoneme etc., so that the order in which they appear in the word is maintained (Figure 10).

As final part of the multimodal inputs the teacher robot motor directives were presented on the 4 motor units (forward, backward, turn right and turn left) one for each of the possible actions with only one active at a time. The activation values in all three input areas were between 0 and 1.

During training the student robot received all the inputs, however when testing, either the language area for recognition or the motor inputs for production were omitted. Recognition was verified by comparing the units which are activated on the language area via W^{td} (Figure 9) with the activation pattern belonging to the verbal description of the corresponding behaviour. For action production the robot continuously received its own current x, y and φ coordinates and the language instruction of the behaviour to be performed. Without motor input it had to produce the appropriate motor activations via W^{td} which it had learnt from observing the teacher to produce the required behaviour. The associator model based robot recognised the 'go'-, 'pick'- and the 'lift' behaviour, while the teacher robot was looping between the three behaviours as done during training. For testing the recognition of a behaviour, the teacher robot was placed at random start positions and performed a corresponding action. Furthermore, the trained student robot could successfully recreate a behaviour based on a language input as can be seen from Figure 11.

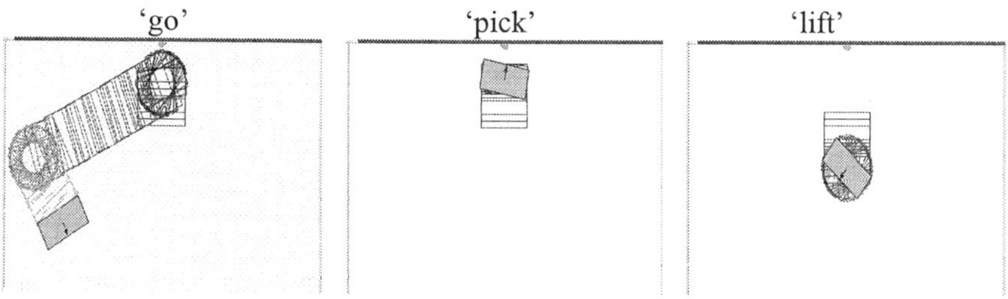

Figure 11 The simulated trained student robot performance when positioned at the same point in the environment but instructed with different language input.

The robot's ability to both recognise an observed behaviour and perform the behaviour that it has learnt by imitating a teacher shows the model was able to recreate some concepts of the mirror neuron system. The student robot produces similar regional unit activation patterns when observing the behaviour and performing it, as seen in Figure 12. In doing so it combined features of the individual models outlined in this paper to achieve the grounding of language and visual information in action in a multimodal approach.

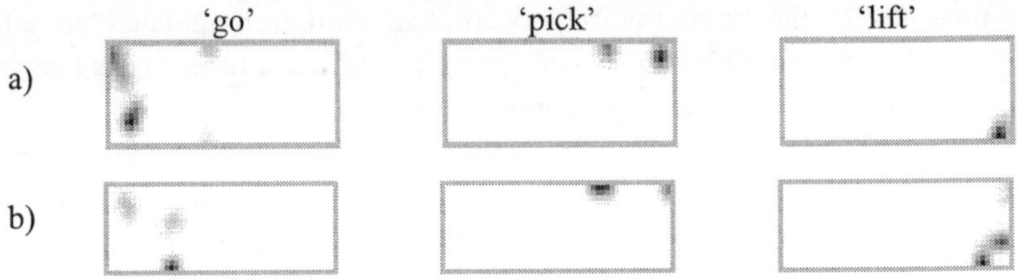

Figure 12. Hidden layer activations for the associator network summed up during short phases while the student robot a) correctly recognises the behaviours and b) performs them based on a language instruction.

6. Discussion

First, learning was done in simulation, then the trained models were transferred to the robot, where the performance of the docking has also been demonstrated. The lower-level vision network (W^m and W^c in Figure 6) was trained with "real" natural images, i.e. small patches randomly cut out from a set of colour images depicting natural scenes. We have used simulation primarily to ease the generation of data material for training but then transferred the simulation to a real robot environment. The "what"-"where" association network (W^{lat} in Figure 6) was trained supervised and therefore requires controlled conditions. We have used a simple orange coloured disc, pasted into a natural image to denote the target object during learning. During performance, the network localises real orange fruit sufficiently robustly. The actor-critic network (W^m and W^c in Figure 6) was trained with a simulator. Due to the robustness of the docking (if the robot should turn left, then it should do so in a large area), the learnt weights can also control the real robot. Actor-critic learning advances relatively quickly, and reasonable performance is achieved after a few successful trials of the simulated agent. Therefore, learning using real hardware is also realistic. Finally, the performance of the associator architecture for multimodal integration (all weights in Figure 9) increases slowly compared to the actor-critic network. While it requires this useful control algorithm to be already present, it can perform learning whenever the robot is active.

7. Conclusion

By combining various neuroscience inspired models for a biomimetic learning in a mirror neuron-based robot we have produced robot behaviours based on visual object localization, language grounding and docking action. The associator architecture was able to combine multimodal inputs of vision, language and motor in order to recognise and produce three behaviours. In doing so we have been able to reproduce an important property of the mirror neuron system and action verb processing. In conclusion we believe mirror-neuron based learning holds a lot of potential for learning robots in the future.

Acknowledgement

This work is part of the MirrorBot project supported by the EU in the FET-IST programme under grant IST-2001-35282.

References

Elshaw M., Weber C., Zochios A., Wermter S. (2004) An Associator Network Approach to Robot Learning by Imitation through Vision, Motor Control and Language. *Proceedings of the International Joint Conference on Neural Networks*, Budapest, Hungary, pp. 591-596.

Elshaw M., Wermter S., Watt P. Self-organisation of Language Instruction for Robot Action. *Proceedings of the International Joint Conference on Neural Networks*. Oregon, USA, pp. 22-27, July 2003.

Gallese, V. and Goldman, A. (1998) Mirror neurons and the simulation theory of mind-reading. *Trends in Cognitive Science*, 2(12), 493-501.

Foster, D., Morris, R. and Dayan, P. (2000) A model of hippocampally dependent navigation, using the temporal difference learning rule. *Hippocampus,* 10, 1-16.

Hauk, O., Johnsrude, I. and Pulvermüller, F. (2004) Somatotopic representation of action of action words in human motor and premotor cortex. *Neuron*, 41, 301-307.

Hauk O. and Pulvermüller F., Neurophysiological Distinction of Action Words in the Fronto-Central Cortex (2004) *Human Brain Mapping*, 21, 191-201.

Hinton, G. E., Dayan, P., Frey, B. J. and Neal, R. (1995) The wake-sleep algorithm for unsupervised neural networks. *Science,* 268, 1158-1161.

Keysers, C., Kohler, E., Umiltà, M.A. Fogassi, L., Nanetti, L. and Gallese, V. (2003) Audio-visual mirror neurones and action recognition. *Exp. Brain Res.*, 153, 628-636.

Kohler, E., Keysers, C., Umilta, M., Fogassi, L., Gallese, V. and Rizzolatti, G. (2002) Hearing sounds, understanding actions: Action representation in mirror neurons. *Science*, 297, 846-848.

Pulvermüller, F. (1999) Words in the brain's language. *Cognitive Neuroscience*, 22(2), 253-336.

Pulvermüller, F. (2003) *The neuroscience of language: On brain circuits of words and serial order*, Cambridge, UK: Cambridge University Press.

Pulvermüller, F. (2001) Brain reflections of words and their meaning. *Trends in Cognitive Science*, 5(12), 517-524.

Rizzolatti, G. and Arbib, M. 1998, Language within our grasp. *Trends in Neuroscience*, 21(5), 188-194.

Weber C., Wermter S. (2003) Object Localisation using laterally connected "What" and "Where" associator networks. *Proceedings of the International Conference on Artificial Neural Networks,* Istanbul, Turkey, pp. 813-820.

Weber C., Elshaw M., Zochios A., Wermter S. (2004a) A Multimodal Hierarchical Approach to Robot Learning by Imitation. *Fourth International Workshop on Epigenetic Robotics: Modeling Cognitive Development in Robotic Systems,* Genoa, Italy, pp. 131-134.

Weber C., Wermter S., Zochios A. (2004b) Robot Docking with Neural Vision and Reinforcement. *Knowledge Based Systems*, 12(2-4), 165-72.

Wermter S., Austin J., Willshaw D., Elshaw M. (2001) Towards novel neuroscience-inspired computing. In Wermter S., Austin J. and Willshaw D. (Eds) Emergent Neural Computational Architectures based on Neuroscience. Springer, Heidelberg, Germany. pp. 1-19.

Wermter S., Elshaw M. (2003) Learning robot actions based on self-organising Language Memory. *Neural Networks,* 16(5-6), 691-699.

Wermter S., Weber C., Elshaw M., Panchev C., Erwin H., Pulvermüller F., (2004) Towards Multimodal Neural Robot Learning. *Robotics and Autonomous Systems Journal,* 47(2-3), 171-175.

GROUNDING LANGUAGE IN PERCEPTION: A CONNECTIONIST MODEL OF SPATIAL TERMS AND VAGUE QUANTIFIERS

ANGELO CANGELOSI[A] KENNY R. COVENTRY[B] ROHANA RAJAPAKSE[A]
DAN JOYCE[C] ALISON BACON[B] LYNN RICHARDS[B] STEVEN N. NEWSTEAD[B]

[A] *Adaptive Behaviour and Cognition Group*
School of Computing, Communications and Electronics, University of Plymouth,
Drake Circus, Plymouth PL4 8AA, UK
{acangelosi, rrajapakse}@plymouth.ac.uk

[B] *School of Psychology, University of Plymouth*
{kcoventry, ambacon, lvrichards }@plymouth.ac.uk

[C] *Sobell Dept. of Motor Neuroscience, University College London,*
dan.joyce@virgin.net

This paper presents a new connectionist model of spatial language based on real psycholinguistic data. It puts together various constraints on object knowledge ("what") and on object localisation ("where") in order to influence the comprehension of a range of linguistic terms, mirroring what participants do in experiments. The computational model consists of a vision processing module for input scenes, an Elman network module for the representation of object dynamics, and a dual-route network for the production of object names and linguistic prepositions describing the scene. Preliminary simulations on the prediction of spatial term ratings are presented, and extensions of the model to vague quantifiers and other syntactic categories are considered.

1. Introduction

Spatial language and cognition is concerned with the understanding of the cognitive and linguistic mechanisms affecting the appraisal of spatial tasks. One very important issue is the relative extent to which various constraints on object knowledge ("what" information) and on object localisation ("where" information) influence the comprehension of a range of linguistic terms. Extensive research has been dedicated to the understanding of the effects of "where" information, that is geometric factors related to the relative position and orientation of the objects, in the use of spatial prepositions. For example, Logan and Sadler (1996) have proposed the existence of individual spatial templates for prepositions such as *above* and *under*. These templates specify the level of appropriateness of each spatial term for the different positions (in a 7x7 grid) of a located object with respect to a reference point. Regier (1996) developed a constrained connectionist model of spatial language able to deal with the effects of geometric factors on various spatial terms. More recently, Regier and Carlson (2001) have studied the combined effects of attentional and geometric factors in their Attention Vector Sum (AVS) model. In parallel with these studies on geometric factors, significant evidence has been gathered on the role of a range of "extra-geometric" factors on spatial language comprehension. These extra-geometric factors include the knowledge of object properties and function, and general knowledge of object interaction dynamics. For example, Coventry et al. (2001) have investigated the relative importance of object function and geometric position on the comprehension of *over*, *under*, *above* and *below*. Whether an object is depicted as fulfilling its function or not (e.g. rain shown to fall on an umbrella protecting a man from getting wet, or shown to miss the umbrella and therefore failing to protect the man) is a better predictor of the acceptability of *over* and *under* to describe such scenes than the relative positions of umbrella and man, for example, while conversely the relative positions of umbrella and man are a better predictor the acceptability of *above* and *below* than function.

More recently, Coventry and Garrod (2004) have developed the "functional geometric framework" to explain the integration of the "what" and "where" factors in spatial language. They argue that the application of geometric and extra-geometric routines underlie the comprehension of spatial prepositions. The application of such routines is driven by knowledge of the objects involved in the scene and how those objects typically interact in past learned interactions between those objects. Such a framework is also consistent with growing theoretical arguments and experimental evidence on the role of grounding language in action and perception (Barsalou 1999; Glenberg & Kashak 2002; Glenberg, this volume; Cangelosi, in press). For example, the idea that meaning is constructed as a result of putting together multiple constraints fits with recent work by Glenberg and Kaschak. They have proposed that the meaning of a sentence is constructed by indexing words or phrases to real objects or perceptual analog symbols for those objects, deriving affordances from the objects and symbols and then meshing the affordances under the guidance of syntax. Barsalou (1999) places similar emphasis on perceptual representation for objects and nouns in his perceptual symbol systems account. For Barsalou, words are associated with schematic memories extracted from perceptual states which become integrated into what Barsalou terms simulators. Cangelosi (in press) uses the Cognitive Symbol Grounding framework based on the hypothesis that symbols are directly grounded in internal categorical representations, whilst at the same time having logical (e.g. syntactic) relationships with other symbols. Some symbols, those corresponding to the basic vocabulary, need to be learned and directly grounded, through experience, in the objects they refer to (and the categorical representations that they activate). This is the case, for example, of words learned during early lexical development. Other symbols can then be grounded in representations of categorical entities constructed by the individuals, not necessarily through direct experience and interaction (e.g. when new concepts are learned through deduction). The internal categorical representations, constituting the meanings upon which symbols are grounded, include perceptual, sensorimotor, and social categories, as well as internal state representations. He considers two modelling approaches to symbol grounding: (i) the connectionist approach, based on artificial neural networks for category learning and naming tasks, and (ii) the embodied modelling approach, based on adaptive agent simulations and cognitive robots. These models provide an integrative view of cognitive systems and help our understanding of the relationships between vision, action and language.

In this paper we present a new computational model for spatial language, in which visual scenes are described by selecting the spatial terms that most appropriately describe them (i.e. consistent with human subjects' acceptability ratings). Due to the fact that the model is able to ground the selection of the spatial terms directly into the visual scene, this work helps bridge the gap between theories of meaning which capture meaning in terms of symbol-symbol relations (Landauer & Dumais, 1997) versus those which "ground" language directly in perceptual representation (Regier and Carlson, 2001). In particular, the new model will be able to directly ground the names of objects in visual scenes. Spatial terms, such as the prepositions *over*, *under*, *above* and *below* will also be grounded in information on objects' locations and interaction provided in the input scenes. This will also create a system in which symbol-symbol relationships (e.g. between prepositions and nouns) also permit a prediction of the interaction between objects.

2. The Computational Model

The computational model has a hybrid vision-connectionist architecture (Figure 1) with three main modules: (1) Visual Routines, (2) Elman Networks, (3) Dual-Route Network.

The Visual Routine module uses a series of Ullman-type vision processing routines (Joyce et al. 2002) to identify the constituent objects of a visual scene. It is directly inspired by recent findings and theories of visual object processing, such as Edelman's (1999) feedforward chorus model. The input to the Visual routine module consists of seven frames from 60-second movies (one frame every ten seconds). The scene involves three objects: a located object (e.g. teapot), a liquid substance (e.g. water) and a reference object (e.g. a container such as a cup). The frames are presented to the model, which processes them at a variety of spatial scales and resolutions for object form and motion features yielding a visual buffer. In addition to the basic scale representation, texture, edge and region boundary features are extracted. The processing of each frame results in three arrays of 9x12 activations, representing retinotopically organised and isotropic receptive fields for each of the three objects.

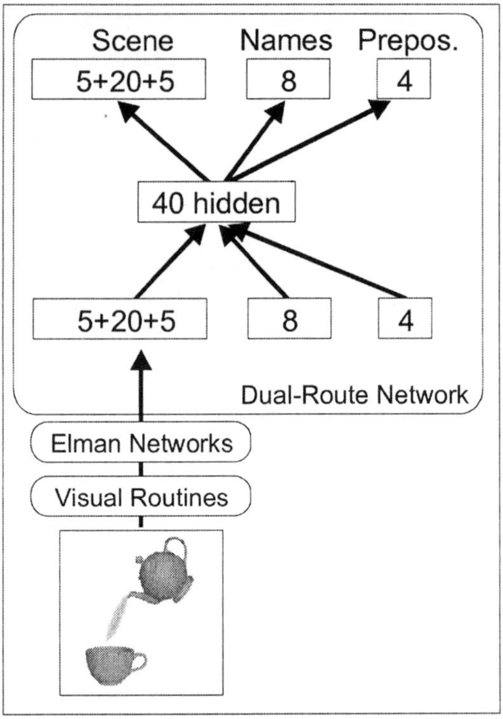

Figure 1 – Dual-route network with input from Visual Routines and the Elman Networks modules.

The Elman Network module utilises the output information from the vision module to produce a compressed neural representation of the dynamics of the scene (e.g. movement of liquid flow between the located and reference objects). Three separate Elman networks are used, respectively for each of the three object types in the scene (located object, liquid, reference objects). Three different networks are currently needed, because the vision module is not currently able to segment and identify objects autonomously. All networks have 108 input and output units. The networks perform a typical prediction task, by producing in output the matrix of the next frame in the 7-frame sequence. The Elman network predicting the liquid flow has 20 hidden units and 20 context (memory) units. The two networks for the located and reference objects only require 5 hidden and context units, because the prediction task is trivial (objects are static). The scope of the Elman network module is to compress the temporal and dynamic information in the scene. This is achieved in the hidden units' activation at the last frame presentation, when the networks have been able to predict all the frames of the scene. The network training protocol consists of a collection of sequences shown to the network in random order (but with fixed sequential order for the seven frames of each scene). Networks are able to correctly learn the prediction task for both training and generalization scenes (Joyce et al., 2003).

The third module consists of a Dual-Route neural network. This architecture combines visual and linguistic information for both linguistic production and comprehension tasks (Plunkett et al., 1992; Cangelosi et a., 2000). This is the core component of the model, as it integrates visual and linguistic knowledge to produce a description of the visual scene. The network receives in input information on the scene through the activation values of the Elman networks' hidden units. It will then produce in output a judgment regarding the appropriate spatial terms describing the visual scene and the names of the objects involved in the scene. The activation values of the linguistic output nodes correspond to rating values given by subjects for the spatial prepositions.

The network used in this simulation had 30 input visual units and 12 input linguistic nodes. The 30 visual nodes corresponded to the 30 hidden units of the three Elman networks (respectively 5 for the network processing the teapot, 20 the liquid and 5 the containers).The linguistic units consisted of the 8 names of objects and the 4 prepositions *over*, *under*, *above* and *below*. The dual route network has 40 hidden nodes. The output layer had the same number and type of units as those in input. As a matter of fact, the dual route network can be considered to be an autoassociator.

3. Simulation and Results

3.1. *Experimental Data and Training of the Network*

Preliminary simulations of the model focussed on the spatial prepositions *over*, *under*, *above* and *below*. For the model training, experimental data from a series of experiments on spatial language were used (Coventry et al., submitted). The visual scenes, used for both the experiment and the model, involved: 3 different reference objects (containers: a plate, a dish and a bowl), 2 levels of closure of container (lid on/closed and lid off/open), 6 different positions for the located objects (in a 3x2 grid position "higher" than the other objects), 2 directions of the located object (left and rightward facing), and 3 functional conditions (the liquid ends in the contained, or misses it, or no liquid is present). This constitutes a stimulus set of 216 movies in total (i.e. 3x2x6x2x3 experimental design). Figure 2 shows the initial and final frames of a sample scene.

The methodology used for these experiments involved the presentation of pictures together with sentences of the form "*The teapot is over the cup*". Participants had to rate the appropriateness of each sentence to describe the movies using a Lickert scale (range from 1 = totally unacceptable to 9 = totally acceptable). Typically, these experiments show effects of geometry and function, together with interactions between these variables and *over/under* versus *above/below* (see also Coventry et al., 2001).

Figure 2. Example of training stimuli. First (left) and last (right) frame of the 60 second movie. This represents a functional scene, i.e. when the liquid finishes in the container.

The subjects' ratings were used to train the dual-route neural network through error backpropagation. An innovative method for backpropagation training was used, where ratings were converted into probability of scene-preposition pairings. This is to avoid a training regime in which the network is assumed to be giving four simultaneous judgments for each scene. Instead human subjects, especially during developmental learning, tend to choose only one preposition to describe a scene (although the choice of a preposition may depend on various contextual factors). To simulate such a learning strategy, the original ratings of each scene-preposition pair were converted into frequency of presentation of a stimulus with an associated localist teaching input. At each learning cycle, only one output preposition unit is set to 1 (maximum activation/choice), whilst the other 3 prepositions units are trained to be inactive. To obtain such a frequency, the original average ratings were scaled and normalized within each scene and within the whole training set. For example, two quite extreme ratings of 7.12 and 3.96 (range 1-10) respectively correspond to presentation frequencies of 28 and 7.

The conversion of ratings into preposition frequencies resulted in a training epoch of 21611 stimuli (scene-preposition pairs) for the dual-route network. Three networks were trained using different initial random weights and different random sets of generalisation test stimuli (10% of scenes that were never used during training). The training parameters included a learning rate of 0.01 and momentum of 0.8, and a total number of training epochs of 500.

3.2. *Results*

The average final error (RMS) for the 30 vision units was 0.008 for both training and testing data, and 0.003 for the 6 output units of the object names. More importantly, for the 4 spatial preposition output units, the error was 0.044 with training data and 0.05 with generalisation data. The error values in the preposition units were calculated off-line by comparing the actual output of the 4 preposition units and the rating data converted to produce the stimulus frequencies (the actual error values used for the weight correction are always higher because they use localist teaching input).

These results clearly indicate that the networks produce rating values similar to that of experimental subjects. The networks' ability to correctly generalize the use (ratings) of the four prepositions with the novel scenes of the generalization set indicates that the model has learned the relationships between the objects involved in their scene, their geometrical and functional properties and the linguistic terms. They also indicate that the training algorithm based on presentation frequency, instead of rating teaching input, works well and provides a psychologically-plausible learning regime.

Additional simulations (Coventry et al., in submission) have also shown that the model can accurately predict new experimental rating data for new scenes. For example, this is the case where only the initial frames are shown and the networks (i.e. the Elman nets) must "replay" the scene and predict its end frame (i.e. where the liquid ends). The Elman network hidden activation values were then passed to the dual route network to generate new rating for the four spatial prepositions. To compare the model's prediction with the performance of real subjects, a new experiment was conducted. Subjects had to predict the end states of the initial frames of movie and rate the appropriateness of the linguistic descriptions. The acceptability ratings for both networks and subjects were overall lower for the predicted scenes rather than the end state scenes. The dual rout network error for the four preposition units was below 5%. In addition, in both networks and subjects the effects of geometry, function and interactions between these variables and *over/under* versus *above/below* were still present, indicating that participants do predict where the liquid will go in order to ascertain the appropriateness of these prepositions.

4. Discussions and Future Work

The preliminary results of this hybrid vision and connectionist model of spatial prepositions demonstrated the feasibility of building a language processing model directly grounded in perception. In addition, the model's ability to replicate accurately subjects' performance in the liquid-flow prediction simulation (and experiment) further supports its psychological validity.

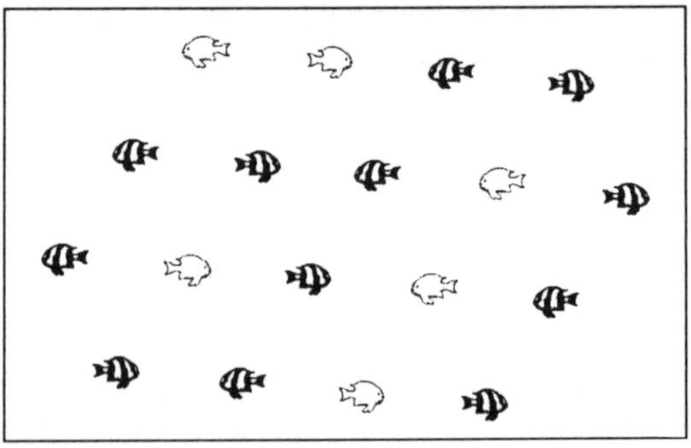

Figure 3. Example of stimuli for quantifier experiments. Subjects (and networks) have to rate sentences such as "There are *several* striped fish.

This model is currently being extended to deal with further linguistic terms, namely vague quantifiers such as *some*, *few*, *a few*, *lots of* (Figure 3). The hypothesis is that this grounded connectionist approach will permit the identification of the main mechanisms responsible for quantification judgment and their linguistic expression. Vague quantifiers like *a few* and *several* exhibit many of the same context effects that have been observed for spatial prepositions. For example, relative size of objects and their expected frequency (e.g. Hormann 1983; Newstead & Coventry 2000) have both been shown to affect the comprehension of quantifiers. "A few cars" is associated with a smaller number than "a few crumbs". In addition, new experiments (Coventry et al., in preparation) have demonstrated that the rating of vague quantifiers is affected by the extent to which objects are grouped together and the degree of spacing between objects. The issue we are exploring with the new model is that these context effects originate from visual processing constraints such that information regarding specific numbers of objects in a scene cannot be derived very easily from visual processing of that scene.

We also hope to be able to extend the model further by considering the direct interaction of the model (e.g. a cognitive agent) with objects in its environment. This is in contrast with the use of a "passive" model that observes interactions between objects through the presentations of movies. This new model might involve the use of a robotic arm (Massera et al., this volume; Cangelosi in press) that builds categorical and linguistic representation of the world by learning to manipulate objects and interact with them. This approach is consistent with the embodied framework in cognitive psychology (e.g. Barsalou's and Glenberg's grounded theories of cognition) and in cognitive systems studies (e.g. Steels, 2003).

Acknowledgments

This research was supported by the UK Engineering and Physical Research Sciences Council (EPSRC Grants GR/N38145 and GR/S26569)

References

Barsalou, L. W. (1999). Perceptual symbol systems. Behavioral and Brain Sciences, 22(4), 577-660.

Cangelosi (in press). Grounding symbols in perceptual and sensorimotor categories: Connectionist and embodied approaches. In H. Cohen & C. Lefebvre (Eds), *Categorization in Cognitive Science*, Elsevier

Cangelosi A., Greco A. & Harnad S. (2000). From robotic toil to symbolic theft: Grounding transfer from entry-level to higher-level categories. *Connection Science*, 12(2), 143-162.

Coventry, K. R. & Garrod, S. C. (2004). *Saying, Seeing and Acting: The Psychological Semantics of Spatial Prepositions*. Psychology Press, Hove.

Coventry K.R., Prat-Sala M., Richards L.V. (2001). The interplay between geometry and function in the comprehension of 'over', 'under', 'above' and 'below'. *Journal of Memory and Language*, 44, 376-398.

Coventry K.R. et al. (in submission). Spatial language and perceptual symbol symbols: Implementing the functional geometric framework.

Edelman S. (1999). *Representation and Recognition in Vision*. MIT Press.

Glenberg A.M., Kaschak M. (2002). Grounding language in action. *Psychonomic Bulletin and Review*, 9(3), 558-565.

Joyce D., Richards L., Cangelosi A., Coventry K.R. (2002), Object representation-by-fragments in the visual system: A neurocomputational model. In L. Wang et al. (Eds), *Proceedings of the 9th International Conference on Neural Information Processing (ICONP02)* IEEE Press.

Joyce. D. W., Richards, L. V., Cangelosi, A. & Coventry, K. R. (2003). On the foundations of perceptual symbol systems: Specifying embodied representations via connectionism. In F. Dretje et al. (Eds.), *The Logic of Cognitive Systems. Proceedings of the Fifth International Conference on Cognitive Modelling*, pp147-152. Universitats-Verlag Bamberg, Germany.

Hormann H. (1983). Then calculating listener, or how many are einige, mehrere and ein paar (some, several and a few). In R. Bauerle, C. Schwarze, & A, von Stechow (Eds.), *Meaning, use and interpretation of language*. Berlin.

Landauer, T., & Dumais, S. (1997). A solution to Plato's problem: the latent semantic analysis theory of acquisition, induction and representation of knowledge. *Psychological Review*, 104, 211-240.

Logan G.D., Sadler D.D. (1996). A computational analysis of the apprehension of spatial relations. In P. Bloom, M.A. Peterson, L. Nadel, M.F. Garrett (Eds.), *Language and Space* (pp. 493-530). Cambridge, Mass.: MIT Press.

Massera G., Nolfi S., Cangelosi A. (in press), Evolving a simulated robotic arm able to grasp objects. In A. Cangelosi A., G. Bugmann & R. Borisyuk (Eds.), *Modelling Language, Cognition and Action: Proceedings of the 9th Neural Computation and Psychology Workshop*. Singapore: World Scientific

Newstead S.E., Coventry K.R. (2000). The role of context and functionality in the interpretation of quantifiers. *European Journal of Cognitive Psychology*, 12(2), 243-259.

Plunkett, K., Sinha, C., Moller, M.F & Strandsry, O. (1992). Symbol grounding or the emergence of symbols? Vocabulary growth in children and a connectionist net. *Connection Science*, 4(3-4), 293-312.

Regier T. (1996). *The human semantic potential: Spatial language and constrained connectionism*. Cambridge Mass.: MIT Press.

Regier T., Carlson L.A. (2001) Grounding spatial language in perception: An empirical and computational investigation. *Journal of Experimental Psychology: General*, 130(2), 273-298.

Steels L. (2003). Evolving grounded communication for robots. *Trends in Cognitive Science*. 7(7), 308-312.

STRUCTURED CONNECTIONIST MODELS OF LANGUAGE, COGNITION AND ACTION

NANCY CHANG, JEROME FELDMAN AND SRINI NARAYANAN
International Computer Science Institute, 1947 Center Street, Suite 600
Berkeley, CA 94704, USA
{nchang,jfeldman,snarayan}@icsi.berkeley.edu

The Neural Theory of Language project aims to build structured connectionist models of language and cognition consistent with constraints from all domains and at all levels. These constraints include recent experimental evidence that details of neural computation and brain architecture play a crucial role in language processing. We focus in this paper on the computational level and explore the role of embodied representations and simulative inference in language understanding.

1. Introduction

The original promise of neural computation derived from its potential to serve as a mechanism for formulating and testing theories that link brain and behavior. Although subsequent work has addressed a variety of specific technical problems, it has devoted relatively less attention toward building general models of intelligent behavior, especially in the domain of language. In this paper we summarize some progress made on this front by the Neural Theory of Language (NTL) project[1], an interdisciplinary group at ICSI and UC Berkeley. We focus on our efforts to build models of language understanding consistent with biological, psychological, linguistic and computational constraints.

The intellectual base for the NTL enterprise is a synthesis of findings from cognitively motivated approaches to linguistics (Lakoff 1987, Lakoff & Johnson 1980) and structured connectionist modeling (Feldman & Ballard 1982), linked by a three-part *Embodiment Hypothesis*:

1. Many concepts are directly embodied in motor, perceptual and other neural structures.
2. All other concepts derive their inferential structure via mappings to these embodied structures.

---

[1] Previously the L_0 group; see also Feldman *et al.* (1996) (a summary of early work), Feldman & Narayanan (2003), and http://www.icsi.berkeley.edu/NTL.

3. Structured connectionist models provide a suitable computational formalism for such neurally grounded representations and mappings.

These ideas have been extended with a *Simulation Hypothesis*—that language understanding exploits many of the same structures used for action, perception, imagination, memory and other neurally grounded processes, and that language provides parameters for simulations using such embodied structures. We explore these hypotheses using methods and convergent constraints from several related disciplines, including direct imaging and behavioral experiments, theoretical developments in linguistics and computation, and system implementations that illustrate important behaviors and may have practical applications as well.

To our knowledge, structured connectionist models (SCMs) comprise the best formalism for capturing the computational relationships between neural activity and complex language and thought. In some cases, the connection between language and neural computation has been direct: Regier's (1996) model of the acquisition of spatial relations terms captured cognitive linguistic phenomena directly with SCMs incorporating aspects of the human visual system. In other cases we have exploited an intermediate computational modeling level as a bridge between descriptions of behavior (cognitive and linguistic phenomena) and the SCM level, which can in turn be mapped to detailed neural architectures. This approach to describing language using neural computation is analogous to other layered abstractions in science, such as using the intermediate level(s) of chemistry to link biology and physics.

As exemplified by the models discussed below, the computational level of description makes use of various formalisms (e.g., feature structures, unification, probabilistic belief networks, Petri nets). But crucially, these formalisms have principled realizations using SCMs. Bailey's (1997) model of the acquisition of hand action verbs, for example, used feature structures encoding parameters for *executing schemas* (or *x-schemas*, described below) capturing many features of motor control; both feature structures and x-schemas also have implementations using SCMs (Bailey 1997, Shastri *et al.* 1998). The intermediate computational level thus allows a convenient level of abstraction while grounding linguistic and cognitive theories in more detailed and biologically inspired models.

This paper focuses on recent efforts to explore the Simulation Hypothesis at the computational level. The basic model is shown in Figure 1. Crucial to our model is the notion of the *construction* as the basic unit of linguistic representation, adopted from cognitive and constructional approaches to grammar (Goldberg 1995, Fillmore & Kay 1999). Constructions are mappings between schematic representations of *form* (phonological schemas) and *meaning* (conceptual schemas), as depicted in the figure. For our purposes, constructions are further designed to supply parameters for simulations.

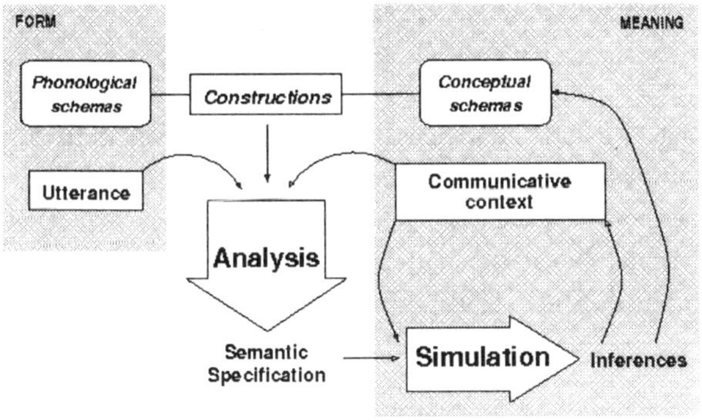

Figure 1. Simulation-based language understanding, consisting of an *analysis* process that draws on *constructions* (pairings of phonological schemas in the form domain with conceptual schemas in the meaning domain) to interpret an utterance in a communicative context. The result of analysis is a *semantic specification* that provides parameters for an active simulation.

Language understanding in this model proceeds in two phases. The *analysis* process determines which constructions are instantiated by a given utterance, drawing on linguistic knowledge (the set of known constructions and their associated forms and meanings), conceptual knowledge (entity and event types), and the current communicative context. The result of analysis is a *semantic specification* that provides the parameters for a *simulation* process that activates embodied conceptual structures represented as x-schemas, resulting in inferences that are then used to update the current context.

The paper is structured as follows. We begin with some linguistic and biological motivation for simulation (Section 2). We then elaborate on each component above: the construction-based grammatical formalism (Section 3); the analysis process linking surface forms with simulations (Section 4); the simulation process and our mechanism for simulation-based inference (Section 5), first proposed as part of a model of metaphorical inference in news stories (Narayanan 1999). Together these models show how embodied conceptual and linguistic structures can be integrated within a simulation-based framework to provide a common representational toolkit for language, cognition and action.

2. Evidence for simulation in language understanding

The idea that language draws on embodied structures is not new. Cognitive linguists (Lakoff 1987, Talmy 2000) have noted that patterns of sensorimotor experience, or *image schemas,* play a central role in semantic representation crosslinguistically. (For example, the prepositions *to* and *into* both involve motion of a trajector relative to a landmark, but *into* also involves containment.) This observation applies not only to literal language but also to abstract and metaphorical language (Lakoff & Johnson 1980). Recent neurobiological and behavioral findings also support the notion that perceptual and motor systems are activated during language understanding. (See also Bergen *et al.* 2004.)

Neurobiological evidence centers on experiments showing that areas of motor and pre-motor cortex associated with specific body parts are activated in response to motor language referring to those body parts. Other studies (Pulvermüller *et al.* 2001, Hauk *et al.* 2004) found that verbs associated with different effectors (e.g., mouth/*chew*, leg/*kick*, hand/*grab*) are processed at different rates and in different regions of motor cortex (i.e., areas responsible for the appropriate mouth/leg/hand motion display more activation). Tettamanti *et al.* (in press) have also shown that passive listening to sentences describing mouth/leg/hand motions activates corresponding parts of pre-motor cortex.

Several psycholinguistic experiments offer behavioral evidence for the automatic and unconscious use of perceptual and motor systems during language use. Richardson *et al.* (2003) found that sentences with visual semantic components can result in selective interference with visual processing. For example, subjects processing sentences encoding upwards motion (e.g., *The ant climbed*) take longer to perform a visual categorization task in the upper part of their visual field. Another experiment showed that subjects performing a physical action in response to a sentence take longer to perform the action if it is incompatible with the motor actions described in the sentence (Glenberg & Kashak 2002). These experiments, along with the imaging results above, provide suggestive evidence for an integrated, multimodal action representation that serves as a common substrate for action, perception and language.

3. Embodied Construction Grammar

The analysis process described in Section 1 relies on Embodied Construction Grammar (ECG) (Bergen & Chang, in press; Chang *et al.* 2002), a computationally precise formalism for representing constructions. As in other construction-based grammars (Kay & Fillmore 1999; Goldberg 1995; Croft 2001), constructions express generalizations linking the domains of form and meaning. Constructions vary in size (from morphemes and lexical items to larger phrasal and clausal units) and specificity (from frozen and partially frozen idioms to more abstract grammatical constructions); and they encompass information that crosscuts traditional levels of linguistic analysis (e.g., phonological, morphological, syntactic, semantic and pragmatic).

ECG is designed to serve as an interface between language and simulation. It includes representations of meaning in terms of embodied structures (called *embodied schemas*) that specify parameters for simulation, and construction representations linking to embodied schemas. Some constructions directly specify which perceptual and motor schemas to deploy (e.g., *to* and *into* from Section 2, or action words like *walk* or *run*), while others specify how to combine subsidiary parameterized representations. The schema and construction formalisms include mechanisms for expressing type constraints, *identification* (or unification) constraints, self-reference, constituency and dependency relations. Computationally, both constructions and schemas are implemented using typed feature structures with unification constraints, organized in a typed inheritance hierarchy. Figure 2 shows some simple example ECG constructions, including the lexical HARRY and RAN constructions and a more complex DIRECTED-MOTION construction. We highlight some of their key properties; see Bergen & Chang (in press) for a more detailed description.

```
construction HARRY
  subcase of REF-EXPR
  form : "Harry"
  meaning : Harry
```

```
construction RAN
  subcase of MOTION-VERB
  form : "ran"
  meaning : Run-Action
    evokes Predication as p
    p.setting.time ← past
```

```
construction DIRECTED-MOTION
  constituents
    x : REF-EXPR
    y : MOTION-VERB
    z : DIR-EXPR
  form
    x_f before y_f
    y_f before z_f
  meaning
    evokes Predication as p
    evokes Dir-Motion as d
    p ↔ y_m.p
    p.schema ↔ d
    d.mover ↔ x_m
    d.means ↔ y_m
    d.direction ↔ z_m
```

Figure 2. ECG example constructions:.lexical constructions linking a specific form with embodied meanings and constraints, and a clausal DIRECTED-MOTION construction associating word order relations with a directed motion scene.

Each construction has **form** and **meaning** components (or *poles)*, and potentially a set of constructional **constituents**. It may be specified as a **subcase of** another construction (thus inheriting constraints). The HARRY construction is a subcase of the REF-EXPR (referring expression) construction, a general construction similar to the traditional NP. HARRY links its form (simplified as the orthographic form "Harry") to a referent whose type is constrained (using a colon) to instantiate the category Harry (of people known by the name "Harry"). Similarly, RAN is defined as a subcase of MOTION-VERB that links its form to the Run-Action schema, corresponding to an x-schema capturing perceptual and motor knowledge about running. It is this association that grounds the RAN construction in terms of its underlying motor-perceptual representations. RAN also constrains its associated evoked predication to take place in the past. The flexible **evokes as** relation allows schemas to be activated as part of the meaning pole without requiring inheritance or constituency.

The DIRECTED-MOTION construction corresponds to expressions describing the motion of some entity in some direction. The construction has internal **constituents** of types REF-EXPR, MOTION-VERB and DIR-EXPR (direction expressions, not defined here but including directional particles like *out* as well as phrases like *to the store*). The construction pairs word order constraints over its constituents' form poles with identification constraints (using a double-headed arrow) over its constituents' meaning poles, using a subscripted f or m on the relevant constituent names to access their respective form and meaning poles. The meaning pole constraints specify the role fillers of the evoked Dir-Motion (directed motion) schema and link these to its associated predication.

These examples represent a particularly simple subset of English, but the formalism has also been applied to more complex crosslinguistic phenomena. Overall, the formalism provides means of linking linguistic structures with embodied simulations, through the processes to which we now turn.

4. Constructional analysis using embodied constructions

Constructional analysis is the process of determining which constructions are instantiated by a particular utterance in a situational context. It is thus analogous to parsing in traditional systems, but also incorporates meaning throughout, and its output is not merely a set of structures (analogous to a parse tree) but also a semantic specification (or *semspec*) that indicates which embodied schemas are evoked by the constructions and how they are related. This semspec serves as input to the simulation process to be described in Section 5.

We briefly illustrate the constructional analysis process for the sentence *Harry ran home* (Figure 3). The individual input words trigger lexical constructions HARRY, RAN and HOME, shown in the center column of the figure as linking their forms (left) with their accompanying semantics (right). (We assume HOME is defined as a special subcase of DIR-EXPR, whose referent is determined during a reference resolution process not described here.) These in turn trigger a search for a larger construction that can account for these items appearing in the specified order; the clausal DIRECTED-MOTION construction (from Figure 2) successfully matches both form and meaning constraints of the input and effects additional bindings among its constituents. The resulting semspec (schemas and bindings on the right side of the figure) indicates which perceptual and motor structures should be activated and how they are related, thus supplying the parameterization needed for the mental simulation of the described scene, i.e., a directed motion event in which the mover (Harry) moves by means of running in the direction of (his) home.

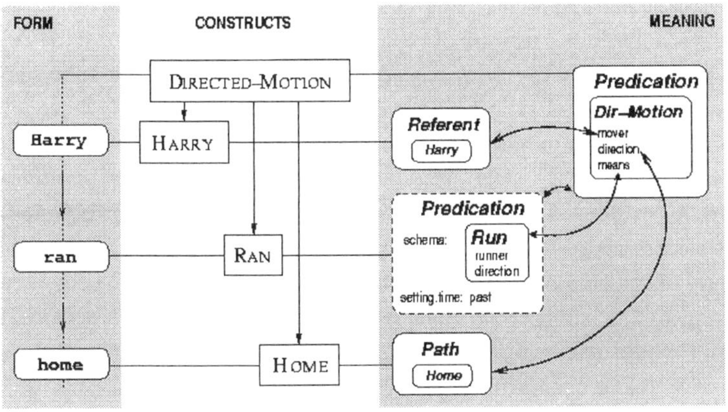

Figure 3: A simplified analysis of *Harry ran home*. The central boxes depict the constructs involved, each linking form (left) and meaning (right). The DIRECTED-MOTION construction asserts an ordering on its constituents' forms and bindings (curved arrows) on their meanings.

A construction analyzer has been implemented and tested on a corpus of English child-directed utterances involving simple motion events (Bryant 2003), including intransitive motion, directed motion, caused motion and dative constructions (Goldberg 1995). The analyzer extends partial parsing and unification-based chart parsing methods to check semantic constraints. It can thus find partial interpretations for sentences not covered by its grammar, and include only semantically coherent analyses in the chart. The best-matching set of constructions is chosen based on coverage of the utterance and semantic fit, favoring analyses with semspecs that are more complete.

5. Simulation-based inference

The semspec produced by constructional analysis specifies parameters for simulations. The model assumes that embodied representations of events and actions can be activated with respect to structures not linked directly to the body in its current physical context. Instead of being carried out directly, these actions can trigger simulations of what would happen in the imagined situation. This ability to simulate or imagine situations is a core component of human intelligence and is central to our model of language.

The computational formalism used in simulation is the *executing schema*, or *x-schema* (Narayanan 1999), a graph-based, token-passing formalism based on stochastic Petri nets that has an SCM interpretation (Shastri *et al.* 1998). X-schemas capture hierarchical structure, sequential flow, concurrency, resource consumption and other aspects of motor control and event structure. X-schemas can be *parameterized* (by the semspec) to execute with variable values and can represent both planned actions and events in the physical world. X-schemas are connected via input/output links; during execution they may make control transitions that modify state, leading to dynamic *activation*, *inhibition* and *interruption* of other x-schemas. X-schema inference consists of changes to the current system state, which is captured in the current graph marking and thus inherently distributed over the entire network of x-schemas and its parameters. A simulation of the semspec in Figure 3 would activate a RUN x-schema whose execution would result in the consumption of tokens representing the runner's energy as well as a change in the runner's location.

Narayanan (1999) describes how the simulation-based model can be applied to metaphorical language as well, which is understood as mapping to underlying embodied meaning, as hypothesized by Lakoff & Johnson (1980). In the computational implementation, the *source* domain of embodied spatial motion is encoded using the *x-schema* formalism described above. This basic model is extended so that simulation-based inferences in the source domain (e.g., resulting from executing a FALL x-schema) are projected via metaphorical mappings (e.g., FALLING IS FAILURE) to license inferences in more abstract *target* domains (e.g., international economics); the sentence "France fell into a recession" may thus be understood to involve an economic failure, where the state of recession is understood as a hole that contains France. The target domain is represented using a temporal belief network, with probabilistic belief update yielding additional inferences; this computational mechanism also plausibly models neural inference by spreading activation.

The system has been tested on narratives from the domain of international economics, using a source domain model of about 100 linked x-schemas in the domains of health and spatial motion and about 50 metaphor maps, all developed using a database of 30 discourse fragments from newspaper stories. The system made a surprising variety of subtle inferences related to abstract plans and actions, involving *goals* (accomplishment, modification, subsumption, concordance, thwarting), *resources* (consumption, production, depletion, level), *aspect* (temporal structure of events), *frame-based* inferences, *perspective*, and *communicative intent*. These demonstrate how embodied representations facilitate inference in both concrete physical and more abstract domains.

6. Conclusions

Our explorations of embodied models of language and cognition demonstrate how a layered methodology has allowed us to reap the insights of linguistic theory while imposing the formal rigor of computational modeling. While we have focused on the computational requirements for realizing a simulation-based model of language understanding, the computational mechanisms employed also have plausible connectionist interpretations, as required by the NTL research paradigm. Additional support for the model's cognitive plausibility comes from two related projects: (1) a model of how children learn their earliest constructions, based on the ECG formalism and analysis process used here (Chang 2005); and (2) a Bayesian model of human online sentence processing assuming construction-based representations (Narayanan & Jurafsky 1998). Overall, we hope that both the methodological approach we have taken and the specific realizations of simulation-based language understanding we have described will bring us closer to fulfilling the promise of neural computation for illuminating links between brain and behavior.

Acknowledgments

We thank the Klaus Tschira Foundation for support of this work and are happy to acknowledge that it describes the collective effort of the NTL group. We also thank an anonymous reviewer for helpful suggestions.

References

Bailey D. (1997). *When push comes to shove: A computational model of the role of motor control in the acquisition of action verbs*. PhD thesis, UC Berkeley.

Bergen B., Chang N., Narayan S. (2004). Simulated Action in an Embodied Construction Grammar. *Proc. 26th Cognitive Science Society Conference.* (pp. 108-113)

Bergen B., Chang N. (in press). Embodied Construction Grammar in Simulation-Based Language Understanding. In Östman J.-O. and Fried M. (Eds.), *Construction Grammar(s): Cognitive and Cross-language dimensions.* Amsterdam: John Benjamins.

Bryant J. (2003). *Constructional Analysis.* M.S. report, UC Berkeley.

Chang N. (2005). *Constructing Grammar: A Computational Model of the Emergence of Early Constructions.* Ph.D. thesis, UC Berkeley.

Chang N., Feldman J., Porzel R., Sanders K. (2002). Scaling Cognitive Linguistics: Formalisms for Language Understanding. *Proc. SCANALU.*

Croft W. (2001). *Radical Construction Grammar.* Oxford: Oxford Univ. Press.

Feldman J.A., Ballard D.H. (1982). Connectionist models and their properties. *Cognitive Science*, 6, 205-254

Feldman J., Lakoff G., Bailey D., Narayanan S, Regier T, Stolcke A. (1996). L_0—The First Five Years of an Automated Language Acquisition Project. *AI Review,* 10, 103-129

Feldman J., Narayanan S. (2003). Embodied meaning in a neural theory of language. *Brain and Language,* 89, 385-392

Glenberg A.M., Kaschak M.P. (2002). Grounding language in action. *Psychonomic Bulletin & Review*, 9, 558-565

Goldberg A. (1995). *Constructions: A Construction Grammar Approach to Argument Structure.* Chicago: Univ. of Chicago Press.

Hauk O., Johnsrude I., Pulvermüller F. (2004). Somatotopic representation of action words in human motor and premotor cortex. *Neuron*, 41(2), 301-7

Kay P., Fillmore C.J. (1999). Grammatical constructions and linguistic generalizations: The What's X doing Y? construction. *Language,* 75/1, 1-33.

Lakoff G. (1987). *Women, fire, and dangerous things.* Chicago: Univ. of Chicago Press.

Lakoff G., Johnson M. (1980). *Metaphors We Live By.* Chicago: Univ. of Chicago Press.

Narayanan S. (1999). Moving Right Along: A Computational Model of Metaphoric Reasoning about Events. *Proc. AAAI.* (pp. 121-128)

Narayanan S., Jurafsky D. (1998). Bayesian Models of Human Sentence Processing. *Proc. 20th Cognitive Science Society Conference.* (pp. 84-90)

Pulvermüller F., Haerle M., Hummel F. (2001). Walking or Talking?: Behavioral and Neurophysiological Correlates of Action Verb Processing. *Brain and Language* 78, 143–168

Regier T. (1996). *The Human Semantic Potential.* Chicago: Univ. of Chicago Press.

Richardson D.C., Spivey M.J., McRae K.,, Barsalou L.W. (2003). Spatial representations activated during real-time comprehension of verbs. *Cognitive Science*, 27, 767-780

Shastri, L., Grannes D., Narayanan S., Feldman J. (1998). A connectionist encoding of parameterized schemas and reactive plans. In Kraetzschmar G.K. and Palm G. (Eds.), *Hybrid Information Processing in Adaptive Autonomous vehicles*. Berlin: Springer-Verlag.

Talmy, L. (2000). *Toward a Cognitive Semantics*. Cambridge, MA: MIT Press.

Tettamanti M., Buccino G., Saccuman M.C., Gallese V., Danna M., Perani D., Cappa S.F., Fazio F., Rizzolatti G. (in press). Sentences describing actions activate visuomotor execution and observation systems.

PREDICTING THE EFFECT OF SLIGHTLY COMPLEX UTTERANCES

MICHAEL KLEIN[A] KENJI DOYA[B]

[A] *Institute for Natural Language Processing, Stuttgart University, Azenbergstr.. 12 Stuttgart,Germany 70174, Germany*
kleinml@ims.uni-stuttgart.de

[B] *Initial Research Project, Okinawa Inst.of Science and Technology, 12-22,Suzaki Gushikawa, Okinawa 904-2234, Japan*
doya@irp.oist.jp

This article introduces a neural architecture which is capable of learning to use language in accordance with desires and to understand the intentions expressed by utterances of other people. We explain the architecture and describe simulation experiments we carried out in multi-agent game environments. We then discuss how compositionality is accomplished in our system. Finally, we relate the components of our architecture with the real brain by pointing out functional imaging and other experiments in which subjects had to carry out tasks related to the capabilities of our model.

1. Introduction

1.1. *Motivation*

It is generally agreed, that language is not - in the first place - to describe the world (Wittgenstein, 1953, Austin 1961). Language - like actions - is used to accomplish certain goals. Speaking involves conceiving of an intention and selecting the relevant information to be expressed for the realization of this purpose (Levelt, 1989). It is, further, generally recognized, that the utterance of a speaker achieves its effects by way of getting the addressee recognize the speaker's intention to achieve it (Grice, 1975). In other words, to really understand an utterance is to understand the speaker's communicative intention.

Based on these observations, we believe that if we really want to understand communication, it is important to understand the relation between the utterances and the goals and, therefore, to start the modeling of language production with the goals and to end with the goals in comprehension. Such a neural approach to language use needs to answer several questions. From a language production point of view such a theory needs to explain: (i) how we know, in what situations, given certain desires, it is useful to speak instead of performing another action or simply doing/saying nothing and when do we say what to whom and (ii) how we transform the goal into an utterance (or a sequence of utterances). From the view point of language comprehension, the standard Grician examples should be explained. But such an approach has to explain even more. To understand the use of language means to know what a certain utterance can achieve and also what kind of goals an agent is likely to have in a certain situation. Therefore, such a framework has to offer some sort of *probability filter* helping to process utterances based on the question: What can this person possibly want from me? Such a mechanism would be useful to process pragmatic, semantic, syntactical, and even phonetic and phonological information in a different manner.

Our paper deals with the question of how the human brain acquires, represents, and processes this aspect of language. Regarding the brain as a goal-directed system (with the strength of such systems being the selection of actions with respect to goals), the goal-directedness of language can be treated as a special case of goal-directed actions and language and meaning are treated as one means of achieving its goals. Thus, the learning of language becomes an optimization process, where the performance of the language learner in handling the environment increases with a higher level of linguistic capacity: the more precise we communicate the more likely are our goals to be achieved.

To accomplish complex learning tasks, such as language, the brain uses subsystems, which differ especially with respect to their learning strategies (Doya,1999), but interact to achieve the global goals of the system. In particular, supervised learning can be used to train forward models (Jordan and Rummelhart, 1992) for prediction, and reinforcement learning can be used to train a value function (Sutton, 1998). In our approach, the forward model is used to predict context dependent utterance effects. It is trained by observing the changes in the environment that utterances of other agents bring about. A special feature of our approach is the use of the forward model in language understanding to compute the communicative intention of the speaker. A value function, which determines how desirable states of the world are, is used along with the forward model for action or utterance selection. Verbal and non-verbal actions can be selected, depending on whether speaking or manipulating the environment directly is more likely to bring about the desired change.

In contrast to the simple forward model described in Klein et al. (2004), in this study, we use a more complex model, which enables the prediction of the effects of compositional utterances.

We present the results of simulation experiments in multi-agent language games. Previous work in multi-agent modeling of communication (Steels, 2001;

Cangelosi et al., 2002; Smith et al., 2003) generally investigate what might be called the descriptive aspects of language. These approaches deal with problems such as how object categories emerge, how these categories are given linguistic labels, or how such labels are shared within a language community. These aspects are essential for understanding how a successful act of reference can evolve within a language community. However, to go beyond the descriptive use of language, as it is the goal of the studies described in this chapter, it is necessary to investigate the relation between utterances and goals. Of course (in a certain sense), the agents in some of those previous studies also have goals – the goal of a successful act of reference or the goal of successfully transmitting a correct description of some state of affairs. But to get a better picture of communication, it is important to move on to the general phenomena of goal-directed communication.

1.2. *Communicative and Other Intentions*

In this article, instead of using the term goal, we would like to distinguish desires and intentions. By intentions we understand a set of properties of states which an agent would like the world to have (or not to have). By intentions we understand the states of the world, which we are actively trying to bring about. [a] There is a wide range of reasons to communicate. Some of these reasons we want to hearer to understand, some of them we do not. The intentions, a speaker want the addressee to understand is called the *communicative intention*.

To make agents generate utterances with communicative intentions and to make agents understand the communicative intentions of another agents, we need agents with intentions, and, in the framework described above, this means we need agents with desires. To create agents with desires we given them a reward. This is because agents in an environment (trying to survive) have a genetically build-in reward system. Certain actions lead to states (helping to survive) which are rewarded. An agent can learn which states are rewarded. Furthermore, they can learn which states lead to those states which are rewarded. By learning this, they gain an internal representation of which states are desirable.

---

[a] Please note that we use the terms intention and desire as technical terms. Also language in this context is used as a technical term meaning no more than acquired symbolic means of communication.

Mathematically this can be approached by a *value function*, i.e. a function mapping every state to a (positive or negative) numerical value:

$$V_\pi(s) = E_\pi\{\sum_{k=0}^{\infty} \gamma_k r_{t+k+1}\} \qquad (1)$$

$$V(s_t) \leftarrow V(s_t) + \alpha[r_{t+1} + \gamma V(s_{t+1}) - V(s_t)] \qquad (2)$$

A speaker uses a certain verbal or non-verbal action, because he desires the effect he expects the action to produce in the present context (according to his experience). The action / utterance which will lead to the state of the world which is desired the most is chosen. We can define the intention of the utterance as the desired state of the world which triggered the action.

1.3. *Internal Models*

To select verbal and non-actions / utterances with respect to how much their effect is desired, agents need some function to predict these effects. Such a function, which predicts a sensation s_{t+1}* (effect) based on the state s_t (context) and the action u_t (utterance) has been be called a *forward model* (Jordan, 1992).

$$s_{t+1} = F(s_t, a_t) \qquad (3)$$

$$a_t = \arg\max_u V(F(s_t, u_t)) \qquad (4)$$

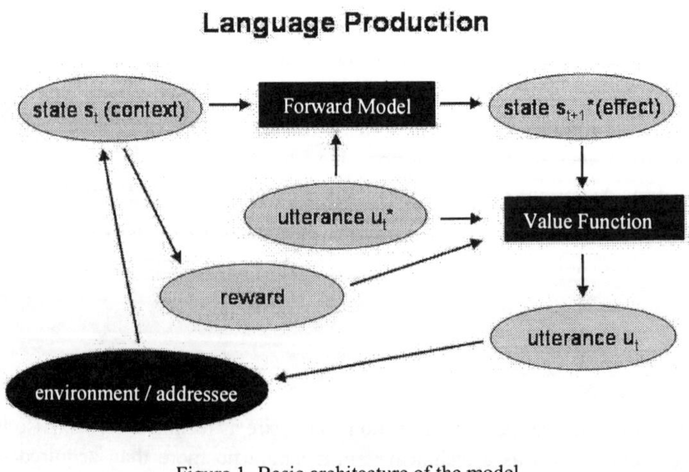

Figure 1. Basic architecture of the model.

2. Training the Value Function

During the training of the value functions, the forward model is rule-based for all possible actions. The task for the agent during these simulations was to learn to behave in an optimal way, employing language and other actions whenever appropriate.

At the very early stages of the training, agents selected *no action*, or senseless actions very often (such as donating objects to other players without being asked). Sensible, but suboptimal actions did occur in the intermediate stage of training (figure 2). No more suboptimal action could be detected after training. Agents used language and selected other actions in an optimal way.

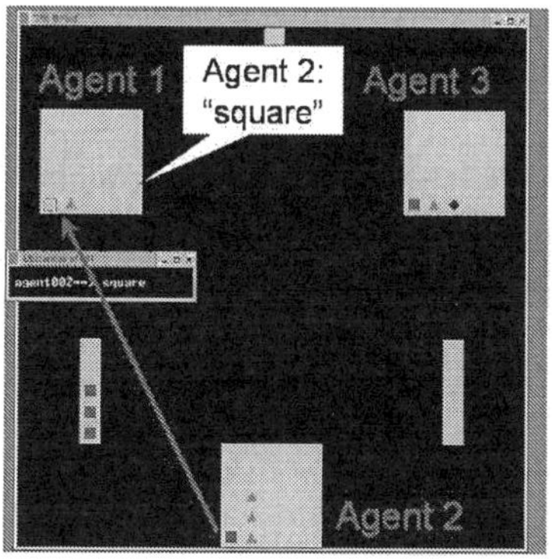

Figure 2. Suboptimal action in early stages of training the value function. The agent 1 asks for the square, which is, of course, useful for him. But to harvest the square tree would be the optimal action in this situation.

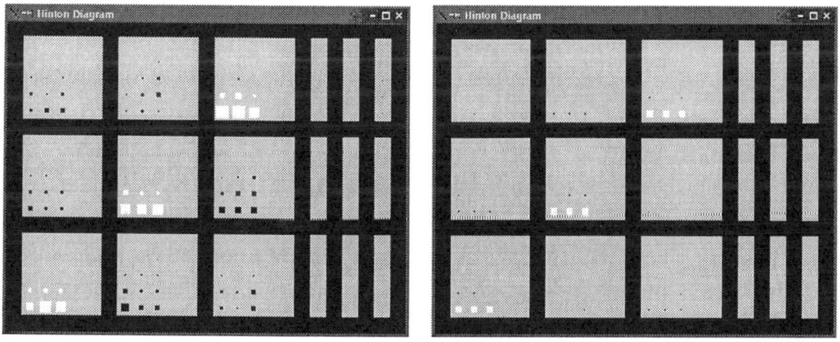

Figure 3. Hinton diagram of the weights of the value function after a training of 20 Million time steps. The left graph shows the weight for $\gamma = 0.9$. The right graph shows the weights for $\gamma = 0.1$. The weights allow optimal action in the first, but not in the second case.

3. Training the Forward Model

3.1. *Non-Compositional Utterances*

To train the value functions, so far, we used rule-based forward models for verbal actions as well, but, ultimately, our goal is to have a model which acquires the use of language. Therefore, the forward model needs to learn the context-dependent consequences of utterances (the context of each utterance is the game-state). Forward models for language use can be trained, because utterances used by others speakers are experienced to have effects on the observable world. These effects of utterances are dependent on the context. Effects on the observable world are indirect and are achieved by a more direct effect on unobservable states, such as the mental states of the addressees. Experienced context-dependent effects of utterances are used to train the forward model. Such a forward model can then act as a function, mapping context configurations and utterances to effects. The forward-model is implemented by a single-layer perceptron trained with supervised learning (Eq. (5)-(7)). It maps verbal utterances and game-states onto game-states.

$$e_k = y_k * - y_k \tag{5}$$

$$\delta w_{ik} = \alpha e_k x_i \tag{6}$$

$$w_{ik} \leftarrow w_{ik} + \delta w_{ik} \tag{7}$$

An addressee understands an utterance by using his forward model applied to the context and the utterance with which he was addressed, in other words, he maps the utterance and the context on the intention of the speaker.

During training of the forward model, besides the language learner, there are also teachers agents needed. Teacher-agents use a rule-based dialogue system to produce and understand utterances. Teachers have no interest in the learner-agents learning of language: they do not perform any actions or utterance with the goal to teach language to the learners, but learner-agents can learn language by observing the language use of the teacher-agents

In such an environment, an agent, equipped with an optimal value function and a rule-based model for non-verbal actions, can learn to use language to achieve his goals by expressing his desires and to understand the desires of other agents.

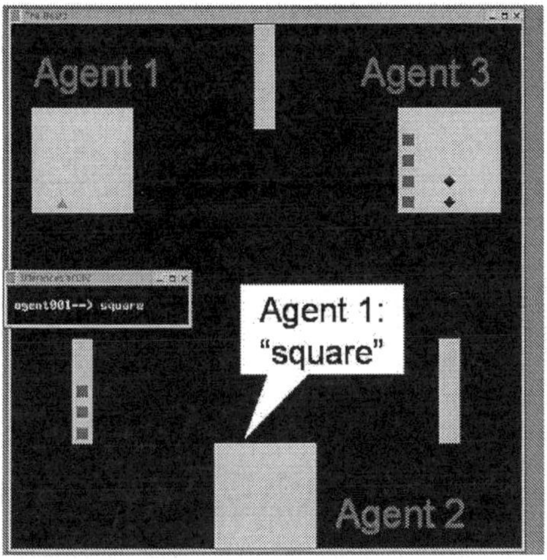

Figure 4. Common error in early stages of language learning. Agent 2 (the language learner) asks agent 1 for the square, which agent 1 does not have. Obviously, the language learner has not yet learned the conditions for successful use of the utterance "square".

We could demonstrate that after training the reactions of the language are appropriate for the utterances the learner is addressed with. Also, the learner's utterances are in accordance with the situation and the desire of the agent. No difference between teachers and learners could be detected.

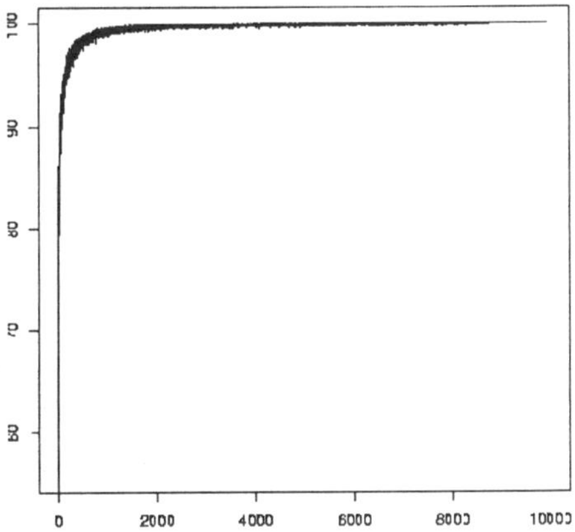

Figure 5. This is the time course of the prediction error for the prediction of the effect of verbal actions. The graph shows the percentage of correct predictions for 5 runs and 10 Million time steps.

4. Complex Utterances

4.1. *Method*

So far, the model is restricted to one word holo-phrases. In human language very few utterances might actually be used a holo-phrases, but most expressions are compositional in nature.

To be able to process compositional utterances we designed a more complex forward model in the following way: In a first step, a feature representation (color, form, movement) is extracted from the utterance, which is used to identify and locate the relevant object in space. Attention is directed on this particular object and a change of this object in the context is predicted. With this forward model, agents are able to understand and produce sentences which they have never heard before, i.e. they derive the meaning of the complex utterance from the meaning of its parts.

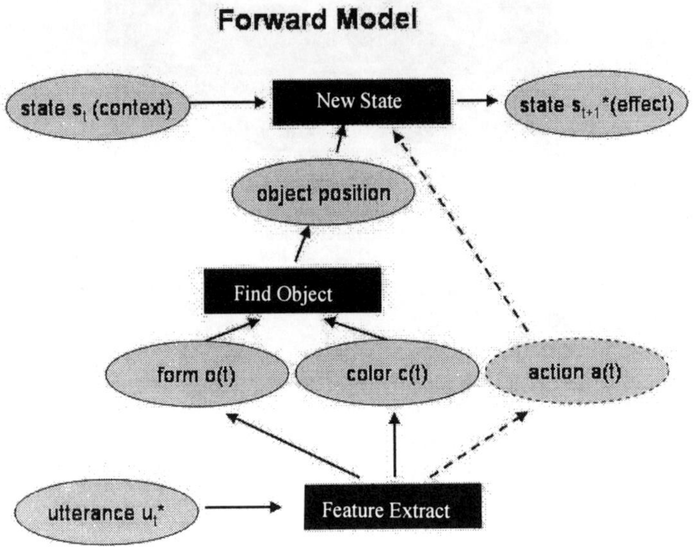

Figure 6. This is the compositional forward model. Like the simple forward model it maps a state (context) and an utterance on the next state.

We could show that agents learn to use the compositional utterances to express their desires. They also learn to understand the desires of other agents expressed by compositional utterance. Although, learning takes significantly longer than in the simulation with the non-compositional model, after training the agents show correct reaction when they are addressed and produce the appropriate utterance in every situation.

5. Discussion

Brain activity correlating to reward prediction (i.e. the value function) was found in the striatum in fMRI studies (e.g. Tanaka, 2004). Also, reward predictive modulation of striatal neuron firing was shown (e.g. Kawagoe, 2004). Neurophysiological and imaging data supports the idea of the cerebellum functioning as a forward model (Kawato, 1999). However, the internal models found in these studies, so far, are internal models of motor control. An internal model required for appropriate language use (predicting context-dependent utterance effects) is very unlikely to be represented in the cerebellum alone, but is likely to involve cortical language and theory of mind areas.

The architecture we chose for the forward model which has to deal with the compositional utterances is designed for one type of utterance only: the request for a certain object. The only compositional aspect so far is the description of the object ("red triangle", "green square"). The action, which is performed as a consequence of the utterance is always to give this object away. However, any other kind of action could easily be integrated in the framework. An word describing the requested actions would have to be added to the utterance ("give red diamond", "eat red square", etc.). From this action-denoting word, a feature representation could be extracted in a manner similar to the way in which object features are extracted in the current model.

A more difficult, but also more interesting extension of this model would be to make parts of the agent's states (or other aspect of the simulation environment) unobservable for the other agents. Then two more types of speech acts (questions and answers) could be introduced into the framework. Such a simulation would possibly shed some more light in the representation of the relation between referential and usage aspects of meaning in the human brain.

6. Conclusion

Inspired by the learning mechanisms of the brain, we were able to develop a model of language learning, meaning representation and communication which is capable of learning language, producing simple utterances in accordance with its goals, and understanding the intentions expressed by utterances of other agents. We believe, that the value function / forward model approach offers an interesting perspective on communication and the representation of meaning and linguistic knowledge in the human brain.

Acknowledgments

This work was supported by the German Academic Exchange Service (DAAD), the German Research Foundation (DFG), and the National Institute of Information and Communications Technology of Japan (NICT).

References

Austin J.L. (1961). *Philosophical Papers*. Oxford University Press.

Cangelosi A., Greco A., Harnad S. (2002). Symbolic Grounding and the Symbolic Theft Hypothesis. In Cangelosi A. and Parisi D., *Simulating the Evolution of Language.* London: Springer (pp. 191-210).

Doya K. (1999). What are the computations of the cerebellum, the basal ganglia and the cerebral cortex? *Neural Networks*, 12, 961-974.

Grice H.P. (1975). Logic and Conversation. In Cole P., Morgan J. (Eds.), *Speech Acts*. New York: Academic Press (pp. 41-58).

Jordan M., Rummelhart D.E. (1992). Forward models: Supervised learning with a distal teacher. *Cognitive Science*, 16, 307-354.

Kawagoe R., Takikawa Y., Hikosaka O.(2004).Reward-predicting activity of dopamine and caudate neurons-a possible mechanism of motivational control of saccadic eye movement. *Journal of Neurophysiology*, 91(2), 1013-24.

Kawato M. (1999). Internal Models for motor control and trajectory planning. *Current Opinion in Neurobiology,* 9, 718-727.

Smith K., Kirby S., and Brighton H. (2003) Iterated Learning: a framework for the emergence of language. *Artificial Life*, 9(4):371--386.

Klein M., Kamp H., Palm G., Doya K. (2004). Expressing and Understanding Desires in Language Games. In Pollack J., Bedau M., Husbands P., Ikegami T. and Watson R.A. (Eds.), *Proc. of the Ninth International Conference on the Sim. and Synthesis of Living Systems*. Boston: MIT Press (pp. 170-175).

Levelt W.J.M. (1989). Speaking - From intention to articulation. MIT Press.

Steels L. (2001). Language games for autonomous robots. *IEEE Intelligent Systems*, 16 (5), 16-22.

Sutton R.S. (1988). Learning to Predict by the Methods of Temporal Differences. *Machine Learning*, 3, 9-44.

Tanaka S.C., Doya K., Okada G., Ueda K., Okamoto Y., Yamawaki S. (2004). Prediction of immediate and future rewards differentially recruits cortico-basal ganglia loops. *Nature Neuroscience*, 7(8), 887-893.

Tomasello M. (2003). *Constructing a language: A usage-based theory of language acquisition*. Cambridge, MA: Harvard University Press.

Wittgenstein L. (1953). *Philosophical Investigations*. Blackwell.

AN ASSOCIATIVE MODEL OF CORTICAL LANGUAGE AND ACTION PROCESSING

ANDREAS KNOBLAUCH[BA] HEINER MARKERT[A] GUENTHER PALM[A]

[A] *Department of Neural Information Processing, University of Ulm*
Oberer Eselsberg O27, 89069 Ulm, Germany
{knoblauch, markert, palm}@neuro.informatik.uni-ulm.de

[B] *Speech and Language Group, MRC Cognition and Brain Sciences Unit*
15 Chaucer Road, Cambridge CB2 2EF, England
andreas.knoblauch@mrc-cbu.cam.ac.uk

The brain correlates of words and their referent actions and objects appear to be strongly coupled neuron ensembles or assemblies distributed over defined cortical areas. In this work we describe the implementation of a cell assembly-based model of several visual, language, planning, and motor areas to enable a robot to understand and react to simple spoken commands. The essential idea is that different cortical areas represent different aspects of the same entity, and that the long-range cortico-cortical projections represent hetero-associative memories that translate between these aspects or representations.

1. Introduction

The brain correlates of words and their referent actions and objects appear to be strongly coupled neuron ensembles in defined cortical areas (Pulvermuller, 1999). Being one of the most promising theoretical frameworks for modelling and understanding the brain, the theory of cell assemblies (Hebb, 1949; Palm, 1990) suggests that entities of the outside world (and also internal states) are coded in overlapping neuron assemblies rather than in single ("grandmother") cells, and that such cell assemblies are generated by Hebbian coincidence or correlation learning. One of our long-term goals is to build a multimodal internal representation using several cortical areas or neuronal maps, which will serve as a basis for the emergence of action semantics, and to compare simulations of these areas to physiological activation of real cortical areas.

In this work we describe a cell assembly-based associative model of several visual, language, planning, and motor areas implemented on a robot to understand and react to simple spoken commands (cf. Fay et al., 2004). The task is to find certain fruits in a complex visual scene according to spoken or typed commands. This involves parsing and understanding of simple sentences, relating the nouns to concrete objects sensed by the camera, and coordinating motor output with planning and sensory processing.

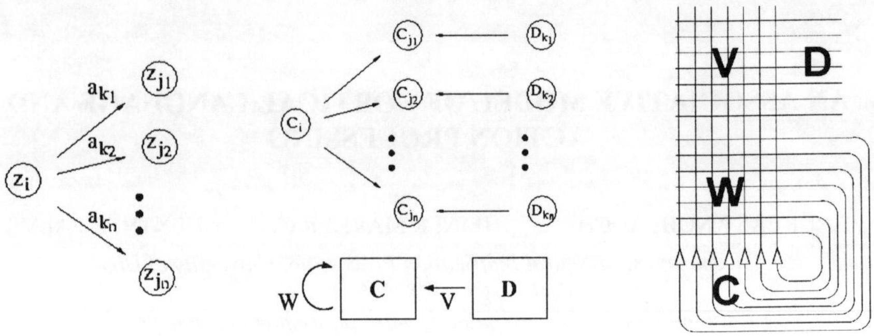

Figure 1. DFA (left), neural network (mid), and cell assemblies (right).

2. Language, finite automata, neural networks and cell assemblies

In this section we briefly review the relation between regular grammars, finite automata and neural networks. Regular grammars can be expressed by generative rules $A \rightarrow a$ or $B \rightarrow bC$ where upper case letters are variables and lower case letters are terminal symbols from an alphabet Σ.

Regular grammars are equivalent to deterministic finite automata (DFA). A DFA can be specified by $M=(Z,\Sigma,\delta,z_0,E)$ where Z is the set of states, Σ is the alphabet, $z_0 \in Z$ is the starting state, $E \subseteq Z$ contains the terminal states, and the function $\delta:(Z,E) \rightarrow Z$ defines the state transitions. A sentence $s=a_1a_2...a_n \in \Sigma^*$ is well formed with respect to the grammar if $\delta(...\delta(\delta(z_0,a_1),a_2),...,a_n) \in E$.

DFAs can be simulated by neural networks: For this it is sufficient to specify a simple model of recurrent binary neurons by $N=(C,D,W,V,c_0)$, where C contains the local cells of the network, D is the set of external input cells, W and V are binary matrices specifying the local recurrent and the input connections (Fig.2). The network evolves in discrete steps, where a unit is activated ($c_i(t)=1$) if its potential $x_i(t)=(Wc(t-1) + Vd(t-1))_i$ exceeds threshold Θ_i, and deactivated ($c_i(t)=0$) otherwise. A simple emulation of the DFA requires one neuron c_i for each state z_i, one neuron d_k for each input symbol a_k, synaptic connections $w_{ij}=d_{kj}=1$ for each state transition $(z_i,a_k) \rightarrow z_j$ and thresholds $\Theta_i=1.5$. If at the beginning only neuron c_0 is active the network obviously simulates the DFA. A biologically more realistic model would interpret the nodes in Fig.1 not as *single* neurons but as groups of nearby strongly interconnected neurons, i.e., local cell assemblies. This architecture has two additional advantages: First, it enables *fault tolerance* since incomplete input can be completed to the whole assembly. Second, overlaps between different assemblies can be used to express hierarchical relations between represented entities.

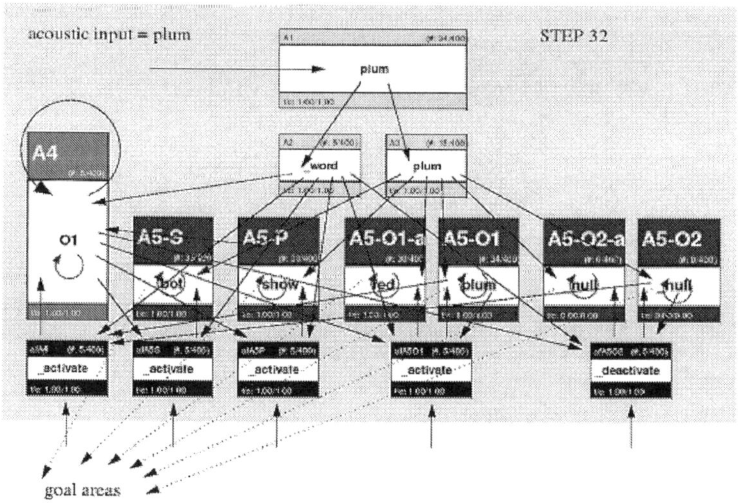

Figure 2. Architecture of cortical language areas.

3. Cortical language model

The language system consists of a standard HMM-based speech recognition system on the word level, and a cortical language system which can analyse streams of words detected with respect to simple regular grammars (Knoblauch et al., 2004). Fig. 2 shows the 15 areas of the language system. Each area is modelled as a spiking associative memory of 400 neurons (Knoblauch & Palm, 2001). Binary patterns constituting the neural assemblies are stored auto-associatively in the local synaptic connections by Hebbian learning. The model can roughly be divided into three parts. (1) Auditory cortical areas A1, A2 and A3. (2) Grammatical areas A4 and A5-X. (3) Simple "activation fields" af-X that coordinate the activation or deactivation of the grammar areas.

When processing language, first auditory input is represented in area A1 by primary linguistic features (such as phonemes), and subsequently classified with respect to function in A2 and content in A3. The main purpose of area A4 is to emulate a DFA in a similar way as the neural network in Fig. 1. Each node corresponds to an assembly representing a grammatical state, and each edge corresponds to a state transition stored in delayed recurrent hetero-associative connections of area A4. E.g., processing of a sentence "Bot show red plum" would activate the state sequence S→Pp→OA1→O1→ok_SPO corresponding to expectation of processing of a subject, a predicate, an object or attribute, and finally an object. If the sentence was not well formed with respect to the grammar, then the sequence terminates in an error state.

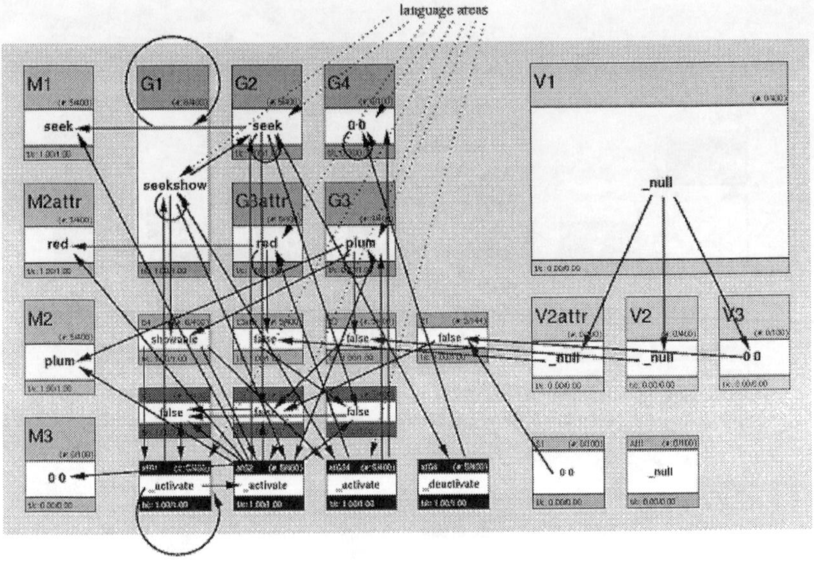

Figure 3. Architecture of the cortical action model.

Finally, the robot has to "understand" the sentence by transforming the word stream into an action representation. This is the purpose of areas A5-X which correspond to different grammatical roles. A5-S represents the subject "bot", A5-P the predicate "show" and A5-O1a/O1 the object "red plum" (Fig. 2).

4. Action processing

Our system for planning, action, and motor processing can be divided into three parts (Fig 3). (1) The planning/goal areas represent the robot's goal after processing a spoken command. Sequence assemblies in area G1 represent lists of actions necessary to complete a task, e.g., seek→point for the command "show object". Area G2 represents the current subgoal, and the remaining GX areas properties of involved objects. (2) The "motor" areas MX represent the motor command necessary to perform the current subgoal, and also control the low level attentional system. (3) Again there are "activation" fields (afX) and additionally "evaluation fields" that compare representations of different areas.

For illustration we describe the task „bot show red plum", where the robot has to point onto a red plum in the vicinity. First the robot has to understand the command as described in section 3 which activates the A5-representations. This information is routed to the goal areas where the first part of the sequence (seek→point) gets activated in G1. Similarly, object information is routed to areas G2, G3, G3attr. Since the plum's location is unknown, there is no activity in area G4. After checking semantics, corresponding assemblies in the motor areas are activated. This activates the attentional system and initiates the robot to seek the plum. Finding the plum activates corresponding visual assemblies in areas VX. The evaluation fields detect this and initiate areas G1, G2 to switch to the next state "point" of the action sequence. The robot will then adjust its "finger position" represented in area S1 in order to point to the plum. The matching of the positions will be detected by the evaluation fields and this eventually activates the final state in G1.

5. Conclusion

We have described the implementation of a cell assembly-based model of cortical language and action processing on a robot (cf. Knoblauch et al., 2004; Fay et al., 2004). The model consists of about 40 neuron populations each modelled as a spiking associative memory containing many "local" cell assemblies stored in local auto-associative connections (Knoblauch & Palm, 2001). The neuron populations can be interpreted as different cortical and subcortical areas, where it is a long term goal of this project to establish a mapping of our "areas" into real cortex (Pulvermuller, 1999).

Acknowledgments

This work has been supported by the MirrorBot project of the European Union.

References

Hebb D.O. (1949). *The organization of behavior. A neuopsychological theory.* New York: Wiley.

Fay R., Kaufmann U., Knoblauch A., Markert H., Palm G. (2004). Integrating object recognition, visual attention, language and action processing on a robot using a neurobiologically plausible associative architecture. In: *Proceedings of the AI-Workshop on NeuroBotics 9/04*, University of Ulm

Knoblauch A., Palm G. (2001). Pattern separation and synchronization in spiking associative memories and visual cortical areas. *Neural Networks, 14, 763-780*

Knoblauch A., Fay R., Markert H., Kaufmann U., Palm G. (2004) Associating words to visually recognized objects. In Coradeschi S. and Saffiotti A. (Eds.), *Anchoring symbols to sensor data. Papers from the AAAI workshop, Technical Report WS-04/03.* Menlo Park, CA: AAAI Press. (pp. 10-16)

Palm G. (1990). Cell assemblies as a guideline for brain research. *Concepts in Neuroscience, 1, 133-148*

Pulvermuller F. (1999). Words in the brain's language. *Behavioral and Brain Sciences, 22, 253-336*

Categorization and Language

CARVING NATURE AT ITS JOINTS AND CARVING JOINTS INTO NATURE: HOW LABELS AUGMENT CATEGORY REPRESENTATIONS

GARY LUPYAN

Carnegie Mellon University Department of Psychology and
Center for the Neural Basis of Cognition
Pittsburgh, PA 15213 USA
glupyan@cnbc.cmu.edu

Using words to label categories is a true human universal. In addition to their public function in communication, labels may also serve private functions in shaping how concepts are represented. The present work explored the effects of assigning category labels on perceptual representations. A connectionist simulation is presented that examines the effects of labels on learning different types of categories. It is found that labels can augment perceptual information, and play an especially important role in shaping representations of entities whose perceptual features alone are insufficient for reliable classification.

1. Introduction

The phrase carving nature at its joints goes back to the ancient Greeks, who noticed that properties in the world are not equally distributed, but rather come in clusters. For instance, things that have beaks also tend to fly, so the property [has beak] is more likely to occur with [flies] than not. In forming a concept [bird], we pay heed to this correlation. But concept formation is not just a passive process of observing correlations. As the only species to evolve language, the human process of concept formation is bi-directional—not only do we learn to represent the natural joints of nature, we continuously carve our own joints using words. The present work explores through a connectionist simulation what happens when, in learning to form a concept, we are presented with not just the correlated perceptual properties of a concept like [bird], but also with the corresponding category label "bird."

The most obvious function of words is to communicate information between individuals. One provocative consequence is that once the meaning of words is shared in a community, their use allows for the learning of novel combinations of concepts through manipulation of labels rather than trial-and-error learning (Cangelosi, Greco, & Harnad, 2000). But in addition to this public function of labels, they may also have a private function. Words may help to cohere concepts, reinforcing the correlated perceptual features that comprise them, and aid in contrasting similar, but functionally different concepts. The goal of the first simulation was to serve as a broad testing platform for the argument that labels have an effect on representation of categories and that the effect depends on how well-formed the categories are. Specifically, the presence of labels may help in categorizing exemplars in high-variability categories, but not in low-variability ones. Consider categorizing apples versus strawberries on the one hand, and tables versus chairs on the other. Apples and strawberries are members of what can justifiably be called *low-variability* categories. These categories have high internal coherence (Gentner, 1981), having both a high within-category similarity and a high between-category contrast. On the other hand, categorizing objects like chairs and tables is made more difficult by both their high perceptual within-category variability (tables and chairs can take a wide variety of shapes) and the fuzzy boundary between the categories (some tables are quite chair-like, and vice-versa). Including labels in one's experience of such *high-variability* categories may facilitate accurate identification of exemplars. On the other hand, including labels as part of training on low-variability categories is not predicted to make a difference due to the perceptual well-formedness of such categories. If the effect of labels is to increase the internal coherence of a category (Kotovsky & Gentner, 1996; Kersten & Smith, 2002), then in doing so, representations of exemplars from one category will be made more similar, while representations of items from separate categories will be made more distinct.

A secondary aim of this simulation was to further explore the consequences of introducing labels: if labels improve categorization accuracy, does this improvement come at a cost? A basic tenet of categorization is that placing an item into a category involves highlighting some properties while abstracting over others (e.g., Pothos, 2004). For instance, color is probably unimportant to the category *chair*, and so should be suppressed in the representation of a particular chair. However, color is important for the category *banana*, and so should be highlighted in its representation. If resources are abundant and each item can be memorized, abstraction and highlighting of particular dimensions may speed responses, but is otherwise not crucial. If one can afford to encode accurately every observable property of an object, categorization is not necessarily compromised by representing all the dimensions. However, when resources are limited, preventing memorization of individual exemplars (as was the case in the current simulation), the properties that are important to a category will be represented at the expense of properties irrelevant to the category. Making a representation more categorical (i.e., moving it closer to the prototype) will therefore lead to better categorization (smaller categorization errors), but at the same time, greater difficulty with reconstructing the exact item with all its original features (i.e., larger reconstruction error). Furthermore, this effect should be larger for high-variability categories since it is for these categories that labels are predicted to have the greatest effects.

2. Method

2.1. *Network Architecture*

The simulations used a standard feedforward network configured as an auto-associator. This particular framework is used in the present model for several reasons. First, architectures of this type are quite familiar, commonly used to model category learning (e.g., Plunkett, Sinha, Moller, & Strandsby, 1992; Mareschal, French, & Quinn, 2000). Auto-associators perform a type of principal component analysis, first picking up on the broad distinctions between exemplars, and then on the finer differences. With training, representations in the hidden layer come to reflect the category structure present in the training set. Because these networks are self-supervised, no external teaching signal is necessary—the structure of the training set itself guides learning.

The network was configured with the following set of connections: A 30-unit input layer connected to a 6-unit hidden layer that connected to a second 6-unit hidden layer, which in turn connected to a 30-unit output layer.[i] In the labeling conditions, the second hidden layer also connected to a 4-unit labeling output layer. Because labels were only presented as outputs, they could have an effect only during training. Labels were not included as inputs because it was discovered that doing so did not change the results qualitatively, while greatly complicating the training procedure. If labels were to have an effect on representations, it was having to represent the labels as outputs that would make the difference.

Due to the difference between the size of the two output layers (30 versus 4 units), the error derivatives of the labeling output layer were multiplied by a constant factor to 10. This approximately equated the contributions of the two layers.

2.2. *Stimuli*

The training corpora consisted of 30-unit binary patterns. First, four category prototypes were generated with each one overlapping on 24 out of 30 bits with the other prototypes. Training items were generated by randomly flipping a set number of bits: 5 out of 30 for the *low-variability* training set and 10 out of 30 for the *high-variability* set. Each category consisted of 8 exemplars for a total of 32 (4*8) exemplars per training set. The category labels were modeled by 4-bit orthogonal vectors (e.g., 1000 for all examples from the first category, 0100 for the second category, etc.).

2.3. *Training procedure*

The networks were trained in Lens v.2.4 with cross entropy error using "Doug's Momentum," a modified version of steepest gradient descent (Rohde, 1999). The learning rate was set to 0.1, and momentum to 0.9. Training was done in batch mode—weights were updated only after presenting all the examples.

On each training trial, the outputs of the input layer were clamped to a randomly selected example. Since the network's task was simply to auto-associate the inputs, the targets of the main output layer were identical to the network inputs. In the labeling conditions, the second output layer provided the additional targets corresponding to the category labels. Twenty randomly-initialized networks were run in each of the four condition (*labeled low-variability*, *unlabeled low-variability*, *labeled high-variability,* and *unlabeled high-variability*).

2.4. *Testing procedure*

Testing was conducted by running a forward pass through the network for each item after training. Results are reported for 200 epochs of training. Three measures were computed to gauge the networks' performance. First, an accuracy measure was calculated to answer the question "is the network's response closest to the correct category?" This was done by computing the Euclidean distance between the produced output and the 4 category prototypes. The network's response was marked correct if the minimum distance corresponded to the correct category. The responses were translated to a percentage correct measure over the 32 testing items (8 for each category). The second measure was the cross-entropy error of the training items, and answers the broad question "how good is the network at representing the trained patterns?" Since the networks' hidden layers were kept deliberately small to prevent the network from memorizing the exact patterns, this error is not expected to be close to 0. The third measure corresponded to the error of the mapping from the trained or novel items to the category prototype from which the item was derived. This answers the question "how categorical is the network's representation?" In other words, how close to the correct category is the network's response?

3. Results

The simulation results confirmed both predictions. Labels did not affect categorization performance for the low-variability corpus, $t(38) = 1.16$, ns, with performance very close to 100% regardless of labels. Labels did affect categorization for the fuzzier category structure of the high-variability corpus $t(38)=22.57$, $p<.0001$, creating a highly significant interaction between labeling condition and category structure as a predictor of categorization accuracy, $F(3,76)=487.18$, $p<.0001$. Performance was very close to 100% for the high-variability corpus when trained with labels, but dropped to an average of 55.6% when trained without labels (chance=25%; see Figure 1).

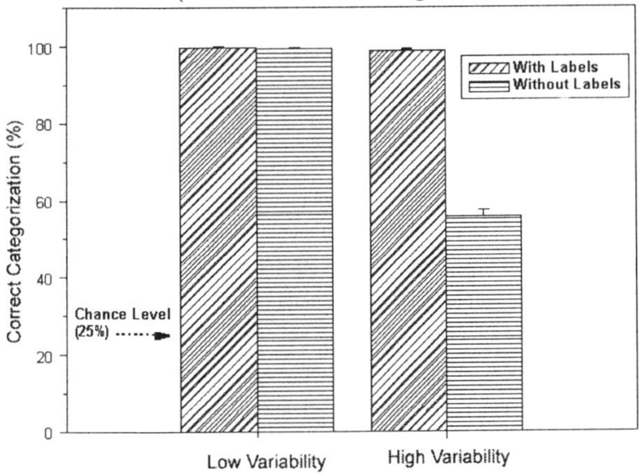

Figure 1: Labels improve categorization accuracy for high-variability categories, but not low-variability ones.

A parallel pattern of performance was observed when the network was tested on novel items. When tested on previously unseen items, performance was predictably lower overall, but labels still did not affect categorization accuracy for the low-variability corpus $t(38)=0.6$, while significantly aiding performance in the high-variability corpus $t(38)=7.06$, $p<.0001$, with correct categorization averages of 54.75% and 37.20% with and without labels, respectively. The interaction remained highly significant: $F(3,76)=368.27$, $p<.0001$. As a way to ensure that the null effect of labels on the low-variability corpus was not due to a ceiling effect, an additional analysis examined performance over time. Figure 2 shows that for the low-variability condition, labels did not affect performance at any point during training $t(24)=.65$. For the high-variability corpus, labels improved categorization accuracy throughout training both for the trained exemplars $t(24)=7.30$, $p<.0001$) and for the novel items $t(24)=8.03$, $p<.0001$.

The second measure looked at reconstruction error—how efficient was the network at auto-associating the input patterns. Figure 3 shows that exemplars from the high-variability categories were more difficult to reconstruct accurately $F(1,78)=553.24$, $p<.0001$. More importantly, there was a differential effect of labels on the two corpora $F(3,76)=550$, $p<.0001$. The presence of labels did not affect reconstruction accuracy for the low-variability set, $t(38)=-.15$, but led to a significant increase in the error for the high-variability set $t(38)=14$, $p<.0001$.

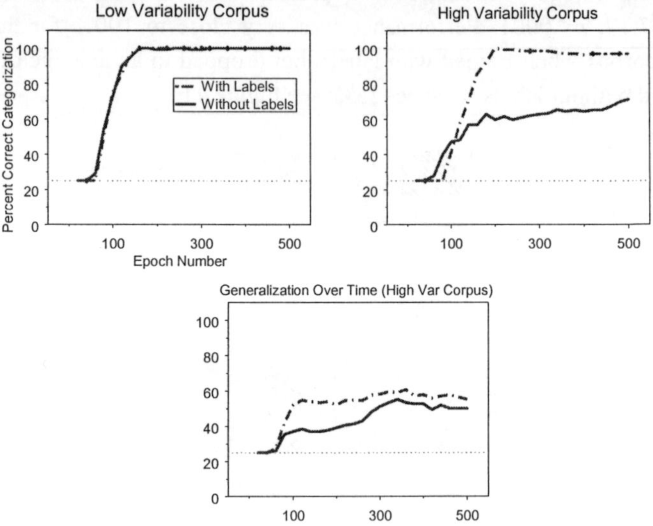

Figure 2: Averaged training profiles for the high and low variability corpora. At no point in training, do labels improve categorization performance for the low-variability categories.

To examine what caused this increase in reconstruction error, the final measure looked at the degree to which the network's representations were clustered around the prototype. This measure was the error between the network's output and the category prototype. Here, labels led to a smaller error (more categorical responses) for both training corpora $F(1,78)=224.38$, $p < .0001$; however, there was also a significant interaction that indicated that labels had more effect for the high-variability corpus than the low-variability one $F(3,76)=2087$, $p<.0001$. The effect of labels on the degree of clustering can be observed more directly by looking at principal component analysis (PCA) plots of the hidden unit activations of the trained network. Figure 4 shows the effect of labels on forming more categorical representations for the high- and low- variability sets. The degree to which exemplar representations were categorical was enhanced for both corpora, but since low-variability unlabeled representations were already quite categorical, the inclusion of labels did not alter the response accuracy.

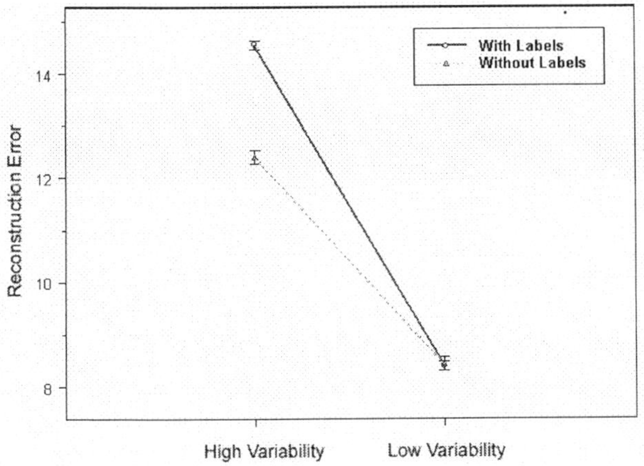

Figure 3: The improved categorization accuracy comes at a cost to accurate representation.

4. Discussion

As predicted, labels improved categorization accuracy of items organized in high-variability, but not low-variability categories. Although training with labels resulted in more categorical representations regardless of the training set, the effect was much greater for the fuzzier categories, suggesting that it is these categories that benefit most from having some of their features abstracted. By associating a single label with numerous exemplars, the resulting category representations of initially ill-formed categories became less idiosyncratic, resulting in improved categorization, and better generalization to unseen exemplars.[ii] In accord with the present findings, Tijsseling and Harnad (1997) found that category labels had no effect on representations when categories were initially very discriminable (e.g., apples and cars). This does not mean, however, that such labels play no role in shaping these category representation. While perceptual properties are quite sufficient to discriminate apples and cars, calling different-looking cars by the same name may help cohere the category, helping individuals to find the features common to its members. The present simulation provides evidence that when presented with a category containing highly-variable exemplars, labels help to highlight common features at the cost of abstracting irrelevant ones.

An intuitive interpretation of these results is that labels made the representations more categorical because training with labels meant that networks were performing a categorization task in addition to the auto-encoding one. One may

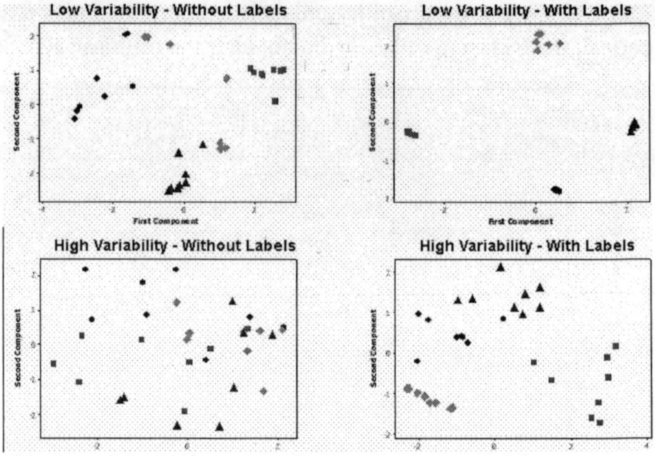

Figure 4: PCA plots of the hidden unit activations for the two traininng corpora and labeling conditions. The different symbols represent the four categories used in training.

therefore object to my referring to the category output nodes in the simulation as labels. Although the function of nodes in connectionist simulations is largely a matter of interpretation, the act of assigning a discrete code to a range of exemplars as done here and in other similar situations (Harnad, Hanson, & Lubin, 1991; Tijsseling & Harnad, 1997) may be particularly characteristic of linguistic labels. While categories can be formed without labels, as evidenced by a common response of an organism to a range of stimuli (e.g., Cangelosi, Greco, & Harnad, 2002), such common behavior does not imply a shared representation among the category members. For instance, a well-behaved cat may be said to have formed a category [surfaces in the house that may be scratched]. However, one cannot infer from the cat's behavior that it is not treating each instance of the category on an individual basis. The learning of a unique category label encourages the formation of a shared (rather than idiosyncratic) representation of the category. Words also allow the combining of concepts. Cangelosi et al. (2000) provide a compelling illustration of the advantage of combining existing grounded symbols into propositions through "symbolic theft" over learning the perceptual properties through "sensorimotor toil." For instance, knowing [horse] and [stripes] permits a working definition of [zebra] (horse+stripes) without having to experience one. The present work suggests that even in the absence of the combinatorial powers of words, providing discrete category labels facilitates the highlighting of relevant features and abstracting of irrelevant ones.

The reason experience with labels led to more categorical representations in the model is simply that each act of labeling was also an instance of categorization. Merely perceiving an object does not require categorizing it. In contrast, naming an object (whether to communicate to another individual or for your own benefit) does require placing it into a category. The repeated experience with categorizing exemplars through labeling has the effect of gradually altering the category representations to be more in line with the named categories. This does not mean that words and language are required to entertain certain thoughts or to hold particular representations (this is an open question, e.g., Hermer-Vazquez, Spelke, & Katsnelson, 1999). Indeed, not making the labels available to the models at test was done expressly to observe the process by which labels may affect representations of exemplars as distinct from using them during recognition.

This reasoning relates directly to work on Categorical Perception (CP) (Harnad, 1987), a phenomenon in which the repeated placement of exemplars into categories exaggerates the perceptual difference between categories (acquired distinctiveness) and collapses perceptual differences within a category (acquired similarity). Categorical perception, unlike language, is not unique to humans (e.g., categorical perception of phonemes in chinchillas, Kuhl & Miller, 1978), and so would appear to not require language. Indeed, the literature on CP rarely makes any mention of the role of words or language. While numerous experiments by Goldstone and colleagues (e.g., Goldstone, 1998 for review) provide evidence that categorization in the laboratory can alter perceptual representations of both familiar and novel dimensions, the role of linguistic labels in this process is not fully specified. However, recent studies have shown that under some circumstances verbal interference can extinguish CP of colors and facial expressions (Roberson & Davidoff, 2000), implicating verbal coding of some kind in the functioning of CP in human adults. Özgen and Davies (2002) provided evidence of how experience with categorization can create new boundaries in the domain of color perception, arguing that lifetime experience with color words heavily influences the resulting color categories. While CP can occur in the absence of language, our ubiquitous experience with named categories suggests that further study of the role of language in category formation will be a useful and productive line of research.

Acknowledgments

Sincere gratitude to Jay McClelland and David Rakison for useful discussion. This material is based upon work supported by a National Science Foundation Graduate Research Fellowship.

References

Cangelosi, A., Greco, A., & Harnad, S. (2000). From robotic toil to symbolic theft: grounding transfer from entry-level to higher-level categories. *Connection Science, 12,* 143-162.

Cangelosi, A., Greco, A., & Harnad, S. (2002). Symbol Grounding and the Symbolic Theft Hypothesis. In A.Cangelosi & D. Parisi (Eds.), *Simulating the Evolution of Language* (London: Springer.
Gentner, D. (1981). Some Interesting Difference between Nouns and Verbs. *Cognition and Brain Theory, 4,* 161-177.
Goldstone, R. L. (1998). Perceptual learning. *Annual Review of Psychology, 49,* 585-612.
Harnad, S. (1987). *Categorical Perception: The Groundwork of Cognition.* New York: Cambridge University Press.
Harnad, S., Hanson, S. J., & Lubin, J. (1991). Categorical Perception and the Evolution of Supervised Learning in Neural Nets. In D.W.Powers & L. Reeker (Eds.), *Working Papers of the AAI Spring Symposium on Machine Learning of Natural Language and Ontology* (pp. 65-74). Stanford University.
Hermer-Vazquez, L., Spelke, E. S., & Katsnelson, A. S. (1999). Sources of flexibility in human cognition: Dual-task studies of space and language. *Cognitive Psychology, 39,* 3-36.
Homa, D. & Cultice, J. (1984). Role of Feedback, Category Size, and Stimulus Distortion on the Acquisition and Utilization of Ill-Defined Categories. *Journal of Experimental Psychology-Learning Memory and Cognition, 10,* 83-94.
Kersten, A. W. & Smith, L. B. (2002). Attention to novel objects during verb learning. *Child Development, 73,* 93-109.
Kotovsky, L. & Gentner, D. (1996). Comparison and categorization in the development of relational similarity. *Child Development, 67,* 2797-2822.
Kuhl, P. K. & Miller, J. D. (1978). Speech perception by the chinchilla: Identification functions for synthetic VOT stim. *The Journal of the Acoustical Society of America, 63,* 905-917.
Mareschal, D., French, R. M., & Quinn, P. C. (2000). A connectionist account of asymmetric category learning in early infancy. *Developmental Psychology, 36,* 635-645.
Ozgen, E. & Davies, I. R. L. (2002). Acquisition of categorical color perception: A perceptual learning approach to the linguistic relativity hypothesis. *Journal of Experimental Psychology-General, 131,* 477-493.
Plunkett, K., Sinha, C., Moller, M. F., & Strandsby, O. (1992). Symbol Grounding or the Emergence of Symbols? *Connection Science, 4,* 293-312.
Pothos, E. M. The Rules versus Similarity Distinction. *Behavioral and Brain Sciences,* (in press).
Roberson, D. & Davidoff, J. (2000). The categorical perception of colors and facial expressions: The effect of verbal interference. *Memory & Cognition, 28,* 977-986.
Rohde, D. L. T. (1999). *Lens: The light, efficient network simulator* (Rep. No. CMU-CS-99-164). School of Computer Science, Carnegie Mellon University.
Tijsseling, A. & Harnad, S. (1997). Warping Similarity Space in Category Learning by BackProp Nets. In M.Ramscar, U. Hahn, E. Cambouropolos, & H. Pain (Eds.), *Proceedings of SimCat 1997: Interdisciplinary Workshop on Similarity and Categorization* (pp. 263-269). Edinburgh University.

---

[i] The reported pattern of results does not change if only one hidden layer is used. The architecture made use of two hidden layers in order to investigate the effects of selective lesioning of one layer versus the other. These results are not discussed here.

[ii] Interestingly, in a study the author was not aware of at the time, Homa and Cultice (1984) reported a strikingly similar pattern of results in a behavioral experiment using adult participants. In investigating the role of feedback on learning categories of varying levels of well-formedness, Homa and Cultice (1984) found that while feedback provided little benefit for well-formed categories, learning ill-defined categories was only possible with category feedback.

LANGUAGE AS AN AID TO CATEGORIZATION: A NEURAL NETWORK MODEL OF EARLY LANGUAGE ACQUISITION

MARCO MIROLLI[AB] DOMENICO PARISI[A]
[A] *Institute of Cognitive Sciences and Technologies, National Research Council, 15 Viale Marx, 00137, Rome, Italy*
d.parisi@istc.cnr.it

[B] *Philosophy and Social Sciences Department, University of Siena, 47 Via Roma, 53100, Siena, Italy*
mirolli2@unisi.it

The paper describes a neural network model of early language acquisition with an emphasis on how language positively influences the categories with which the child categorizes reality. Language begins when the two separate networks that are responsible for nonlinguistic sensory-motor mappings and for recognizing and repeating linguistic sounds become connected together at 1 year of age. Language makes more similar the internal representations of different inputs that must be responded to with the same action and more different the internal representations of inputs that must be responded to with different actions.

1. Introduction

Understanding the relationship between language and cognition is one of the most difficult and important challenges facing cognitive science. Language has not only a communicative (social) function, but also a cognitive (individual) function (Vygotsky 1962, Jackendoff 1996, Clark 1998, Carruthers 2002). The view of language as something that transforms all human psychological processes dates back as early as the 1930s, with the work of Russian scholar Lev Vygotsky (1962, 1978), but it is a view which has been rather neglected in mainstream cognitive science. Nonetheless, the idea has not completely disappeared and recently has been raising increasing interest in philosophy and cognitive science (Carruthers & Boucher, 1998; Gentner & Goldwin-Meadow, 2003). Indeed, in recent years, language has been proposed to have an influence on, and to improve, many human psychological functions, among them categorization (Clark, 1998; Gentner, 2003), learning (Waxman & Markov, 1995; Nazzi & Gopnik, 2001), selective attention (Jackendoff, 1996; Clark, 1998), memory (Gruber & Goschke, 2004), voluntary control (Diaz & Berk 1992), perspective taking (Tomasello, 2003), analogy making (Gentner 2003), and reflexive thinking (Dennett 1991, 1996; Carruthers 2003). This paper describes some very simple neural network simulations that (a) model the very earliest stages of language acquisition and (b) investigate the effects of language on categorization.

During her first year of life, the child learns to control her movements, to make appropriate sensory-motor mappings, to categorize perceptual experiences, and to reproduce her own sounds and the linguistic sounds of her environment. Notwithstanding all these progresses, this phase of child development is called 'pre-linguistic' because in her first 10-12 months the child does not show any strictly linguistic competence, that is, she is able neither to understand nor to meaningfully produce words. It is only around the end of her first year that the child learns to connect the linguistic sounds that have become familiar to her with their meanings as indicated by the fact that she reacts correctly to linguistic stimuli and she produces words in the appropriate circumstances.

Language acquisition can be considered to involve three sub-tasks(Kit, 2002): the acquisition of linguistic forms, the acquisition of non-linguistic sensory-motor mappings, and the association between linguistic forms and specific sensory-motor mappings, which become the meanings of the linguistic forms. Behavioral evidence (Bloom, 1994) suggests that the acquisition of linguistic forms and the acquisition of sensory-motor mappings run quite independently until the end of the first year and that only after the child has acquired a certain ability to map sensory inputs into motor outputs and to categorize experiences, on one side, and to recognize and produce linguistic forms, on the other side, the third task, the association of linguistic forms with specific sensory-motor mappings, can begin. The model of language learning presented here is based on this kind of behavioral evidence.

2. Method

2.1. *The neural network and the environment*

The neural network used in our simulations is modular. It is constituted by two sub-networks with three layer each, which we call the sensory-motor sub-network and the linguistic sub-network. The hidden layers of the two sub-networks are reciprocally connected by two matrices of connection weights so that the linguistic and the sensory-motor systems can interact with the each other (figure 1).

The sensory-motor sub-network has 16 bipolar input units (each unit's activation can be either 1 or –1) which encode the properties of perceived objects, 2 hidden units (with continuous activation in the interval [–1; 1]), and 2 output units which encode the action performed by the network in response to each object. The activation of the two output units is thresholded to be either 1 or – 1, so that there are only four possible 'actions': <-1; -1>, <-1; 1>, <1; -1>, <1; 1>. The network's environment consists of 480 objects, belonging to 4 categories of 120 exemplars each. There are four prototype vectors, one for each category[a], and the perceptual properties of objects are generated by flipping 4 bites of the prototype to which the object belongs.

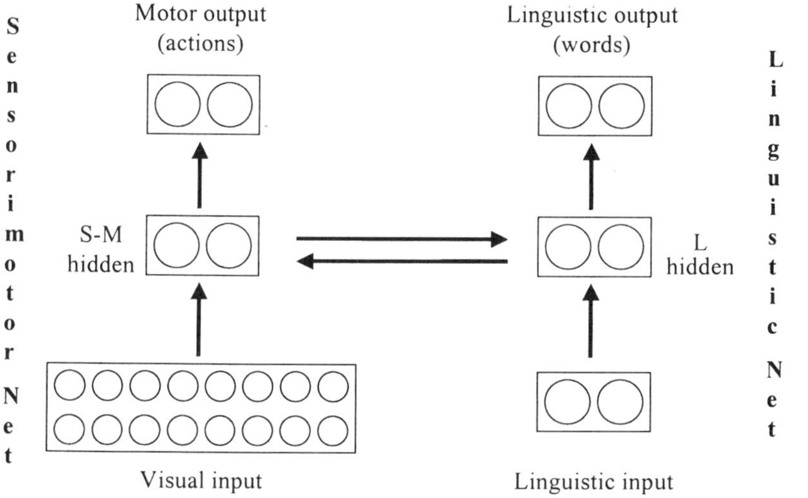

Figure 1. The neural network.

The linguistic sub-network has 2 input units encoding incoming linguistic signals, 2 hidden units, and 2 output units which represent emitted sounds. All the units of the linguistic sub-network have continuous activations in the interval [– 1; 1]. The linguistic environment is constituted by 4 words, which can be interpreted as the names of the four kinds of objects or of the appropriate actions to be performed upon them. Since words are pronounced in different ways by different persons and by the same person at different times, the acoustic inputs are created by changing slightly 4 prototype vectors, one for each word (<-0.5; -0.5>, <-0.5; 0.5>, <0.5; -0.5>, <0.5; 0.5>). There are 120 instances of each word and each instance is produced by changing both values of the corresponding prototype vector by a amount randomly chosen in the range [–0.25; 0.25] .

---

[a] The prototype vectors are: <-1; -1; -1; -1; -1; -1; -1; -1; 1; 1; 1; 1; 1; 1; 1; 1>; <-1; -1; -1; -1; 1; 1; 1; 1; -1; -1; -1; -1; 1; 1; 1; 1>; <-1; -1; 1; 1; -1; -1; 1; 1; -1; -1; 1; 1; -1; -1; 1; 1> and <-1; 1; -1; 1; -1; 1; -1; 1; -1; 1; -1; 1; -1; 1; -1; 1>

2.2. The two stages of learning

The entire network goes through two successive stages of learning which are meant to correspond approximately to child's learning from birth to 1 year and to her learning from 1 year on, respectively. In the first stage of learning the two sub-networks are trained independently to accomplish two different tasks. The sensory-motor network learns to categorize objects and the linguistic network learns to repeat or imitate words. At the end of this first stage the connections between the hidden layers of the two sub-networks become functional and in the second stage of learning the network learns to associate words with the categories of objects to which they refer.

At the beginning the network is initialised with random connection weights in the range [-0.5; 0.5]. Then, 1000 cycles of backpropagation learning (with a learning rate of 0.2) are run in the following way. For the sensory-motor sub-network, one of the 480 different objects is randomly chosen, its perceptual properties are encoded in the input units of the sensory-motor network, the response of the network is calculated, and the correct action is provided to the network as teaching input (correct responses are <-1; -1>, <-1; 1>, <1; -1> and <1; 1> for objects belonging to category A, B, C and D, respectively). For the linguistic sub-network, one of the 480 possible instances of words is randomly chosen and encoded in the linguistic input units, the sound produced by the linguistic network in response to this heard sound is calculated, and the same sound heard as input is given as teaching input (in other words, the linguistic network has to accomplish an autoassociative task).

During this first stage of learning, the connections that link the hidden layers of the two sub-networks are non-functional. After 1000 cycles of back-propagation learning, the inter-network connections become functional and in the second stage of learning their weights are modified so that the network learns to associate the internal representations of objects (the vectors of activation of the sensory-motor hidden layer) with the internal representations of the appropriate words (the vectors of activation of the linguistic hidden layer), and vice versa. This second phase of learning runs as follow. One of the 480 objects is chosen randomly, together with one of the 120 instantiations of the word that designates the category to which the object belongs. The object and the word are given as input to the sensory-motor network and to the linguistic network, respectively. Then, the vectors of activation of the two hidden layers are calculated and two cycles of the delta rule learning algorithm are applied by considering the two hidden layers with the connection weights in between them as two distinct perceptrons: one perceptron has the sensory-motor hidden layer has its input layer and the linguistic hidden layer as its output layer, and the other perceptron has the linguistic hidden layer as its input layer and the sensory-motor hidden layer as its output layer. The delta rule is applied by using the vector of activation observed in the linguistic hidden layer as teaching input for the first perceptron and the vector of activation observed in the sensory-motor hidden layer as teaching input for the second perceptron. The procedure, like that of the first stage of learning, is applied 1000 times with a learning rate of 0.2.

3. Results

We have run two kind of tests, behavioral tests and internal representations tests. The behavioral tests have the function of assessing the performance of the network, while with the internal representation tests we analyse the effects of language on the categorization abilities of the network.

3.1. *Behavioral tests*

There are 4 behavioral tests: categorization, linguistic imitation, naming, and comprehension. In the categorization test the sensory-motor network is given an object as input and its performance is calculated as the difference between its action response and the right one (according to the object's category). In the linguistic imitation test the linguistic network is given the instantiation of a word as input and performance is how much the network is able to reproduce the heard sound in its sound output. In the naming test we give the network one of the 480 objects as input, activation spreads to the sensory-motor hidden units, then to the linguistic hidden units, and finally to the linguistic output units. We calculate the network's performance with respect to the vector prototype of the word that designates the category of the perceived object. The comprehension test is symmetrical to the naming one. In this test, we give the network an instantiation of a word as input and let the activation spread from the linguistic input units to the linguistic hidden units, then to the sensory-motor hidden units and, finally to the motor output units of the sensory-motor networks. The network's performance is calculated with respect to the response that is appropriate to the meaning of the word, that is, to the action to be produced in response to objects of the category to which the heard word refers.

All the behavioral tests are conducted three times: at the beginning of the simulation (cycle 0), at the end of the first stage of learning (cycle 1000), and at the end of the second stage (cycle 2000). Figure 2a reports the percentage of errors in the four behavioral tests averaged on all the 480 inputs (objects in the categorization and naming tests and instances of words in the linguistic imitation and comprehension tests)[b]. The network learns both to perform the correct action in response to a perceived object and to imitate the linguistic sounds it hears: for both tasks the percentage of error decreases from around the 75% at the beginning of the simulation to 0%, that is, no error at all, at the end of the first stage of learning. Even the second stage of learning is very successful: while at the end of the first stage of learning the network gives wrong responses almost 90% of the time in the naming test and around 70% of the time in the comprehension test, at the end of the second stage the network has acquired a perfect linguistic competence in that it always names objects correctly and responds appropriately to all the linguistic signals present in its environment.

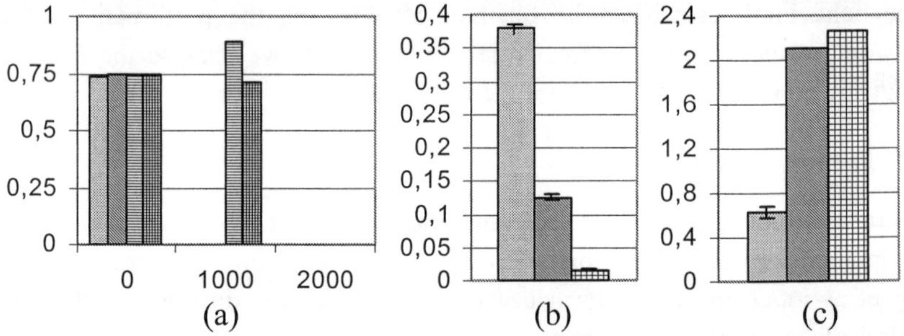

Figure 2. (a): Percentage of errors in the categorization (left), linguistic imitation (middle-left), naming (middle-right) and comprehension (right) tests with random weights (0) and after the first (1000) and the second stage of learning (2000). (b): Average clouds' dimension before learning (left) and after learning in the 'vision' (middle), and 'language' (right) conditions. (c): Average distance between clouds before learning (left) and after learning in the 'vision' (middle), and 'language' (right) conditions. See text for details.

3.2. *Internal representations tests*

Given the perceptual properties of an object as input, we can look at the activation vector of the two sensory-motor hidden units (in the range [-1; 1]) and consider this activation vector as the network's internal representation of that object. Therefore, the internal representation of an object can be viewed as a point in a bi-dimensional space, with each dimension of the space corresponding to the activation state of one hidden node; consequently, the internal representation of one category of objects is the "cloud" of points which represent all the objects belonging to that category (see figure 3). A good categorization is one in which (1) objects belonging to the same category are represented similarly, that is, the size of their cloud is small, and (2) objects belonging to

[b] All the results are averaged over 10 replication of the simulation with different initial conditions (random connection weights).

different categories are represented differently, that is, the different clouds are distant from each other. So, for assessing goodness of categorization we use two measures: (1) cloud size, calculated as the average distance between all the points belonging to a cloud and the cloud's geometric center, averaged over the four clouds, and (2) distance between clouds, calculated as the average distance between the centers of the four clouds. We analyse the categorization performance of the trained neural network in two conditions: 'vision' and 'language'.

The vision condition corresponds to the situation described above, in which the clouds are the network's internal representations given the objects as visual input. In the language condition the network perceives the objects together with their names (random instances of the four words corresponding to the four categories). In this condition we present both kinds of input, then we calculate the linguistic hidden unit activations given the linguistic input, and then we measure the sensory-motor hidden unit activations given both the visual input and the activation arriving from the linguistic hidden units.

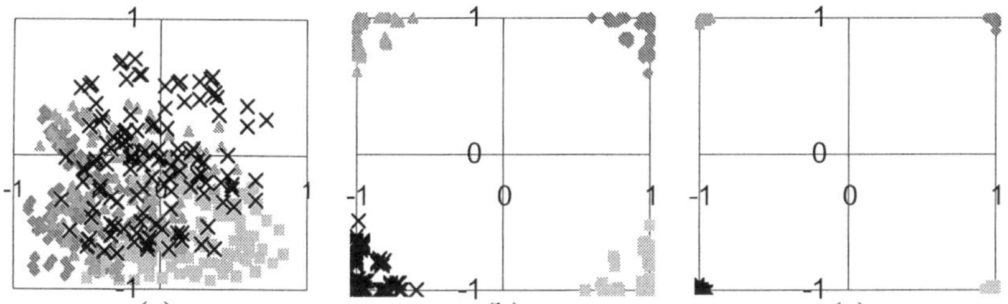

Figure 3. Network's internal representations (clouds) before learning (a) and after learning in the vision (b) and language (c) conditions. See text for details.

Figure 3 shows the internal representations of objects before learning (a), and after learning in the vision (b) and language conditions (c). While before learning the four clouds are big and highly overlapping and so the network cannot discriminate between different kinds of objects (fig. 3a), after learning objects' categories are clearly distinguished in that each class of objects occupies one of the four corners of the network's representational space (fig. 3b). In other words, during training the network has learnt to categorize objects by decreasing the average size of the clouds (fig. 2b, left and middle bars) and by increasing the average distance between the clouds (fig 2c, left and middle bars). This is just another example of the well-known categorical perception effect (Harnad et al.1995). What is more interesting is the effect that language has upon categorization: as you can see comparing the internal representations in the vision (fig. 3b) and the language (fig. 3c) conditions, categorization is consistently facilitated when objects are perceived together with their 'names'. In fact, the four clouds are both significantly smaller (fig. 2b middle and right bars) and more distant from each other (fig. 2c middle and right bars).

4. Discussion and conclusions

We have described a neural network model of early language acquisition. The model assumes that during her first year of life the child separately learns various sensory-motor mappings (reaching, manipulating, categorizing objects) and various sound-related abilities (recognizing sounds, repeating her own sounds and the sounds produced by others). It is only at the end of the first year that language learning begins. The two separate networks that are responsible, respectively, for the sensory-motor mappings and for the sound-related abilities become functionally connected and language learning consists in learning the connection weights linking the two networks. These weights allow the child to produce linguistic sounds in the appropriate circumstances and to react appropriately to the linguistic sounds produced by other individuals.

But language also changes the way in which the child categorizes reality. Input-output mappings require categorization. Categorization is to make the internal representations (patterns of activation in the network's internal units) of different inputs more similar if the different inputs must be responded to with the same action, and more different if the inputs must be responded to with different actions (Di Ferdinando & Parisi 2004). All organisms categorize reality. And infants categorize reality during their first year in order to generate the appropriate input/output mappings. But our model shows that language can influence cognition by inducing better categorization of non-linguistic inputs.

Our model can also be considered as a general model of the integration of multiple sensory-motor mappings in the brain. Much activity in the brain consists in exploiting the co-variations observed in experience to establish correlations between different sensory-motor mappings. For example, visual-motor mappings are correlated with proprioceptive-motor and tactile-motor mappings in such a way that given one type of input the brain can predict the other one. Language is but one example of this type of integration between various non-linguistic sensory-motor mappings and the sensory-motor mapping from heard sound to pronounced sounds. Given a visual input from an object, the brain generates the internal representation of the word (sound) that designates the object. Or, given a word (sound), the brain generates the internal representation of the visually or haptically perceived object designated by the word.

However, language is somewhat different from other types of integration between different sensory-motor mappings. While other correlations between different sensory-motor mappings are not arbitrary, language is arbitrary - and this may require a species-specific genetically inherited basis for language learning. Given a visual input from a new object, the brain can predict how the visually perceived object would feel if touched. In contrast, given a new object the brain cannot predict what is the word that designates the object or, given a new word, the brain cannot predict which object is designated by the word. Furthermore, variations in how the same word is pronounced do not predict variations in the object designated by the word. Another important difference between language and other types of integration between different sensory-motor mappings is that the overall space of variation of heard or pronounced linguistic signals appears to be much smaller than the space of variation of the objects or actions that are designated by the linguistic signals. This is particularly important for categorization because it might contribute to explaining why language makes our sensory-motor categories more distinctive and more compact. Finally, language may be different because the sensory-motor mapping from heard to pronounced words can easily become a circular or closed circuit. When one sees an object, the visual input from the object does not only generate the internal representation of the word that designates the object but it can generate the actual sound of the word through private or inner speech.

In fact, unlike animal communication systems, human language is not only used to communicate with other individuals but it is also used to communicate with oneself, either externally (private speech) or internally (inner speech). In the simulations described in the paper, the language which arrives to the neural network is produced by another individual. We intend to further explore how language influences categorization by applying our neural network model to private speech and inner speech, trying to answer the following question: How an individual categorizes reality when he or she is exposed to reality and at the same time to language stimuli that the individual him- or herself generates, either externally or internally?

Acknowledgments

The research presented in this paper has been supported by the ECAGENTS project founded by the Future and Emerging Technologies program (IST-FET) of the European Community under EU R&D contract IST-2003-1940. Thanks also to an anonymous referee for his useful comments.

References

Bloom P. (Ed.) (1994). *Language acquisition: Core readings*. Cambridge, MA: MIT Press.

Cangelosi A., Harnad S. (2000). The adaptive advantage of symbolic theft over sensorimotor toil: Grounding language in perceptual categories. *Evolution of Communication*, 4, 117-142

Carruthers P. (2002). The cognitive functions of language. *Behavioral and Brain Sciences*, 25, 657-726

Carruthers P., Boucher J. (Eds) (1998). *Language and thought: Interdisciplinary themes*. Cambridge: Cambridge University Press

Clark A. (1998). Magic words: How language augments human computation. In Carruthers and Boucher (Eds) (1998).

Dennett D.C. (1991). *Consciousness Explained*. New York: Little Brown & Co.

Dennett D.C. (1996). Reflections on language and mind. In Carruthers and Boucher (Eds) (1998).

Di Ferdinando A., Parisi D. (2004). Internal representations of sensory input reflect the motor output with which organisms respond to the input. In Carsetti A. (Ed), *Seeing, thinking and knowing*. Dordrecht: Kluwer, 115-141

Diaz R., Berk L. (Eds) (1992). *Private speech: From social interaction to self regulation*. New Jersey: Erlbaum.

Gentner D., Goldin-Meadow S. (Eds) (2003). *Language in mind*. Cambridge, MA: MIT Press.

Gruber O., Goschke T. (2004). Executive control emerging from dynamic interactions between brain systems mediating language, working memory and attentional processes. *Acta Psychologica*, 115, 105-121

Harnad S., Hanson S.J., Lubin J. (1995). Learned categorical perception in neural nets: implications for symbol grounding. In: Honavar and Uhr (Eds) *Symbol processors and connectionist network models in artificial intelligence and cognitive modelling: steps toward principled integration*. New York, NY: Academic Press, 191-206

Jackendoff R. (1996). How langauge help us think. *Pragmatics and Cognition*, 4, 1-34

Kit C. (2002). How does lexical acquisition begin? A cognitive perspective. *Cognitive Science*, 1, 1-50

Nazzi T., Gopnik A. (2001). Linguistic and cognitive abilities in infancy: When does language become a tool for categorization? *Cognition*, 80, 303-312

Vygotsky, L.S. (1962). *Thought and language*. Cambridge, MA: MIT Press

Vygotsky, L.S. (1978). *Mind in society: The development of higher psychological processes*. Cambridge, MA: Harvard University Press

Waxman S.R., Markow D.B. (1995). Words as invitations to form Categories: Evidence from 12- to 13-month-old infants. *Cognitive Psychology*, 29, 257-302

Language Processing

A CONNECTIONIST EVALUATION OF SCHEMES TO MEASURE DIFFICULTY OF WORDS BASED ON THEIR PHONOLOGICAL STRUCTURE

JENNIFER HAYES PETER HOWELL

Department of Psychology, University College London
Gower Street, London, WC1E 6BT, UK
Jennifer.hayes@ucl.ac.uk

Two schemes for measuring the complexity of words (the Index of Phonetic Complexity and Analysis of Phonological Structure) were compared with regards to how well they predict which items in a stream of spontaneous speech produced by people who stutter, were likely to be dysfluent. Connectionist models were trained to output 'stuttered' or 'fluent' as a response to each word using either the IPC or the ANOPhS measure for that word. While the model trained using the IPC codes did not learn to discriminate between fluent and dysfluent items the ANOPhS model showed evidence of learning.

1. Theories of Stuttering

1.1. *Introduction*

All speakers, particularly children, experience fluency breakdown at different points in utterances. For speakers who stutter (SWS) 10% or more of their speech consists of dysfluencies that involve pausing and repetition of complete words said previously or production of parts of words (such as prolongation or repetition of the first part of the word). Some theories of stuttering that address how these dysfluencies arise consider that this involves perceptual monitoring and correction of speech output (where the perceptual and the speech production systems are linked) while others argue that production is independent of perception and dysfluencies can be accounted for entirely in the operation of the production system. An example and evaluation of each of these types of theory follows.

1.2. *Covert repair hypothesis*

Kolk and Postma's (1997) covert repair hypothesis (CRH) is a linked model. They argue that errors that arise in the speech production system of both fluent and dysfluent speakers are monitored by a speech perceptual system that interrupts and corrects any errors as they occur. For instance, when a speaker intends to say the word "cat" phonetically related items such as "rat" also become activated. When sufficient time is allowed for words to reach full activation, the target word will have a higher level than its competitors and the intended word is produced. However, one of the co-activated competitors may have higher activation when speech is produced rapidly if activation builds up at similar rates in the early stages of word selection. Words selected incorrectly are detected by the perceptual monitor before they are output (by internal monitoring processes). Speakers correct such errors before output (the covert repair process) that leads to whole and part word repetitions and pauses. Kolk and Postma explain the higher incidence of fluency breakdown of these forms in SWS by proposing that they have slow phonetic systems which increase the chance of errors arising that are repaired covertly.

1.3. *EXPLAN*

The EXPLAN model of Howell (2004) explains why whole and part word repetitions and hesitations occur, without using a perceptual monitoring system so it does not assume that SWS have slow phonetic systems. According to EXPLAN fluency failures occur because words in an utterance take different amounts of time to plan and execute. Word planning (PLAN) is done by the linguistic system independently of word execution (EX) which is organized by the motor system. PLAN and EX occur in parallel. As PLAN and EX are independent, the current word can be executed, while the plan of the next word is being generated. Usually the planning process is rapid (Sternberg, Monsell, Knoll & Wright 1978) resulting in the plan being delivered ahead of when it is needed for execution. Fluency failures occur when the plan of the word about to be produced is not ready at the time the execution of the previous word has concluded. There are two influences that lead to the plan for the current word not being ready in time a) when the prior word/s is/are produced rapidly and b) when the planning time of the current word is slow. Rapid execution of a preceding word calls for the plan for the current word to be ready earlier than when the preceding word is produced more slowly. Planning time increases when words are complex and can also lead to situations in which the plan is not ready when execution of the preceding word has been completed. Research suggests that SWS have motor timing problems (Max, Caruso & Gracco, 2003) suggesting execution rate varies and that motor response timing varies in SWS when linguistic complexity changes suggesting an interaction between these factors in SWS. In the context of EXPLAN, these findings could provide an account of why SWS produce more dysfluencies than fluent speakers. Our long-term intention is to develop a connectionist model to predict where fluency failure will occur in stuttered speech. EXPLAN does not use perceptual monitoring so, if the model is successful, is more parsimonious than CRH (Kolk and Postma,

1997). Full realization of the connectionist version of EXPLAN will need to include a representation of execution time of the previous word and of the phonetic complexity of the current word. Execution time of the previous word is easily measured but an effective measure of word complexity is not so straightforward. The connectionist models reported here assess two schemes which measure the phonetic complexity of words. The complexity measures alone are used to see whether they can predict which items in a stream of naturalistic speech from SWS were stuttered.

2. Schemes of Phonetic Complexity

2.1. *IPC*

The Index of phonetic complexity (IPC), Jakielski, (1998) is based on the idea that the consonant and vowel sounds that children find difficult in the early stages of language development will continue to cause difficulty for adult speakers (properties identified in the IPC metric are summarized in Table 1). Each occurrence in a word of one of the 8 factors marked as difficult adds one point to the IPC score for that word. Subsequent research, however, has shown that only two of the IPC factors (factor 3, word length, and factor 5, consonant by manner) predict stuttering across all age groups and languages (Howell, Au-Yeung, Yaruss and Eldridge, in press).

Table 1. The IPC Metric.

Factor	No score (easy)	One point each (difficult)
Vowel by class	Monothongs /diphthongs	Rhotics
Word shape	Ends with vowel	Ends with consonant
Word length (number of syllables)	Monosyllabic /disyllabic	>three syllables
Consonant by place	Labials/coronals /glottals	Dorsals (k,g,ng sounds)
Consonant by manner	Stops/nasals/glides	Fricatives /affricatives/liquids
Singleton consonants by place	Reduplicated e.g. VC-VC	Variegated e.g. VC-CV
Contiguous consonants	No clusters	Clusters
Cluster by place	Homorganic	Heteroganic

2.2. *ANOPhS*

The Analysis of Phonological Structure (ANOPhS) (Diment, Howell and Harris, in preparation) is also based on consonant and vowel sounds that children find difficult to produce. It is argued to be an improvement on the IPC scheme because it gives emphasis to the parts of words that are particularly problematic for SWS (i.e., the onsets of words). ANOPhS has ten parameters (these are indicated in Table 2). Four of the ten parameters are computed for each syllable of the word. Thus a six syllable word could have a maximum ANOPhS score of 30 (the six parameters that apply once per word plus six (number of syllables) x the four parameters that apply to each syllable). Like IPC, the overall ANOPhS score for a word is computed by adding 1 point whenever a difficult factor is present.

Table 2. ANOPhS scheme.

Parameter	No score (easy)	One point each (hard)
Onset present*	Begins in a consonant	Begins in a vowel
Onset complexity*	Begins with vowel or single consonant	Begins with consonant cluster
Pre-initial /s/*	Does not begin with /s/	Begins with /s/
Rhyme*	Vowel in syllable is followed by a consonant	Vowel in syllable is not followed by a consonant
Word end segment	Ends with vowel	Ends with consonant
Word final suffix	No morphemic suffix present in final syllable	morphemic suffix present in final syllable
Left adjunction	Perfect alignment at left word edge	left adjoined syllable
Right adjunction	Perfect alignment at right word edge	Right adjoined syllable
Manner- obstruents	Plosives (p,b,t,d,k,g sounds)	Fricatives /Affricatives (f/v/s/z/sh/ch/dz/j/h /th/dh sounds)
Manner-sonorant	Glides, Nasals (w,ym,n,ng sounds)	Liquids (l,r)

* calculated for each syllable

3. Connectionist Modeling

3.1. *Training*

Two multi-layer perceptron models were trained on the same data set of 6,743 spontaneously produced words from eight adult male SWS. For one model the words in the training set were rcpresented by their index of phonetic complexity (IPC) codes (Jakielski, 1998) and for the other the input was made up of ANOPhS codes (Diment et al, in preparation). The IPC model had eight input units (one for each IPC factor) and two output units (stuttered or fluent) and four hidden units. The ANOPhS model had 30 input units (one for each possible ANOPhS factor) and two output units (stuttered or fluent) and ten hidden units. The models were required to output a stuttered or a fluent response depending on the phonetic complexity of the word measured by either the IPC or ANOPhS schcme. Both models were trained for 10,000 epochs.

3.2. ***Results***

IPC model

Fluent Items

The IPC model was tested using a further set of 1000 unseen words selected randomly from the speech of the eight adult speakers included in the training set. Of the 1000 items in the test set 900 were produced fluently and 100 were stuttered. Figure 1 shows the output of the IPC model for all 900 fluent items (i.e., where the correct output was “0”) included in the test set. From Figure 1 we can conclude that the model output values were between 0.02 and 0.14 for fluent items in the test set.

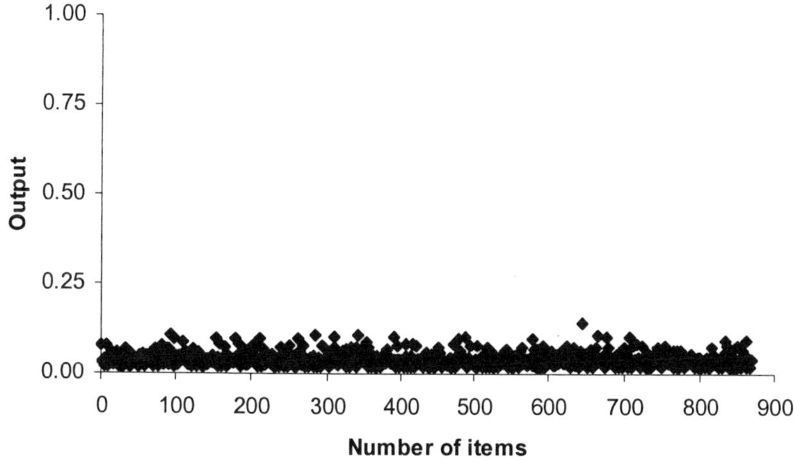

Figure 1. Output of IPC model for fluent items

Stuttered Items

The output of the IPC model for stuttered items (i.e., where the correct output should have been "1") is shown in Figure 2. From Figure 2, it is apparent that the range of values the IPC model outputs in response to stuttered items (i.e., between 0.02 and 0.16) was very similar to the range of values it output in response to fluent items. Thus it seems that IPC model was not able to learn to categorize items as being fluent or stuttered based on their IPC scores.

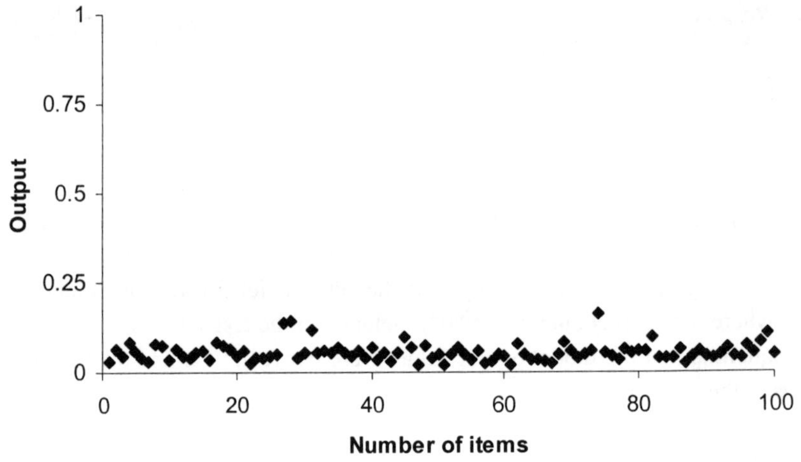

Figure 2. Output of IPC model for stuttered items

ANOPhS model

Fluent Items

The ANOPhS model was also tested using the same set of 1000 words that made up the test set for the IPC model. Again the correct output for a fluent item was "0". Figure 3 shows the output for all 900 fluent items included in the test set. It is evident that the model outputs "0" (the correct output) or an output very close to "0" for all but three items. The model does not output "1" (the correct output for stuttered items) for any fluent items.

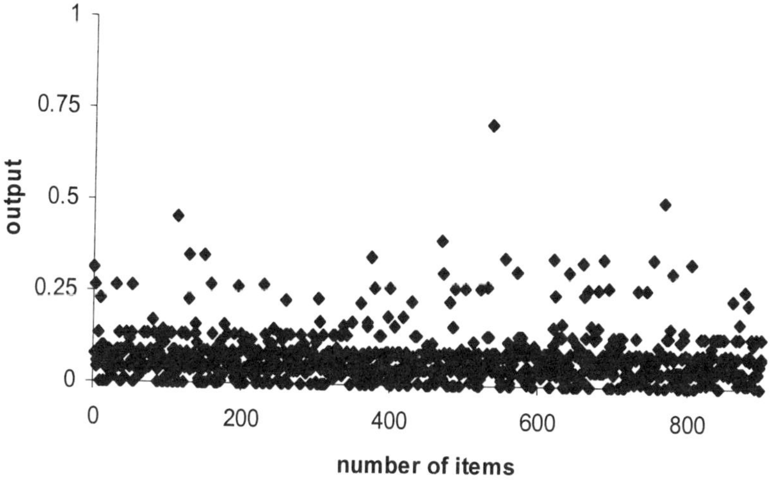

Figure 3. Output of ANOPhS model for fluent items

Stuttered Items

Again the correct output for a stuttered item was "1". Figure 4. shows the output for all 100 stuttered items included in the test set. The model outputs "0" (the output for fluent items) for the majority of stuttered items. By producing an output of "0" for the majority of items (stuttered and fluent) the model is in fact highly successful in performing the categorical task it was set as the correct output for 90% of the input was a fluent response (output"0"). The output graphs for stuttered and fluent items do not, however, have identical distributions. Proportionally more stuttered items have outputs above 0.25 (17% for stuttered 3.5% for fluent) and certainly above 0.5 (16% for stuttered 0.3% for fluent) than was the case for fluent items. It seems that despite the large inequality in the distribution of stuttered items in the training and test sets that the ANOPhS coding scheme provided enough information to enable some discrimination between fluent and stuttered items to take place.

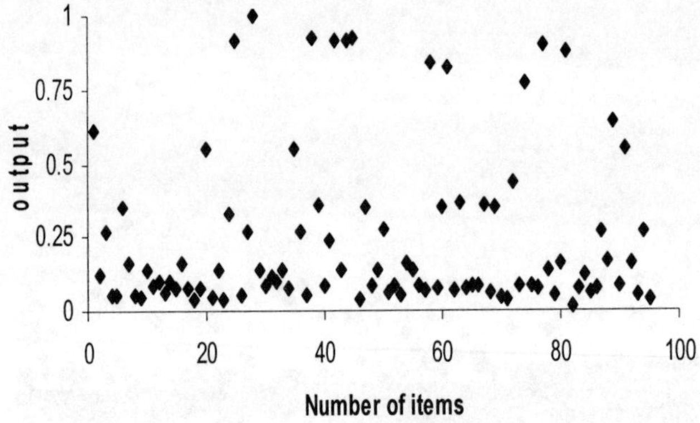

Figure 4. Output of ANOPhS model for stuttered items

Hidden unit representations of stuttered and fluent items in the ANOPhS model.

It is evident that clusters of stuttered items occur in the same locations as clusters of fluent items in all except the top left hand corner of Figure 5. Subsequent analysis has shown that the items clustering in the top left share the ANOPhS parameters 1 (beginning in a consonant), 2 (beginning in a consonant cluster), 3 (having a consonant after the vowel in a syllable) and 4 (ending in a consonant). Consonant clusters at the beginning and end of words or syllables seem to have a role in training the network about which items are likely to be stuttered.

3.3. *Discussion*

The IPC model trained on speech from SWS output a similar range of values for stuttered and fluent items when tested on 1000 unseen items. Thus there was no evidence that the model learned to predict which items would be stuttered and which would be fluent based upon their IPC codes. This failure of IPC codes to predict fluency has also been seen in an analysis of the speech of children who stutter (Weiss & Jakielski, 2001). The ANOPhS model was more promising at learning to categorise items as fluent or stuttered. The model produced the correct output for the majority of fluent items and, while it also produced a fluent output in response to the majority of stuttered items, it produced proportionally more responses in the direction of a stuttered response for stuttered items than it did for fluent items.

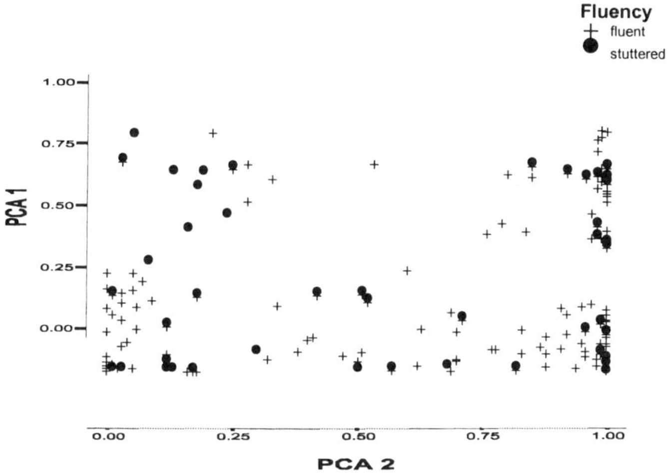

Figure 5. First two principal components of the hidden units in the ANOPhS model.

Thus, although the ANOPhS model appears to output a fluent response in the majority of cases, and in this way was successful in 90% of cases on the categorical task, it does seem to be showing some discrimination between the two types of item. Inspection of the hidden unit representations of the ANOPhS model indicates that to some extent the representations of stuttered items are clustering together and away from those of fluent words. Thus of the two coding schemes ANOPhS would seem to be more successful for future work. Further modelling is intended in which the proportion of stuttered items in the training and test set will be manipulated (over and under sampled) to see if the model still produces a fluent output in the majority of cases. Further analyses also need to be conducted to investigate whether ANOPhS is as successful if fewer parameters are used as has been found for IPC (Howell et al., in press). In the current experiment there was some evidence that the presence of consonant clusters at the beginning and end of a word was permitting some discrimination between words in terms of whether they were likely to be stuttered or not. ANOPhS will be used (possibly in a modified form) as a metric of phonetic complexity in future connectionist models implementing EXPLAN. The next major step in the implementation will be to include execution information about the sequence preceding difficult words (as indexed by ANOPhS). The intention will be to asses whether the speech patterns of SWS can be simulated using an autonomous model in which there is no link between the speech production and perceptual systems. Other architectures, such as support vector machines for classification/ supervised learning purposes and unsupervised learning methods such as self-organizing maps for increased cognitive realism and/or more insightful results concerning the relationships between the features will also be utilised to investigate the potential of the ANOPhS scheme. More specific methods for dealing with the feature selection e.g. based on Bayesian criteria will also be investigated.

Acknowledgments

This research was supported by a grant from the Wellcome Trust.

References

Diment R., Howell, P. & Harris, J. (in preparation). Analysis of phonological structure.

Howell, P. (2004). Assessment of some contemporary theories of stuttering that apply to spontaneous speech. *Contemporary Issues in Communicative Sciences and Disorders.* 39, 122-139.

Howell, P., Au-Yeung, J., Yarus, S. & Eldridge, K. (in Press). Phonetic difficulty and stuttering in English. *Journal of Fluency Disorders.*

Jakielski, K. J. (1998). *Motor organization in the acquisition of consonant clusters.* Dissertation/PhD thesis: University of Texas Austin.

Kolk, H. & Postma, A. (1997). Stuttering as a covert repairs phenomenon. In R. F. Curlee & G. M. Siegel (Eds.), *Nature and treatments of stuttering: New directions* (pp. 182–203). Needham Heights, MA: Allyn & Bacon.

Max, L., Caruso, A.J. & Gracco, V.L. (2003). Kinematic analyses of speech, orofacial nonspeech, and finger movements in stuttering and nonstuttering individuals. *Journal of Speech,Language, and Hearing Research* 46, 215-232.

Sternberg, S., Monsell, S., Knoll, R.L. & C.E. Wright. (1978). The latency and duration of rapid movement sequences:Comparison of speech and typing. In G. E. Stelmach (Ed.), *Information control in motor control and learning,* New York: Academic Press.

Weiss, A.L. & Jakielski, K.J. (2001). *Phonetic coplexity measurement and prediction of children's disfluencies: A preliminary study.* The 4th Nijmegen Conferences on Speech Motor Control and Stuttering. B. Maassen, W. Hulstijn, R. Kent, H. F. M. Peters & P. H. M. M. van Lieshout (Eds.). Nijmegen: Uitgeverij Vantilt.

SENTENCE COMPREHENSION WITHOUT PROPOSITIONAL STRUCTURE

STEFAN L. FRANK
Nijmegen Institute for Cognition and Information
Radboud University Nijmegen, Montessorilaan 3
6525 HR Nijmegen, The Netherlands
E-mail: S.Frank@nici.ru.nl

Comprehending a sentence requires the construction of a mental representation of the situation the sentence describes. Many researchers assume that, apart from such a situational representation, there is a level of representation at which the propositional structure of the sentence is encoded. This paper presents a simple sentence comprehension model, consisting of a neural network that transforms sentences into representations of the events they describe. During training, the network develops internal representations of the sentences. An investigation of these representations reveals that they can encode propositional information without implementing propositional structure.

1. Introduction

The mental representation of discourse has generally been assumed to involve three distinct levels (Kintsch & Van Dijk, 1978; Van Dijk & Kintsch, 1983): the *surface text*, the *textbase*, and the *situation model*. The first of these consists of the text's literal wording. This gives rise to the second level, in which the text's propositional structure is encoded. For instance, the sentence *Bob plays soccer* may be represented in the textbase as a predicate relating two arguments: PLAY(BOB,SOCCER).[a] The highest level of representation, the situation model, includes inferences derived from the reader's background knowledge and experience. In the current example, this may be information about 'what it is like' when Bob plays soccer, for instance, that he is likely to be outside.

Although there is considerable empirical evidence for the existence of a situational representation in the cognitive system (e.g., Zwaan, Madden,

[a]Examples of surface text shall be printed in *italics*, while SMALLCAPS denote propositions. Situation model descriptions are printed in normal font.

Yaxley, & Aveyard, 2004), evidence supporting the existence of a textbase-level representation is weak. Data by Ratcliff and McKoon (1978) and Dell, McKoon, and Ratcliff (1983) has been taken as evidence for a propositional decomposition of text, but as argued by Zwaan (1999) and Frank, Koppen, Noordman, and Vonk (in press), it can also be interpreted differently. Since similar propositions usually describe similar situations, it is difficult to distinguish between propositional and situational representations.

An experiment by Fletcher and Chrysler (1990) is often assumed to indicate that text is indeed represented at three distinct levels. They showed that readers more often confuse two sentences that differ only in surface text than two sentences that also differ propositionally, which in turn are more difficult to tell apart than two sentences that also differ situationally. However, as will be explained in Section 3.3, Frank et al. (in press) have shown that these results do not require distinct levels of representation. They trained a neural network to convert sentences into situational representations of the described events. The network's *single* intermediate level of representation could then account for Fletcher and Chrysler's findings.

The situational representations used by Frank et al. (in press) were taken from the Distributed Situation Space (DSS) model (Frank, Koppen, Noordman, & Vonk, 2003). This paper shall first explain these situational representations. Next, Frank et al.'s (in press) simple sentence comprehension model is explained, and the nature of its intermediate representation of sentences is investigated.

2. Situational representations

The DSS model (Frank et al., 2003) simulates how readers apply their world knowledge to make inferences during story comprehension. The model represents stories at the situational level only: Its representations do not depend on the wording or structure of the text but only on the relation between story events and world knowledge. Since the amount of world knowledge readers have is simply too large to implement even a significant subset for use by a computational model, a simple *microworld* was developed, all knowledge of which is available to the model.

2.1. *The microworld*

In the microworld, there are two characters whose names are Bob and Jilly. All activities they can engage in, as well as all situations they may find themselves in, can be described using just 14 basic events. These include states like 'Bob is tired' and 'Jilly is outside', and actions such as 'they play soccer', 'they play hide-and-seek', 'Bob plays with the dog', 'Jilly plays a computer game', 'Bob wins', and 'Jilly wins'. The basic events can be combined using the Boolean operators of negation, conjunction, and disjunction. However, not all combinations of basic events are as likely to occur. For instance, Bob and Jilly can only play one game at a time, and they never play a computer game outside or soccer inside. Also, they are more likely to be both outside or both not outside (i.e., inside) than to be at different locations.

2.2. *Representing microworld knowledge and situations*

Knowledge about the microworld was not implemented directly. Instead, a sequence of 250 consecutive example situations was constructed, following the microworld constraints. In each of these examples, each of the 14 basic events is either stated to be the case or not the case. That is, each example situation can be represented by a 14-dimensional binary vector, which served as training input to a Self-Organizing Map (SOM; Kohonen, 1995).

During training, the map develops a representation for each basic event, four of which are shown in Fig. 1. For each basic event p, each SOM cell i has a unique value $\mu_i(p)$ between 0 and 1, which is indicated by the cell's grayness in Fig. 1. This value, called the membership value of i for event p, denotes the extent to which the cell is part of the representation of p.

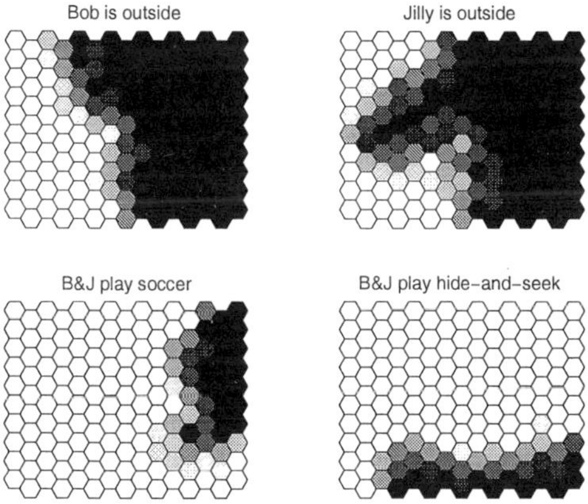

Figure 1. Self-organized maps of four basic events.

A useful property of these representations is that any microworld situation can be represented by combining the basic events using negation ($\neg$), conjunction ($\wedge$), and disjunction ($\vee$):

$$\begin{aligned}\mu_i(\neg p) &= 1 - \mu_i(p)\\ \mu_i(p \wedge q) &= \mu_i(p)\mu_i(q) \qquad (1)\\ \mu_i(p \vee q) &= \mu_i(p) + \mu_i(q) - \mu_i(p)\mu_i(q).\end{aligned}$$

Since the SOM consists of 150 cells, $\mu(p) = (\mu_1(p), \ldots, \mu_{150}(p))$ forms a 150-dimensional vector representation of p. That is, each microworld situation corresponds to a point in 150-dimensional 'situation space'. Since there is no one-to-one relation between dimensions of this space and events in the microworld, this representation is distributed. It is the distributed nature of situation space that gives the DSS model its name.

2.3. *Belief values*

The SOM representations are not arbitrary but encode the probabilistic relations among the events. This is easy to see from the four patterns in Fig. 1. Although they are meaningless by themselves, the relations among them reflect relations among the represented events. The patterns representing 'Bob is outside' and 'Jilly is outside' overlap for a large part, because Bob and Jilly are more likely to be at the same place than to be at different places. Likewise, the representations of 'Bob and Jilly play soccer' and 'Bob and Jilly play hide-and-seek' do not overlap at all, because Bob and Jilly cannot play soccer and hide-and-seek simultaneously. Also, the activation pattern for 'they play soccer' falls completely within those for being outside, since soccer can only be played outside.

Frank et al. (2003) have shown that the conditional probability that event p occurs in the microworld, given that q does, can be estimated quite accurately from their representations. This estimation, denoted $\tau(p|q)$, is called the *belief value* of p in event q, and is a measure for the 'belief' the model has in the occurrence of p. It is computed by

$$\tau(p|q) = \frac{\sum_i \mu_i(p)\mu_i(q)}{\sum_i \mu_i(q)}.$$

Also, the a priori probability (belief value) of event p follows from its representation: $\tau(p) = \frac{1}{150}\sum_i \mu_i(p)$.

Note that Equation 1 preserves the relation between belief values and probabilities in the microworld. Also note that the DSS model's representation of world *knowledge* cannot be separated from the representation of world *situations*. Instead, the vector representations of situations implement world knowledge. This unity of situation and knowledge shows that the DSS model indeed represents story events at the situational level.

3. A simple sentence comprehension model

Frank et al. (in press) argue that the goal of sentence comprehension is the construction of a situation model and that an intermediate, possibly propositional representation may only arise to the extent that it is useful for reaching this goal. They trained a simple recurrent neural network (Elman, 1990) to convert sentences from a *microlanguage* into the DSS model's vector representation of the corresponding events in the microworld. The intermediate representation that developed in the network's hidden layer forms an alternative type of textbase-level representation.

3.1. *The microlanguage*

The microlanguage consists of 15 words: *Bob, Jilly, and, plays, is, wins, loses, soccer, hide-and-seek, a_computer_game, with_the_dog, outside, inside, tired, awake.* Note that both *a_computer_game* and *with_the_dog* are considered one word. By combining the 15 words following the grammar of Table 1, 328 different sentences can be constructed.

Table 1. Grammar of the microlanguage.

<table>
<tr><td>S</td><td>→</td><td>NP VP</td></tr>
<tr><td>NP</td><td>→</td><td>Bob | Jilly | Bob and Jilly | Jilly and Bob</td></tr>
<tr><td>VP</td><td>→</td><td>plays Game [Place | and is State | and Result]</td></tr>
<tr><td></td><td>→</td><td>is Place [and plays Game | and State | and Result]</td></tr>
<tr><td></td><td>→</td><td>is State [and plays Game | and Place | and Result]</td></tr>
<tr><td></td><td>→</td><td>Result [and plays Game | Place | and is State]</td></tr>
<tr><td>Game</td><td>→</td><td>soccer | hide-and-seek | a_computer_game | with_the_dog</td></tr>
<tr><td>Place</td><td>→</td><td>outside | inside</td></tr>
<tr><td>State</td><td>→</td><td>tired | awake</td></tr>
<tr><td>Result</td><td>→</td><td>wins | loses</td></tr>
</table>

3.2. *The model*

A simple recurrent network was trained to transform 290 of the 328 microlanguage sentences into the DSS representation of the described events. The network consisted of 15 localist input elements, one for each word, and processed sentences one word at a time. The hidden layer had six elements and received activation from the input layer as well as a copy of its own previous activation. The output layer consisted of 150 elements, one for each DSS vector element. Seven such networks were trained, each with a different initial random weight setting.

During training, the difference between the desired and the actual output was backpropagated to adjust the connection weights. After training, however, the adequacy of the network's output was measured by using belief values. The idea is that comprehension of a sentence describing event p should result in an output vector $X(p)$ in which p's belief value $\tau(p|X(p))$ is larger than its a priori belief value $\tau(p)$. In the ideal case, when $X(p) = \mu(p)$ so $\tau(p|X(p)) = \tau(p|p)$, the 'amount of comprehension' is defined to equal 1. If $\tau(p|X(p)) = \tau(p)$, the network has not learned anything about the sentence and the amount of comprehension is defined as 0. It is also possible for the sentence to be misunderstood, in which case $\tau(p|X(p)) < \tau(p)$ and the amount of comprehension is negative. Formally, the amount of comprehension of sentence p equals

$$\text{compr}(p) = \frac{\tau(p|X(p)) - \tau(p)}{\tau(p|p) - \tau(p)}.$$

After training, the networks' average comprehension scores lie significantly above 0, both for training and test sentences. This shows that the network learned its task and could extrapolate to novel sentences.

3.3. *Probing the hidden layer*

After processing a sentence, the activations of the hidden-layer elements form an intermediate vector representation of the sentence, called 'the sentence's vector'. Frank et al. (in press) investigated the nature of this representation. It was found that, on average, the euclidean distance between vectors representing sentences that differ at the surface text level but not propositionally is shorter than the distance between vectors representing sentences that do differ propositionally. Also, vectors of sentences that differ propositionally but describe identical situations were, on average, closer together than vectors of sentences describing different situations. Assuming vectors that are closer together are harder to tell apart, these results correspond to Fletcher and Chrysler's (1990), discussed in the introduction.

Here, we shall probe the hidden layer in more detail. In particular, we are interested in finding out if particular sentence vectors encode mostly textual, propositional, or situational information, and how this relates to comprehension scores. This is done by computing distances between vectors of triplets of sentences, such as those labelled A, B, and C_1 in Table 2.

Table 2. Four microlanguage sentences, constituting two test triplets, and their propositional structures.

	sentence	proposition
A	*Bob is tired and inside*	INSIDE(BOB) ∧ TIRED(BOB)
B	*Bob is tired and plays soccer*	PLAY(BOB,SOCCER) ∧ TIRED(BOB)
C_1	*Bob is outside and tired*	OUTSIDE(BOB) ∧ TIRED(BOB)
C_2	*Bob plays a_computer_game and is tired*	PLAY(BOB,COMP) ∧ TIRED(BOB)

Which two of the three sentences A, B, and C_1 are most similar to each other depends on the representational level one looks at. When concerned only with the surface text, A and B are most similar since they have a four-word sequence in common (*Bob is tired and*). At the propositional level, A and C_1 are most similar because they differ by only one predicate (INSIDE vs OUTSIDE). Situationally, however, B and C_1 are most similar since soccer must be played outside which excludes being inside.

Replacing sentence C_1 by C_2 results in another triplet. Again, A and B are most similar textually. For this reason, (A,B) is called the 'textual pair'. Propositionally, B and C_2 have most in common, while A and C_2 describe similar situations. Therefore, pairs (A,C_1) and (B,C_2) are the 'propositional' pairs, while (A,C_2) and (B,C_1) are 'situational'. By computing distances between sentence vectors of these pairs, the triplet's level of representation can be revealed. For instance, a relatively short distance between the textual pair indicates that the triplet's representation strongly depends on surface text. Below, we shall probe the networks' intermediate representations to find out what type of information is encoded and how this relates to comprehension scores.

3.4. *Results*

Two of the test triplets used to probe the intermediate representations are shown in Table 2. Thirty additional triplets were formed by replacing *Bob* by *Jilly*, by swapping the clauses on either side of *and*, and by replacing the clause about being tired by being awake, winning, or losing. The 32 test triplets were processed by the seven trained networks, and distances between all three sentence vectors of each triplet were computed. These were normalized, such that the distance between the textual pair (the so-called 'textual distance'), the propositional distance, and the situational distance sum up to 1 for each triplet. Table 3 shows these distances, averaged for each network, as well as the networks' average comprehension scores for the 64 different sentences from the test triplets.

Table 3. Textual, propositional, and situational distances, averaged per network and sorted by the networks' comprehension scores on the triplets' sentences.

	distance		
compr	text	prop	sit
.327	.357	.379	.264
.310	.378	.350	.272
.293	.329	.354	.317
.273	.339	.350	.311
.272	.367	.373	.260
.270	.231	.403	.366
.238	.257	.384	.358
av.	.323	.370	.307

Sentence vectors mainly encode situational and textual information: The average propositional distance is significantly larger than the textual ($t_{446} = 5.50; p < .0001$) and situational distances ($t_{446} = 8.38; p < .0001$). The difference between textual and situational distances is close to significant ($t_{446} = 1.74; p = .08$). Although propositional distances are larger on average, it was found to be the shortest in 11.6% of the cases. In these cases, the main level of representation was propositional.

The results in Table 3 indicate that situational representations lead to higher comprehension scores than textual or propositional representations. Indeed, there is a large negative correlation of $-.69$ between the networks' comprehension scores and the average situational distance. However, due to the limited number of data points, it is not significant ($t_5 = 2.16; p = .08$). Likewise, the correlation between comprehension scores and textual distances is positive but not significant ($r = .66; t_5 = 1.97; p = .11$). The small correlation between comprehension scores and propositional distances is far from significant.

Defining the comprehension of a triplet as the average of the comprehension scores of its three sentences, the number of data points can be increased by looking at the correlation between comprehension and distances per triplet. Although these correlations are lower than those presented above, they are significant for textual ($r = .19; t_{222} = 2.85; p < .005$) and situational ($r = -.26; t_{222} = 3.99; p < .0001$) distances.

3.5. *Discussion*

The more situationally represented sentence triplets are, the better they are comprehended by the network. This is not surprising since comprehension scores measure the quality of the *situational* output representation. If the intermediate representation already encodes the described situation, the output can be expected to be closer to the desired output compared to sentences whose vector is less situational.

During training, the network learns to produce situational output representations of the sentences. This task will make the network tend towards also developing situational *intermediate* representations, which help produce the desired output. As a result, vectors encode mainly situational information. Also, it does not come as a surprise that propositional information in vectors is rare. During training, the network is provided with both textual input and the desired situational output, while a propositional representation is something it has to develop by itself. It is likely that a more complex microlanguage and microworld, consisting of several entities that can be combined in different ways, will result in more propositional representations, since it may be the need for comprehending such complexities that drives the development of propositional intermediate representations.

4. Conclusion

The mental representation of text is usually assumed to include both a propositional textbase level and a situational level. The DSS model shows how world knowledge and a situational representation of text can be implemented. Frank et al.'s (in press) simple sentence comprehension model is a first attempt to extend the DSS model with a textbase-like level. However, the intermediate representations developed by the model are very different from the propositional structures that are traditionally thought to constitute the textbase.

First of all, there are no predicate-argument structures in the intermediate representation. Second, although sentence vectors encode mainly propositional information in a few cases, they are never purely propositional. Textual and situational information also affects the intermediate representation. Third, the construction of a textbase is not assumed to be a goal of text comprehension. Instead, there is only one comprehension goal: the construction of a situational representation. Even if some propositional information is encoded in the network's hidden layer, it is only there to the extent that it is useful for accomplishing this goal.

References

Dell, G.S., McKoon, G., & Ratcliff, R. (1983). The activation of antecedent information during the processing of anaphoric reference in reading. *Journal of Verbal Learning and Verbal Behavior*, 22, 121-132.

Elman, J.L. (1990). Finding structure in time. *Cognitive Science*, 14, 179-211.

Fletcher, C.R., & Chrysler, S.T. (1990). Surface forms, textbases, and situation models: recognition memory for three types of textual information. *Discourse Processes*, 13, 175-190.

Frank, S.L., Koppen, M., Noordman, L.G.M., & Vonk, W. (2003). Modeling knowledge-based inferences in story comprehension. *Cognitive Science*, 27, 875-910.

Frank, S.L., Koppen, M., Noordman, L.G.M., & Vonk, W. (in press). Modeling multiple levels of text representation. In F. Schmalhofer & C.A. Perfetti (Eds.), *Higher level language processes in the brain: inference and comprehension processes.* Mahwah, NJ: Erlbaum.

Kintsch, W., & Van Dijk, T.A. (1978). Toward a model of text comprehension and production. *Psychological Review*, 85, 363-394.

Kohonen, T. (1995). *Self-Organizing Maps.* Berlin: Springer.

Ratcliff, R., & McKoon, G. (1978). Priming in item recognition: evidence for the propositional structure of sentences. *Journal of Verbal Learning and Verbal Behavior*, 17, 403-417.

Van Dijk, T.A., & Kintsch, W. (1983). *Strategies of discourse comprehension.* New York: Academic Press.

Zwaan, R.A. (1999). Embodied cognition, perceptual symbols, and situation models. *Discourse Processes*, 28, 81-88.

Zwaan, R.A., Madden, C.J., Yaxley, R.H., & Aveyard, M.E. (2004). Moving words: dynamic representations in language comprehension. *Cognitive Science*, 28, 611-619.

Acknowledgements

The research presented here was supported by grant 451-04-043 of the Netherlands Organization for Scientific Research (NWO). I would like to thank Mathieu Koppen and Leo Noordman for their comments on earlier versions of this paper.

EMERGENCE OF LINGUISTIC FEATURES: INDEPENDENT COMPONENT ANALYSIS OF CONTEXTS

TIMO HONKELA[A] AAPO HYVÄRINEN[B] JAAKKO VÄYRYNEN[A]

[A] *Neural Networks Research Centre, Helsinki University of Technology*
P.O.Box 5400, FI-02015 TKK, Finland
timo.honkela@hut.fi

[B] *HIIT Basic Research Unit, Department of Computer Science, University of Helsinki*
P.O.Box 68, FI-00014 University of Helsinki, Finland
aapo.hyvarinen@helsinki.fi

We study how independent component analysis can be used to create automatically syntactic and semantic features based on analyzing words in contexts.

1. Introduction

We show that independent component analysis (ICA) (Hyvärinen et al. 2001) applied on word context data gives distinct features that reflect syntactic and semantic categories. The analysis gives features or categories that are both explicit and can easily be interpreted by humans. This result can be obtained without any human supervision or tagged corpora that would have some predetermined morphological, syntactic or semantic information. The results include both an emergence of clear distinctive categories or features and a distributed representation. This is based on the fact that a word may belong to several categories simultaneously in a graded manner. We wish that our model provides additional understanding on potential cognitive mechanisms in natural language learning and understanding. Our approach attempts to show that it is possible that much of the linguistic knowledge is emergent in nature and based on specific learning mechanisms.

2. Independent Component Analysis

In the following, we consider the use of independent component analysis (ICA) (Comon 1994, Jutten & Hérault 1991, Hyvärinen et al. 2001) in the extraction of linguistic features for expression in due contexts. ICA learns features in an unsupervised manner. Several such features can be present in a word, and ICA gives the explicit values of each feature for each word. We expect the features to coincide with known syntactic and semantic categories: for instance, we expect

ICA to be able to find a feature that is shared by words such as "must", "can" and "may". In earlier studies, independent component analysis has been used for document level analysis of texts (see, e.g., Bingham et al. 2001, Bingham et al. 2002).

We first give a brief outline of the basic theory of independent component analysis (Hyvärinen et al. 2001). In the classic version of the ICA model, each observed random variable is represented as a weighted sum of independent random variables. The weights in the sum (which can be negative as well as positive) can be collected in a matrix, called the mixing matrix. The weights are assumed to be different for each observed variable, so that the mixing matrix can be inverted, and the values of the independent components can be computed as some linear functions of the observed variables.

A classical application for the ICA, blind signal separation (BSS) can be used as an example of the basic idea. In BSS, one studies signals that originate from a number of separate sources, e.g., discussants in a cocktail party who are speaking at the same time[a]. The signals are mixed in different proportions depending on the relative distance of a listener to each sound source. This kind of situation is illustrated in Figure 1 in which five listeners have the task of separating five signal sources from each other.

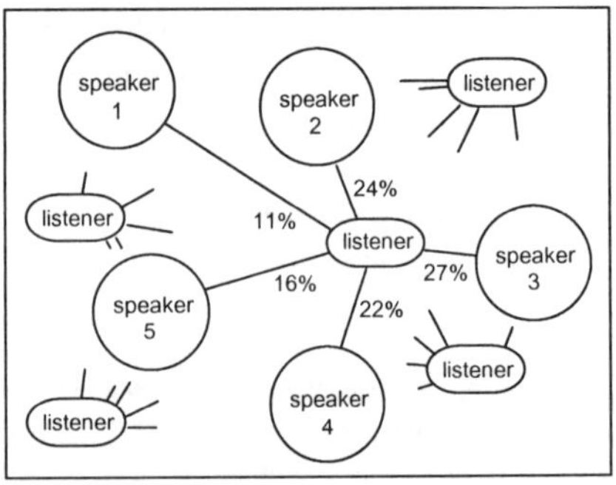

Figure 1. An illustration of a blind signal separation situation and the idea of a mixing matrix. Each listener receives a unique mixture of signals as a combination of five source signals.

[a] The cocktail party should be considered merely as a simple illustration of blind source separation. A real cocktail-party problem is much more complicated than described here due to such factors as the slow propagation of sound in the air, echoes, nonlinearities in the microphones, and noise.

Sources numbered from 1 to 5 refer to the original signals and each listener receives a unique combination of the signals. Based on the distance to each source, the vector that represents the mix of the five sources for one indicated listener in Figure 1 could be [0.11 0.24 0.27 0.22 0.16]. Putting together similar vectors for each listener we would gain a mixing matrix that, in general, would be unknown in a BSS task.

The goal in ICA is to learn the decomposition in an unsupervised manner, which means that we only observe the mixed signals and have no information about the mixing coefficients or the contents of the original signals. For the cocktail party problem, this would mean that ICA would provide an estimate of each original signal as well as of the mixing matrix. It may sound surprising that one could make an estimation of both the original signals and the mixing matrix provided only the observed signals. This challenge could be compared with a situation in which one is asked the sum of which two integers 63 is. The answer could, for instance, be 12+51 or 38+25, or, if negative values are included, there would be an infinite number of options.

The "trick" needed for the successful estimation of the original signals and the mixing matrix is based on is the simple assumption that the original signals are statistically independent. Two variables are independent if information on the value of one variable does not give any information on the value of the other. This does not need to hold for the observed variables. The independence assumption gives ground for finding the estimates the same way as if in summing up two integers to get 63 we would know that one of the numbers would need to be as closely as possible two times larger than the other. Then one could conclude that the sum would be 21+42. Two variables can be independent, for example, if they could be outcomes of two events that have nothing to do with each other, or random signals originating from two different physical processes. In case of two variables, the independence holds if and only if the joint probability of the variables is the same as the product of the two variables considered separately. This definition extends to any number of random variables. Thorough description of the ICA methodology and underlying mathematical and statistical principles is given in (Hyvärinen et al. 2001)

3. Analysis of Word Contexts

We can ask if a mixing structure introduced in the previous section is discernable in a higher level of abstraction than in the process of finding the original speech signals. For the BSS example presented above, the task was to find out what are the original signals and in which proportion each original signal is present in the perceived signals. In a more general level, ICA analysis has been used to find separate underlying components or variables. For linguistic data, the task could be to see whether different underlying categories could be found in an unsupervised manner. For instance, one can study if it is possible to find syntactic and semantic features based on an analysis of words in their contexts. The idea is that maybe words in contexts can be divided into a number of roles that they serve.

Contextual information has widely been used in statistical analysis of natural language corpora (consider, e.g., Church & Hanks 1990, Schütze 1992, Lund et al. 1995, Manning & Schütze 1999). One useful numerical representation for written corpora can be obtained by taking into account the sentential context in which the words occur. First, we represent each word by a vector in an n-dimensional space, and then code each context as an average of the vectors representing the words in that context. In the simplest case, the dimension can be taken equal to the number of different words, and each word is represented by a vector with one element equal to one and others equal to zero. Then the context vector simply gives the frequency of each word in the context. For computational reasons, however, the dimension may be reduced by different methods.

The ICA model in its basic form assumes that the components are independent, which may seem to be a very stringent assumption. However, this assumption need not be taken too seriously, because even if the components are not exactly independent, ICA algorithms can be interpreted as finding the most independent components. Furthermore, it can be shown that ICA estimation can often be interpreted as maximization of the sparseness of the components, i.e., the components should have only a few entries that are significantly different from zero. Sparseness is a very intuitive and useful property that has a long history in sensory coding, see, e.g., (Field 1994), so it seems likely that it it would be useful in this context as well.

For the ICA analyses on the word contexts we applied FastICA[a] software package for Matlab. We formed a context matrix in which the elements consisted of the number of occurrences of one word in the immediate context of another word, i.e, one word followed by another word with no words between them. We used this kind of short context but we are aware that one can use contexts of different sizes (consider, e.g., Redington et al. 1998, McDonald 2000).

We fed the word-context matrix to the FastICA algorithm so that each column was considered one data point, and each row one random variable. The dimension of the data was reduced to 10 by principal component analysis (PCA). The motivation for this operation can be considered, e.g., from the point of view that the ICA method can be divided into three consecutive steps: PCA, normalization of variances and ICA-rotation (Hyvärinen et al. 2001). The dimensionality reduction using PCA was motivated by computational efficiency.

The data used in the experiments consists of collection of e-mails sent to the connectionists mailing list. More details of the approach have been given in (Honkela et al. 2003, Honkela and Hyvärinen 2004).

The results of the ICA analysis corresponded in most cases very well or at least reasonably well with our preliminary intuitions. The system was able to automatically create distributed representations as a meaningful collection of emergent linguistic features; each independent component was one such feature.

In Figure 2, there are three examples of the analysis results[a]. In considering the feature distributions, it is good to keep in mind that the sign of the features is arbitrary. This is because of the ambiguity of the sign in the ICA model: one could multiply a component by -1 without affecting the model. Figure 2 shows how the third component is strong for nouns. For the words "systems" and "psychology" an additional component is strong. Namely, in more thorough analyses it became apparent that plural nouns shared the same pattern of peaked third and fifth component. On the other hand, for nouns that refer to disciplines it became apparent that both third and fourth component is strong.

---

[a] http://www.cis.hut.fi/projects/ica/fastica/

[a] More results can be found in (Honkela et al. 2003) and (Honkela & Hyvärinen 2004). In this paper we focus on the conceptual and modeling aspects of the ICA-based context analysis methodology.

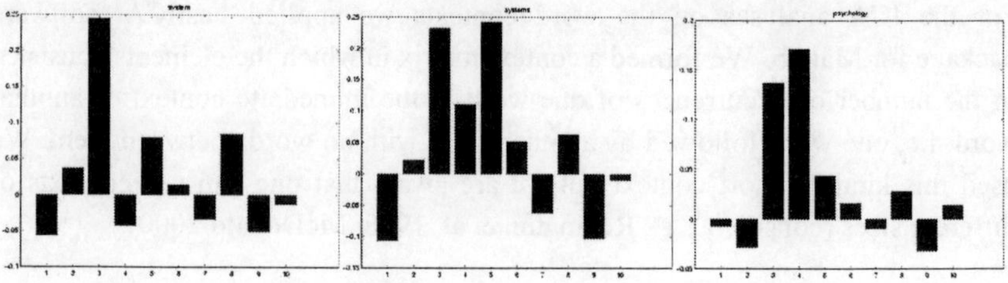

Figure 2. Ten independent components for three words (starting from the left): "system", "systems" and "psychology".

In this analysis we used ten as the number of ICA features which sets a limit on the complexity of the feature encoding. We used this limit in order to demonstrate the power and usefulness of the method in a simple manner. A higher number of features can be used in order to obtain more detailed feature distinctions.

In order to verify the match between ICA-based features and traditional linguistic categories systematically, we have conducted analyses in which the emergent features are compared with the categories provided in tagged corpora. In Figure 3, we show the results of one such analysis for nouns in plural form. Among the ICA-based features the one that has the closest match with the traditional category is first selected. In the study, 1000 most common words for which the traditional linguistic category was known were considered. Each of the points in the diagram corresponds to a word. The position of a point in the y-axis indicates the weight of a particular word in the mixing matrix for the particular feature. Among the words we have circled the ones that are known to belong to the category of plural nouns. Specifically, the comparison has been made using Brown corpus[a] and the corresponding tag is called NNS. The number of those words is 56 among the 1000 most common words. The successful match between the ICA-based feature and the traditional category is rather apparent because the circled points are located in the upper part of the diagram. The majority of the other words have a weight that is close to zero. For illustration, one can list the first 20 words that have the highest weight beginning from the best matching word: "parts", "persons", "places", "conditions", "circumstances", "animals", "times", "forms", "matters", "things", "words", "lines", "houses", "letters", "names", "days", "states", "officers", "books" and "works".

[a] http://helmer.aksis.uib.no/icame/brown/bcm.html

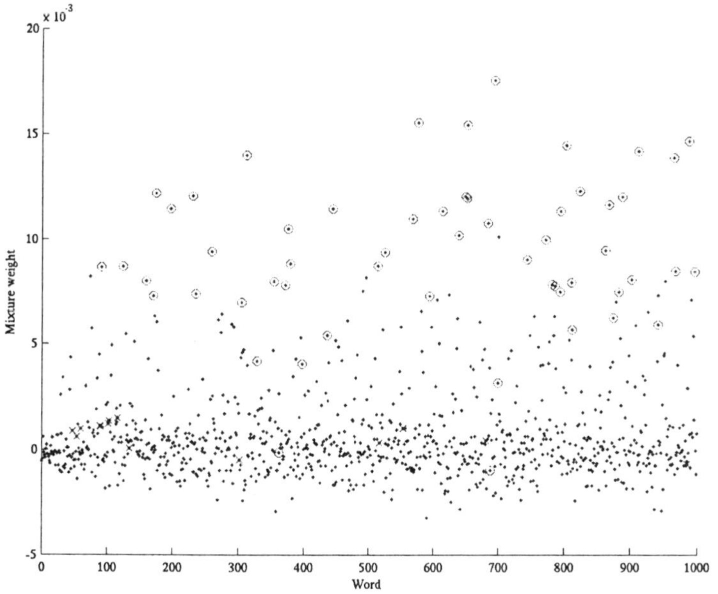

Figure 3. A comparison between an ICA-based emergent feature and a traditional linguistic category in a tagged corpus. Circled words belong to the plural noun (NNS) category. We can see that the words belonging to this particular category are clearly different from most of the other words when the mixture weight (y-axis) is considered.

4. Emergent Representations

In the following, we discuss the potential of independent component analysis for modeling some aspects of language learning and in adaptive conceptual modeling. We compare the merits of the ICA method with some earlier results gained by using the self-organizing maps (SOM). Earlier, the self-organizing map has been used in the analysis of word context data by, e.g., Ritter and Kohonen (1989) (artificially generated short sentences), and Honkela et al. (1995) (Grimm fairy tales). The result can be called a self-organizing map of words, or a word category map. Areas or local regions on a word category map can be considered as implicit categories or classes that have emerged during the learning process. The borderlines for the emergent implicit SOM-based categories have to be determined separately. In the ICA analysis one is able to find the categories in an automated way.

In the SOM-based word category map each word appears in one location. This means, among other things, that one cannot have a map in which several characteristics or categories of one word would be represented unless the categories overlap and accordingly the corresponding areas of the map overlap. On the other hand, the ICA analysis provides a sparse encoding of the words in such a way that there can be a collection of features associated with each word. ICA can provide an arbitrary number of useful features per each word. The features can be syntactic as well as semantic like proposed already by Fillmore (1968). One advantage of using ICA is that the features emerge automatically.

One should also remember that the word context analysis based on written texts is only partially able to reveal the underlying syntactic and semantic feature structures. For instance, it is commonplace that opposites appear in similar contexts in texts. For more realistic language learning simulations it is necessary to include other kinds of contexts including visual perceptions, actions and activities associated with linguistic expressions.

5. Conclusions

We have shown how independent component analysis can find explicit features that characterize words in an intuitively appealing manner. There are various methods such as self-organizing maps (Kohonen 2001, Ritter & Kohonen 1989) and latent semantic analysis (Deerwester et al. 1990) that can be used for automatic statistical methods for linguistic analysis. However, independent component analysis appears to make possible a qualitatively new kind of results that have earlier been obtainable only through hand-made analysis. Moreover, the results indicate that autonomous agents would be able to learn linguistic feature systems in an unsupervised manner.

The analysis results show how the ICA analysis was able to reveal underlying linguistic features based solely on the contextual information. This means that no dictionary or thesaurus was used to guide the ICA analysis and the samples were not labeled. The results include both an emergence of clear distinctive categories or features as well as a distributed representation based on the fact that a word may belong to several categories simultaneously. For illustration purposes in the first experiment we kept the number of features low, i.e., ten. However, similar approach scales well up to higher numbers of dimensions.

Future research directions include analysis of larger corpora for extracting larger number of independent components. On a qualitative level, polysemy will be considered as one particularly challenging research topic. Whether the component values can be applied as degrees of membership for each word in each category is a question of further analysis. The distributed representation can be used as a well-motivated low-dimensional encoding for words. The limited number of dimensions brings computational efficiency whereas the meaningful interpretation of each component provides basis for intelligent processing.

References

Bingham E., Kaban A. and Girolami, M. (2001). Finding topics in dynamical text: application to chat line discussions. *Poster proceedings of the 10th International World Wide Web Conference (WWW10)*, May 1-5, 2001, Hong Kong. (pp. 198-199)

Bingham E., Kuusisto J. and Lagus K. (2002). ICA and SOM in Text Document Analysis. *Proceedings of the 25th ACM SIGIR 2002 International Conference on Research and Development in Information Retrieval*, August 11-15, 2002, Tampere, Finland. (pp. 361-362)

Church K. and Hanks P. (1990). Word association norms, mutual information and lexicography. *Computational Linguistics*, 16, 22-29.

Comon, P. (1994). Independent component analysis - a new concept? *Signal Processing*, 36, 287-314.

Deerwester S., Dumais S., Landauer T., Furnas G. and Harshman R.A. (1990). Indexing by latent semantic analysis. *Journal of the American Society of Information Science*, 41, 391-407.

Field, D.J. (1994). What is the goal of sensory coding? *Neural Computation*, 6, 559-601.

Fillmore C. (1968). *Universals in Linguistic Theory*, chapter The case for case, pp. 1-88. Holt, Rinehart and Winston, Inc.

Honkela T., Pulkki V. and Kohonen T. (1995). Contextual relations of words in Grimm tales analyzed by self-organizing map. In *Proceedings of ICANN-95, International Conference on Artificial Neural Networks*, volume 2, EC2 et Cie. (pp. 3-7)

Honkela T., Hyvärinen A. and Väyrynen J. (2003). *Emergence of linguistic representation by independent component analysis*. Report A72, Espoo, Finland: Helsinki University of Technology, Laboratory of Information and Computer Science.

Honkela T. and Hyvärinen A. (2004). Linguistic Feature Extraction using Independent Component Analysis. In *Proceedings of IJCNN04, International Joint Conference on Neural Networks*.

Hyvärinen A., Karhunen J. and Oja E. (2001). *Independent Component Analysis*. New York: John Wiley & Sons.

Jutten C. and Hérault J. (1991). Blind separation of sources, part I: An adaptive algorithm based on neuromimetic architecture. *Signal Processing*, 24, 1-10.

Kohonen T. (2001). *Self-Organizing Maps*. Berlin, Heidelberg: Springer.

Lund, K., Burgess, C. and Atchley, R.A. (1995). Semantic and associative priming in high-dimensional semantic space. In *Proceedings of the 17th Annual Conference of the Cognitive Science Society*, 660-665.

Manning C. and Schütze H. (1999). *Foundations of Statistical Natural Language Processing.* Cambridge, MA: MIT Press.

McDonald, S. (2000). *Environmental determinants of lexical processing effort.* PhD dissertation, University of Edinburgh.

Redington, M., Chater, N. and Finch, S. (1998). Distributional information: A powerful cue for acquiring syntactic categories. *Cognitive Science*, 22, 425-469.

Ritter H. and Kohonen T. (1989). Self-organizing semantic maps. *Biological Cybernetics*, 61 (4), 241-254.

Schütze H. (1992). Dimensions of meaning. *Proceedings of Supercomputing.* (pp. 787-796)

LATENT LINGUISTIC CODES FOR MORPHEMES USING INDEPENDENT COMPONENT ANALYSIS

KRISTA LAGUS MATHIAS CREUTZ SAMI VIRPIOJA
Neural Networks Research Centre, Helsinki University of Technology,
P.O. Box 5400, 02015 HUT, Finland
krista.lagus@hut.fi

We study properties of morphemes by analyzing their use in a large Finnish text corpus using Independent Component Analysis (ICA). As a result, we obtain emergent linguistic representations for the morphemes. On a coarse level, main syntactic categories are observed. On a more detailed level, the components depict potential thematic roles of the morphemes. An interesting question is whether these discovered lower-dimensional representations could be directly utilized in language processing applications.

1. Introduction

In recent years the use of statistical word-based methods for syntactic and semantic analysis has become increasingly popular. In highly inflecting languages, such as Finnish, a word may have hundreds or even thousands of different inflected forms. Many of the word forms are rare, but they nonetheless make up an important part of the words in a text. Due to the large proportion of rare word forms, statistical analysis methods are of limited use when applied directly to word contexts. Moreover, many cognitively interesting phenomena leave traces *within* words and not only on the sentence level. For example, grammatical relations and to some extent thematic roles (Agent, Patient, Theme, Experiencer, Beneficiary, Location, etc; see e.g., (Saeed, 1997)) are often marked using appropriate inflected forms, not so much using word ordering and function words as in English. We take morphemes as basic units of meaning, and segment every word form into a sequence of morphemes, or more specifically into allomorphs (realization variants). We then apply an unsupervised method called Independent Component Analysis (ICA) (see e.g., Hyvärinen, Karhunen & Oja, 2001) to analyze the contexts of the morphemes to discover latent small-dimensional representations for each morpheme. Ideally each component might code for a single cognitive, syntactic, semantic or phonological property. In a related work (Honkela, Hyvärinen & Väyrynen, 2005), ICA was applied to the analysis of English words based on their averaged contexts.

1.1. *On Finnish Morphology*

Word formation in Finnish mainly takes place using agglutination, that is, the concatenation of morphemes (for a detailed description, see Hakulinen, 1979). A Finnish word typically consists of several morphemes, with an alternation of stems and *suffixes* (and sometimes prefixes):

Finnish word; its segmentation	English translation; literal translation of the segments
kahvinjuojallekin; kahvi + *n* + juo + *ja* + *lle* + *kin*	also for [the] coffee drinker; coffee + of + drink + -er + for + also
tietäisimmeköhän; tietä + *isi* + *mme* + *kö* + *hän*	would we really know?; know + would + we + INTERR + indeed

1.2. *Independent Component Analysis (ICA)*

ICA is a statistical analysis method that attempts to find the latent components (also called sources or underlying factors) that generated a data set. The ICA model (Hyvärinen et al., 2001) assumes that the observations $\mathbf{x}$ are a mixture of the activities of some unknown, independent sources $\mathbf{s}$. In the linear case, $\mathbf{x} = \mathbf{A}\,\mathbf{s}$, where $\mathbf{A}$ is called the mixing matrix. The problem is then to determine $\mathbf{A}$ and $\mathbf{s}$ when only the observations $\mathbf{x}$ are known. A more familiar method may be Principal Component Analysis (PCA) (or Singular Value Decomposition, SVD) which finds *uncorrelated* sources. In contrast, ICA finds *statistically independent* sources, which generally appear to be more interesting since they often correspond more closely to the original underlying processes.

2. Morpheme Context Data and Experiments

We study properties of morphemes by analyzing their use in a large text corpus. For each morpheme i we collect a context vector $\mathbf{x}_i$. The context vectors are then analyzed using ICA.

2.1. *Corpus*

The corpus consists of Finnish newspaper text with 30 million words (tokens) and 1.3 million unique word forms (types). The words were first segmented into morphemes according to the gold-standard segmentation of Hutmegs (Creutz & Lindén, 2004). Hutmegs contains linguistic, semi-automatically produced morpheme segmentations for Finnish (and English) words. On average, a word (token) in the corpus consisted of 2.0 morphemes and the average length of a morpheme was 3.6 letters. As the data morphemes to be studied we chose the 3759 most frequent morphemes.

2.2. *Formation of morpheme context vectors*

As context features we selected the 506 most common morphemes (including the word boundary) occurring in the context of the 3759 chosen data morphemes. For each data morpheme we then formed a 506-dimensional context vector as follows: For each instance of the data morpheme, if one of the context morphemes appeared immediately after the data morpheme within the same word, the corresponding value of the context vector was increased by one. The logarithm of the values was then taken, and the vectors were normalized to unity. Longer contexts could be studied, but our hypothesis was that already a minimal context may lead to interesting results.

2.3. *Application of ICA*

We applied the FastICA algorithm (Hyvärinen, 1999) to obtain the first 50 independent components. As ICA leaves the sign (positive or negative) of the components undetermined, we determined the signs by examining the data distribution for each component. As a result, a significant property is generally shown as a positive value of the component.

3. Results

When looking at the obtained 50 components, 16 seem particular to verbs (8 to inflected and 8 to derived forms), 12 to inflected forms of nouns, 2 to derived forms of nouns (adjectival and caritive, e.g., kodi+ton; home+less), 5 to locations and persons (with some distinction between internal and external locatives). Four components signify adjective stems, of which one is for comparison. Due to vowel harmony many properties appear in the component list twice: once for back vowels and a second time for front vowels. In addition, there are components that are hard to interpret, or are not very specific (e.g. 14 and 30). Figure 1 depicts activations of three morphemes that are country names, along with their interpretations. In Figure 2, three frequent, concrete nouns are displayed.

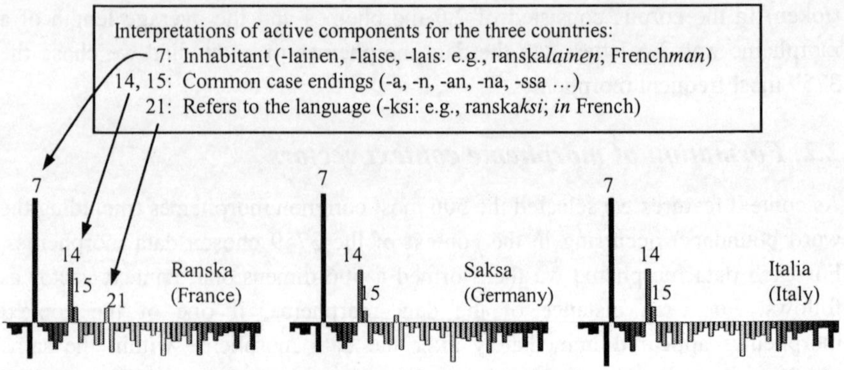

Figure 1: Activations of latent components of three morphemes that are country names: Ranska (France), Saksa (Germany) and Italia (Italy).

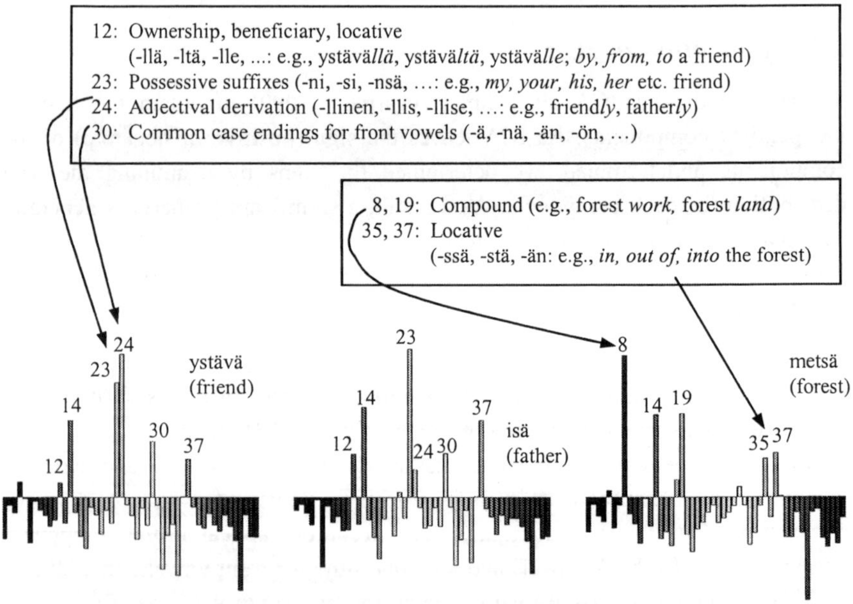

Figure 2: The ICA components for three sample morphemes, ystävä (friend), isä (father) and metsä (forest), along with some interpretations. Component 12 appears to code change and maintenance of the state of ownership and 23 the owner. 24 codes for adjectival role; note that for "friend" this component is strong (friendly), for "father" weaker (fatherly) and for "forest" nonexistent (*forestly).

4. Discussion

Most components appear to signify some interesting property. These include syntactic categories (verb, noun), thematic roles (Location, Beneficiary) cognitive or semantic properties (agent vs. action role), or morpho-phonological properties (e.g., the use of back or front vowels) of the word segments. Typically a morpheme has only a few active components, thus creating *a sparse coding* for the morphemes. An ambiguous morpheme is likely to have a higher number of active components, each reflecting one of the morpheme's central properties or uses. On a coarse level, main syntactic categories are observed. On a more detailed level, the components depict potential thematic roles of the morpheme. This result is interesting since in most cases there is no single context feature that would distinguish such roles, and since many of the context features are themselves ambiguous.

In this experiment we used a linguistic morpheme segmentation for ease of interpretation. It would also be interesting to study what representations ICA discovers for word segments that have been learned in an entirely unsupervised manner, e.g. using the method (Creutz & Lagus, 2004). Questions for future research include discovering suitable context length and optimal number of ICA components. Based on visual inspection, in this case 20 components seemed too few, while 50 was not too many. However, an evaluation method is necessary (such as performance on the TOEFL test of English as a Foreign Language).

The creation of numerical codes for morphemes opens the possibility of using these codes to learn typical patterns of words and to produce new words based on the learned patterns, while operating in a relatively low-dimensional vector space. It is an interesting question whether such codes might be useful for solving a significant problem in statistical language modeling, namely how to estimate characteristics (e.g., probability) of future utterances based on observed utterances. Moreover, one may ask whether the component values could be directly utilized by language understanding applications for decoding the meaning of an utterance.

References

Creutz M., Lagus, K. (2004) Induction of a Simple Morphology for Highly-Inflecting Languages. In *Proceedings of the 7th Meeting of the ACL Interest Group in Computational Phonology (SIGPHON), Barcelona* (pp. 43-51).

Creutz M., Lindén K. (2004) Morpheme Segmentation Gold Standards for Finnish and English. Helsinki University of Technology, Publications in Computer and Information Science, Report A77.

Hakulinen L. (1979) *Suomen kielen rakenne ja kehitys* (The structure and development of the Finnish language). Kustannus-Oy Otava, 4th ed.

Honkela T., Hyvärinen A., Väyrynen J. (2004) Emergence of Linguistic Features: Independent Component Analysis of Context (in this volume).

Hyvärinen A. (1999) Fast and Robust Fixed-Point Algorithms for Independent Component Analysis. *IEEE Transactions on Neural Network,* 10(3):626-634.

Hyvärinen A., Karhunen J., Oja E. (2001) *Independent Component Analysis*. John Wiley & Sons.

Saeed J. I. (1997) *Semantics*. Oxford: Blackwell Publishers.

A NEURAL NETWORK MODEL FOR NATURAL LANGUAGE: THE CASE OF HOMONYMY

ELENI KOUTSOMITOPOULOU*

LexisNexis UK, Information Extraction and Indexing, London, UK
Georgetown University, Washington DC., USA, Computational Linguistics
E-mail: eleni.koutsomitopoulou@lexisnexis.com

The Cognitive Linguistic Adaptive Resonant Network (CLAR-NET, Koutsomitopoulou 2004) is a biologically faithful neural network model of conceptual associations with input from English. Conceptual linguistic associations are analyzed as dynamic resonant patterns represented in terms of neuronal activation. The CLAR-NET model extends the line of research of Loritz 1999 to various linguistic phenomena in the realm of conceptual analysis, among which homonymy. This short presentation outlines the model using homonymy as an illustrative example.

1. Background

In general, formal representation methods give good context-free analyses of noiseless linguistic input by enforcing an abstraction away from the actual properties of real-time natural language (NL). On the other hand, semantic fuzziness and the realities of the linguistic and discourse context are systematically ignored. Natural language is a human neurocognitive phenomenon and as such it can be best represented in a biologically faithful neural network model. Loritz 1991 argues that most connectionist architectures (e.g. Parallel Distributed Processing, PDP) model cerebellar cortex, which he characterizes as "noncognitive" claiming that language learning primarily occurs in cerebral cortex. Grossberg's early Adaptive Resonance Theory (ART) describes cerebral anatomies and processes although without regard to linguistic phenomena. Loritz 1999 first explained a range of phonological, morphological and syntactic NL phenomena within the general early ART framework in a framework he calls "Adaptive Grammar" (AG). Koutsomitopoulou 2004 further extends the AG model to a

*This work is based on research included in the author's doctoral dissertation, Georgetown University, Washington DC, March 2004

number of critical semantic linguistic events. Using lateral inhibition in an ANN model focusing on NL input is a relatively new idea. Even when lateral-inhibition models have been proposed for NL input (as in the PDP model), there is no other provision for the general neurobiological faithfulness of the model. For instance, other models are not sensitive to the inhibition to excitation ratio, which is a neurobiological reality in the brain (e.g. Rubenstein and Merzenich 2003, Mel and Schiller 2004)[a]. Inhibition in STM models (such as that by Cottrell and Small 1983) is recognized as a necessary part of the disambiguation process. However, disambiguation in these models only results in selection among particular (often ad hoc) semantic alternatives of word-level NL input. Similarly, early PDP models like the Interactive Activation model of Rumelhart and McClelland (1981) introduce "interactive connections" to represent top-down cognitive processes. Neither of these models, though, involves any (LTM) learning. Before CLAR-NET there has been no attempt of dynamically modelling high level cognitive/semantic linguistic tasks where lateral inhibition is a ubiquitous local mechanism inherent in a dual, STM and LTM, system[b].

2. Overview of the CLAR-NET Algorithm, Adaptive Resonance Theory and Adaptive Grammar

In the core of the CLAR-NET algorithm lies a natural language parser that feeds the network with discourse. The parse-tree nodes corresponding to the basic argument structure (actor, patient, main verb, etc.) of the sentences and paragraphs of the discourse are directly mapped in the network. This means that the nodes of the network are never hard-wired; they are learned on the fly as new input passes through the parser.

(1) Map input from parser to the network.
(2) Learn input at time t.
(3) Introduce new (phasic) input to the network at time $t + 1$.
(4) Until stabilization, compute overall network resonance.

CLAR-NET calculates both short-term memory (STM) and long-term memory (LTM) by modeling both excitatory and inhibitory forces (corresponding to neurotransmitter release). Roughly, the former facilitates

[a]See Koutsomitopoulou 2004 for an analysis of the inhibition/excitation ratio in the CLAR-NET model.

[b]For more detailed criticism of both PDP and Cottrell's models please refer to Koutsomitopoulou 2004

learning, whereas the latter helps the network classify (disambiguate) a given input. Equations 1 and 2 below were introduced by Grossberg 1972 for the calculation of STM and LTM via excitatory and inhibitory parameters:

$$\dot{x}_j = -Ax_j + Bx_i z_{ij} - Cx_k z_{kj} + I \tag{1}$$

$$\dot{z}_{ij} = -Dz_{ij} + Ex_i x_j \tag{2}$$

Parameter A = natural STM decay (forgetting). *Parameter* B = STM learning rate. *Parameter* C = inhibition rate. *Parameter* D = natural LTM decay. *Parameter* E = LTM learning rate. z_{ij} is the change in the weighted connection between node at site x_j and its excitatory x_i (Hebbian learning). *Parameter* I = type of regulatory exogenous input.

3. Illustrative case: Homonymy

In this paper I briefly present the case of homonymy. For a detailed analysis of this and a range of other linguistic phenomena please refer to Koutsomitopoulou 2004. Consider the following simple natural language sentences:

(1) The bow of this ribbon has two loops.
(2) I used a new bow for shooting those arrows.
(3) Arrow is a weapon.
(4) John has a weapon.

In sentences 1 and 2, *bow* is homonymous. In 1, *bow* belongs in the domain of *ribbons*, whereas in sentence 2 *bow* belongs in the domain of *arrows*. Sentence 3 is a factoid about *arrows*. In psychological terms, "factoids" are akin to "priming stimuli". After Ss 1-3 have been learned, S4 is presented as a "phasic input" or "probe stimulus" at time *t+1*. We seek to understand how the network responds or "learns" in response to this probe stimulus. The network diagram in figure 1 shows how Ss 1-4 and their subordinate (parsed) lexical nodes are mapped into the CLAR-NET system. Notice that sentence S1 has two nominal nodes mapped on the network: Ribbon and Bow. Sentence S2 contains Bow and Arrow. Sentence 3 (S3) is the factoid about Arrow as a Weapon. Sentence 4 (S4) presents phasic input at time *t+1*.

3.1. *STM results and LTM effects of the above experiment*

In order to achieve disambiguation we represent the homonymic concepts by distinctive dynamic resonant configuration patterns that correspond in

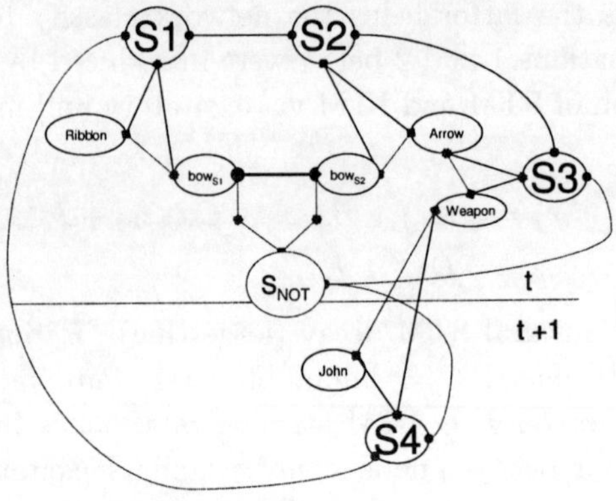

Figure 1. *Bow* homonymy.

our results data to specific neuronal activation patterns. Specifically, the *Bow* node of S2 bearing the connotations of *arrow* and *weapon* achieves higher activation compared to the activation value of *Bow* node of S1, because *Bow* node of S2 and *Bow* node of S1 compete in an inhibitory dipole (represented in the network diagram by the S_{NOT} node).This means that the reading of the homonym *bow* in S2 is preferred in the context of the current network of sentences. The table below presents the x_j values after calculation of the $\Delta\ x_j$ in equation at (1).

Node	t	$t+1$	t_{Stab}
bow_{S_1} - $ribbon_{S_1}$	9.212	8.862	8.082
bow_{S_2} - $arrow_{S_2}$	9.218	8.470	**8.679**

The table below shows the results for the two homonymous *bow* nodes in the same experiment after the calculation of the $\Delta\ z_{ij}$ in equation at (2). In both occasions, *Bow* of S2 shows clear salience over *bow* of S1, which suggests that the network has successfully disambiguated the homonym *bow* term. However, notice that the results in the table below after calculation of LTM show better discrimination between the two terms than when we calculate only STM. Also, extended results (that we cannot for reasons of space present here) show that *arrow* and *weapon* also achieve higher activation when compared to *ribbon*.

Node	t	$t+1$	t_{Stab}
bow_{S_1}	10.051	9.005	8.677
bow_{S_2}	8.715	9.184	**9.912**

Further work should investigate the pattern of rebounds and oscillations of x_j until stabilization, as well as the role of the parameters in the overall pattern of the $\Delta\ z_{ij}$.

4. Conclusions

The CLAR-NET network is a simple but biologically faithful model of language learning where semantics and meaning are treated as "adaptive resonance" phenomena. It is capable of learning to satisfactorily perform conceptual disambiguation and it achieves this goal by representing both (STM) activation and (LTM) learning in a plausible unified fashion.

References

Cottrell G.W., Small S. (1983). A connectionist scheme for modeling word sense disambiguation. *Cognition and Brain Theory,* 6, 89-120.

Grossberg S.(1972a). A neural theory of punishment and avoidance. i: qualitative theory. *Mathematical Biosciences,*15,39-67.

Grossberg S. (1972b). A neural theory of punishment and avoidance. ii: quantitative theory. *Mathematical Biosciences,*15,253-85.

Koutsomitopoulou E. (2004). *A neural network model for the representation of natural language.* PhD thesis, Georgetown University, Washington DC. Ann Arbor: UMI. 65:6, 3137058.

Loritz D. (1991). Cerebral and Cerebellar Models of Language Learning. *Applied Linguistics,* 12:3, 299-318

Loritz D. (1999). *How the Brain Evolved Language.* Oxford University Press.

McClelland J.L., Rumelhart D.E. (1981). An interactive activation model of context effects in letter perception: Part 1. An account of basic finding. *Psychological Review,* 88, 375-407.

Mel B.W., Schiller J. (2004). On the fight between excitation and inhibition: Location is everything. *Sci. STKE2004,* 44.

Rubenstein J.L.R., Merzenich M.M. (2003). Model of autism: increased ratio of excitation/inhibition in key neural systems. *Genes, Brain & Behavior,* vol.2, 5, 255.

Rumelhart D., McClelland, J. (1986). On learning the past tenses of english verbs. In McClelland J. and Rumelhart D. (Eds.), *Parallel Distributed Processing: Explorations in the Microstructure of Cognition.*,vol.2, Cambridge,MA: MIT Press. (pp. 216-271)

[illegible]

4. Conclusions

[illegible]

References

[illegible]

DYNAMIC REPRESENTATIONS OF STACK- AND QUEUE-LIKE SYNTACTIC STRUCTURES

ANDRÉ GRÜNING*

Max-Planck-Institute for Mathematics in the Sciences, Inselstr. 22–25, 04103 Leipzig, Germany

Department of Psychology, University of Warwick, Coventry, CV4 7AL, United Kingdom

We present simulations on how simple recurrent networks (SRNs) learn to represent prototypical centre-embedding and cross-serial languages, and analyse the emerging network dynamics. The networks find solutions for the tasks that can be characterised as simple dynamical systems and yield an explanation why cross-serial languages in an SRN setting are harder to aquire.

1. Introduction

While simple recurrent networks (SRNs) (Elman, 1990) have successfully been used in a wide range of areas within neural computation, they deserve attention in their own right as *non-symbolic* computing devices.

For language processing recursive structures are most interesting. Typical examples—also found in natural languages, see Figure 1—are centre-embeddings and cross-serial dependencies. It is well known how these structures are represented on symbolic automata (Hopcroft and Ullmann, 1979). Due to the different properties of SRNs (instead of a discrete state space, they have e.g. a continuous one), however, is it not clear how they represent these structures. There is a good match of theory and simulation for regular languages (Casey, 1996), but for more complex languages no systematic relationship between emerging dynamics (e.g. Bodén and Blair, 2003, Rodriguez, 2001) and theoretical approaches (e.g. Moore, 1998, Siegelmann and Sontag, 1991) has been established. In particular it is not clear what kind of dynamical system would evolve in a SRN as the representation of a particular language.

*now at *S.I.S.S.A, via Beirut 4, 34014 Trieste, Italy.* E-mail: gruening@sissa.it.

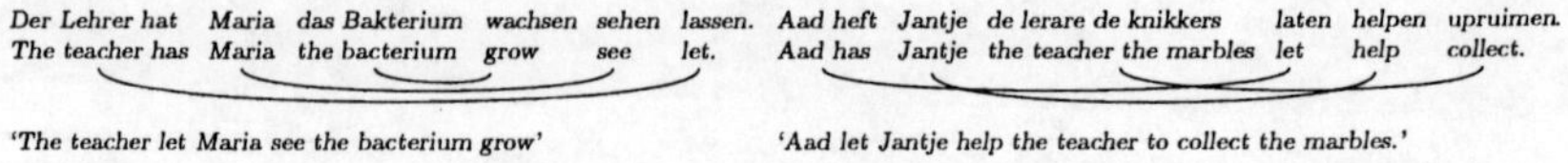

(a) Centre-embeddings in German. (b) Cross-serial dependencies in Dutch.

Figure 1. Natural language examples of recursion. Concentrating on the dependencies between matching nouns and verbs these structures can be described as a palindrome *abccba* for the centre-embedding (a) and a reduplication *abcabc* for the crossed dependencies (b). Dutch example from Bach *et al.* (1986).

2. Simulations

We are especially interested in centre-embedding and cross-serial recursions and concentrate on simple formal languages. Let w be an arbitrary string over the binary alphabet $\{a, b\}$. The centre-embedding palindromic language ww^r is a string w followed by its mirror-reversed copy w^r (e. g. *abba*, *babbbbab* ...). The cross-serial duplicating language ww is just a string w duplicated (e. g. *abab*, *babbbabb*, ...).

For the sake of brevity we will only report on the main results of the simulations and refer to Grüning (2004) for details. We used SRNs with second-order synapses similar to those in Pollack (1991). A suitably adapted version of back-propagation through time was applied.

input:	$b\ a\ b\ b$ ¶¶¶¶¶
target for ww^r:	$- - - - b\ b\ a\ b\ E$
output for ww:	$- - - - b\ a\ b\ b\ E$

Figure 2. Example input and target pairs for two languages. – indicates that the output is irrelevant.

For the training scheme we used a variant of prediction learning. Input symbols a, b are encoded in a one-hot manner, and so are the output symbols a, b. A third input ¶ and output E are provided. The network is trained on input–target pairs as in Figure 2: the second half w^r or w of a string is perfectly predictable from the initial w (the input ¶ indicates the second half is reached) and can be reproduced by the network. Thus it must organise its dynamics so as to implement a stack-like (for ww^r) or queue-like (for ww) memory. The network can then use E to indicate an empty memory.

For each language at least 100 networks with five hidden units were trained on strings of their respective language with $|w| \leq n_{ww^r}$ for ww^r and $\leq n_{ww}$ for ww. The networks were tested on all strings in the training set, and a string counted as an error when its second half was not correctly reproduced by the network. No network could learn ww with $|w| \leq 5$ without error, while 11 out of 100 could for ww with $|w| \leq 4$. For ww^r all networks trained on ww^r with $|w| \leq 4$ got all strings right, and 5 out of

125 did for $|w| \leq 5$. As a first result, we can state that ww is more difficult to acquire in a SRN setting.

However we are not primarily interested in performance but rather in the dynamics. Structured dynamics usually evolve at the edge of performace, thus we visualised the dynamics of those networks that did not make any error for ww^r, $|w| \leq 5$ and ww, $|w| < 4$ while processing 1000 strings. Two-dimensional structures as in Figure 3(a) for ww^r and in Figure 4(a) for ww were recurrent in many of the well-trained networks, and thus we focus on them as essential for the dynamics. We want to find an idealised dynamical system that resembles the actual evolved dynamics. How, then, can we implement ww^r in 2D in a simple way? The key idea is to construct a dynamical system that finds for a given sequence $x_0 x_1 \cdots x_n$ of 0s and 1s, a real number $x \in [0,1)$ with binary expansion $x = 0.x_n x_{n-1} \cdots x_0$, i. e. the reversed sequence. For ww the key idea is similar: find a dynamical system that delivers a real number x with binary expansion $x = 0.x_0 x_1 \cdots x_n$ matching the input $x_0 x_1 \cdots x_n$ sequence itself, see Figure 5. The trajectories for these dynamical systems during input of as and bs are shown in Figures 3(b) and 4(b) and capture the properties of the evolved dynamics quite well.

3. Discussion

Note that for the centre-embedding language the dynamic variables x and y are independent whereas for the cross-serial task the value of y directly influences x. Thus imprecisions in y carry over to x in the latter and accumulate there, while they do not in the former. Hence we need a more precise tuning of the weights for the ww dynamics to achieve the same performance. Thus we could say cross-serial structures are more complicated in an SRNs setting purely for dynamic reasons. Adopting a dynamical systems view of cognitive linguistics, this explains why centre-embedded syntactic structures are more common than cross-serial ones in natural language (Savitch *et al.*, 1987). However cross-serial dependencies seem to be easier to generalise (Bach *et al.*, 1986, Christiansen and Chater, 1999) to additional levels of embedding. In order to reconcile these results we conjecture that cross-serial dependencies are harder to acquire, but once acquired easier to generalise. Work elaborating this point is underway.

Acknowledgements

The author was supported by a reasearch grant of the Max-Planck-Society and by the EC Research Training Network Project "Basic Mechanisms of learning and forgetting".

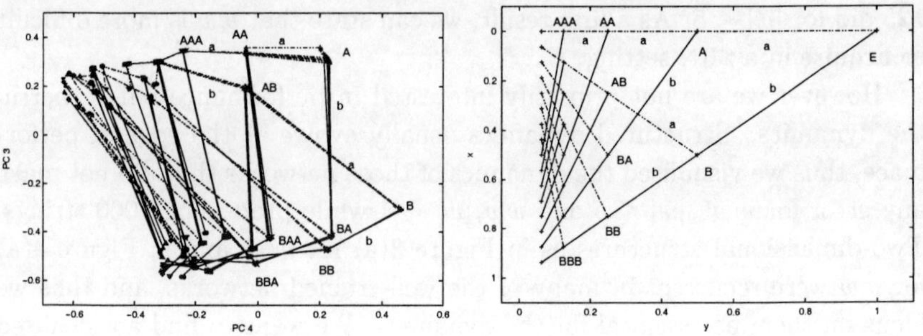

(a) Actual Dynamics.

(b) Essential Dynamics.

Figure 3. Actual and essential dynamics for the centre-embedding language $ww^r, |w| \leq 5$. For the actual dynamics, principle component (PC) analysis helped to find the most interesting 2D projection, and state vectors form clusters. Selected ones are labelled with a string equivalent to the corresponding memory contents. Dash-dotted lines mark trajectories between the clusters caused by an input a, and solid ones correspond to input b. The rather dispersed cluster for empty memory is left out for the sake of clarity. The horizontal axis encodes the number of inputs, and the vertical the contents of the memory: clusters with most recent input symbol a lie in the upper half. For the essential dynamics, imagine the empty stack cluster E smeared out and its solid b trajectory bend to the bottom to get a close fit with the actual dynamics.

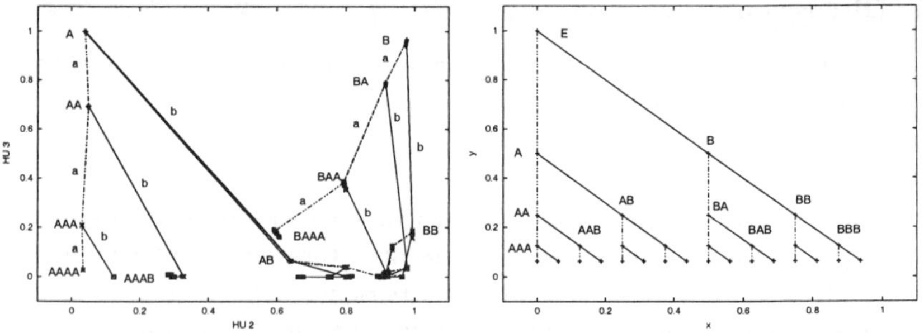

(a) Actual Dynamics.

(b) Essential Dynamics.

Figure 4. Actual and essential dynamics of the cross-serial language $ww, |w| \leq 4$. The same legend as in Figure 3 applies and, for the actual dynamics, the cluster for empty memory has again been left out. The vertical axis is queue depth, the horizontal one the contents: clusters with first input symbol a lie in the left half. The branch for first input symbol b unfortunately suffers from non-linear distortion due to saturation. Otherwise it looks like a twisted version of the essential dynamics.

init:	$x = 0$	$y = 1$
input a:	$x \mapsto (x + \mathbf{0})/2$	$y \mapsto y/2$
input b:	$x \mapsto (x + \mathbf{1})/2$	$y \mapsto y/2$
input ¶:	$x \mapsto 2x \mod 1$	$y \mapsto 2y$

(a) Dynamics for ww^r.

init:	$x := 0$	$y = 1$
input a:	$x \mapsto x + \mathbf{0}$	$y \mapsto y/2$
input b:	$x \mapsto x + \mathbf{y}/2$	$y \mapsto y/2$
input ¶:	$x \mapsto 2x \mod 1$	$y \mapsto 2y$

(b) Dynamics for ww.

Figure 5. The essential dynamical systems in 2D. After applying the corresponding mapping for a sequence of inputs, the binary expansion of x responds to the reversed input sequence of as and bs in (a) and to the sequence itself in (b). For ww^r we manipulate only the easily accessible most significant digit of the binary expansion of x. Dividing by 2 shifts the binary expansion right, and in the vacant first position we can insert a 0 or 1 whichever the input (in fact, one should divide by a number slightly greater than 2 in order to get a gap between the representations of $0.0111\cdots$ and $0.111\cdots$, Siegelmann and Sontag 1991). The second dynamic variable y counts the number of inputs in a multiplicative way. Reading out x, we multiply by 2 and check whether x is greater (output a) or smaller than 1 (output b). The memory is empty (output E), when $y = 1$ again. For ww the difficulty is to directly manipulate a less significant digit of x. y is such that its value corresponds to the least significant digit of x. Thus we can add $y/2$ to x to change the right digit. Reading out proceeds as for ww^r.

References

Bach, E., Brown, C., Marslen-Wilson, W. (1986). Crossed and nested dependencies in German and Dutch: a psycholinguistic study. *Language and Cognitive Processes*, 1, 249–262.

Bodén, M., Blair, A. (2003). Learning the dynamics of embedded clauses. *Applied Intelligence*, 19(1/2), 51–63.

Casey, M. (1996). The dynamics of discrete-time computation, with application to recurrent neural networks and finite state machine extraction. *Neural Computation*, 8, 1135–1178.

Christiansen, M. H., Chater, N. (1999). Toward a connectionist model of recursion in human linguistic performance. *Cognitive Science*, 23(2), 157–205.

Elman, J. L. (1990). Finding structure in time. *Cognitive Science*, 14, 179–211.

Grüning, A. (2004). *Neural Networks and the Complexity of Languages*. Ph.D. thesis, School of Mathematics and Computer Science, University of Leipzig.

Hopcroft, J. E., Ullmann, J. D. (1979). *Introduction to Automata Theory, Languages, and Computation*. Mass.: Addison-Wesley.

Moore, C. (1998). Dynamical recognizers: Real-time language recognition by analog computers. *Theoretical Computer Science*, 201.

Pollack, J. B. (1991). The induction of dynamical recognizers. *Machine Learning*, 7, 227–252.

Rodriguez, P. (2001). Simple recurrent networks learn context-free and context-sensitive languages by counting. *Neural Computation*, 13, 2093–2118.

Savitch, W. J., Bach, E., Marsh, W. (Eds.) (1987). *The Formal Complexity of Natural Language*. Dordrecht: D. Reidel Publishing Company.

Siegelmann, H. T., Sontag, E. D. (1991). On the computational power of neural nets. *Journal of Computer and System Sciences*.

THE ACTIVE ROLE OF PROPER NAMES: EVIDENCE FROM NEURAL NETWORK EXPERIMENTS AND PHILOSOPHY OF LANGUAGE CONSIDERATIONS

BARBARA GIOLITO

University of Eastern Piedmont "A. Avogadro", via Duomo 6,
13100 Vercelli, Italy
bgiolito@lett.unipmn.it

The present work considers the possibility that names play an active role in language, as a *point of attention* around which properties can be clustered. The contribution of neural network experiments of linguistic asks is here discussed in support of such an hypothesis.

1. The nature of proper names

In philosophy of language, the *cluster concept theory* is one of the most influential descriptivist theories of proper names. This was first proposed by Wittgenstein (1953) and then developed by Searle (1958). In this theory, the meaning of names is given by groups of individually not-necessary descriptions.

Such a theory has been criticized by Kripke (1972). He suggests that the use of proper names doesn't usually imply that people are able to associate exhaustive descriptions to names. In many cases, people can associate to a name only a partial description of its referent, unable to univocally individuate such a referent. For example, they can associate to the name 'Goedel' the only description 'a famous logicians that has been living in the XX century'. Alternatively, people can seemingly identify the referent. For example, today we could imagine to identify Goedel as 'the logician that demonstrated the incompleteness of arithmetic', and discover tomorrow that another logician did such a discovery and that Goedel was a cheat. Interesting are also the cases in which the referent of the name is supposed to be a real entity, but the only description we have doesn't refer to it. For example, we can suppose that Giona was a real man, but the only description that we have of it – including that 'he has been living for 3 days inside a whale' – cannot be the a true description of that man. In Kripke's opinion, it is not a description – or a group of descriptions – that can determine the referent of a name, but the act of naming the referent in a certain way. A name can continue to concern the same referent because people intend to keep on using it for the same referent to which it referred the first time.

I accept the Kripkian hypothesis that, sometimes, descriptions do not have a fundamental role in our use of names. Sometimes we use names that we are not able to associate to any description. Nevertheless, in cases like these, we are creating a new entity. That is, we introduce a new speech object represented by the supposition of an entity, supposition obtained by means of the introduction of the name itself. If we are really unable to associate any description to a name, let us call it 'N', then the meaning of this name correspond to a generic description like this: 'the object to which N refers'. The introduction of the name is the fundamental act that causes the use of such a name. This act consists in the introduction of a new object, 'the object named in this way'. Let us consider the following example. Imagine that a friend of mine tells me 'I dislike Bukowski', and that I have no idea about the referent of 'Bukowski'. This would be quite an unusual situation, since it is difficult to hypothesize that we don't know absolutely anything about the entity we are talking about. In this case, we would be probably inclined to suppose that we know, at least, that Bukowski is a man that is doing something that my friend dislikes. But let's consider the rare case in which such a situation is possible, for example when this name has just been introduced. In cases like these, we can just recognize the intention to use the name to refer to a certain entity. If we look at the role of the name, it seems that such a name is a sort of *point* to which we are preparing to associate some data, a *point of attraction à la* Bonomi (Bonomi 1987). In most cases, we are supposed (and we expect) to be able to associate some descriptions to names, and - in a certain way - we can always associate at least a description, the description 'the referent of the name'.

Names can play different roles. A name can be a sort of summary for descriptions we have already ascribed to an object, or it can create the speech object itself. If we don't have any real referent object to use – that is, if we do not know of any link with the object to which the name should refer – our speech object is the supposed entity created by the name itself. We introduce a name and we create a new entity, an entity that - when we have no object able to play such a role - we can use to group the information that we are going to obtain step by step. In several cases, we have some descriptions we can associate to the name, although they are not enough to identify the referent of the name. I could just know that Bukowski has been a famous American writer in the XX century. This information would still be unable, on its own, to univocally identify Bukowski.

There are many situations in which we meaningfully talk about objects we are unable to identify precisely. Also in these cases, names play an active role: they inform us that some data must be referred to the object identified by the name. If I just knew that Bukowski was an American writer, and that he lived in the XX century, when I would use the name 'Bukowski' I would not refer to all American writers, or to an American writer whatsoever that lived in the XX century. This is because the use of the name itself implies my intention to refer to a particular object: the entity – whatever it is – named 'Bukowski'. If the descriptions we can associate to a name 'N' are *per se* insufficient to univocally identify a referent, because they can be applied to many objects, the added description 'the referent of 'N'' makes them gather around the presupposition of a unique object to which they refer. When we don't have any description of Bukowski, the name 'Bukowski' creates the possibility to apply some future descriptions to the same entity (even though that possibility might remain unapplied). When we have insufficient descriptions to individuate the referent, the name groups partial descriptions around the object that is presupposed by the name itself. That is, the new proper name informs us that such an object exists and that next data must be referred to it. In both cases, names draw our attention to a certain point, which becomes a sort of *data identification centre.*

2. The neural network experiments

The theoretical possibility of an active role of names is supported by some language simulation experiments developed by means of neural networks. The simulation presented by Cangelosi, Greco and Harnad (2000) show that the introduction of names modifies the cognitive situation: names are new entities that act on data manipulated by the networks. The networks used in these experiments are 3-layered feed-forward networks. The input and the output layers consist of 2 groups of units (49 visual units and 6 linguistic units), whilst the hidden layer has 5 units. Networks are trained to categorize and name images of circles, ellipses, squares and rectangles received by the visual input units. We suppose that networks are *using names* when they associate an image to a single unit, that can be considered a sort of label for such an image: this interpretation respect the idea of an arbitrary link between names and their referents. At the contrary, the network internal representations of images can be considered a sort of descriptions because they are supposed to maintain a certain analogical relation with the correlated images. The networks must learn two categories of names: the first category names – 'circles', 'ellipses', 'squares' and 'rectangles' – are learnt through direct trial and error experience supervised by corrective feedback (error backpropagation), while the second set of category names – 'symmetric' and 'asymmetric' – is learnt on the basis of strings of symbols alone. The training procedure consists of 3 stages, in which the training error back-propagation algorithm is used. In the first stage, the networks learn to sort the stimuli in the 4 basic categories by producing as output the prototype of each category (that is, the largest instances of the shapes). In the second stage, the networks learn to respond to the stimuli by producing the output visual prototypes and the corresponding names belonging to the first category of names (that is, 'circles', 'ellipses', 'squares' and 'rectangles'). In the third training stage, the networks learn the names 'symmetric' and 'asymmetric'. They receive as input the conjunction of two names, one belonging to the first category – for example, 'square' or 'circle' – and the new name belonging to the second category – in this case 'symmetric'. Networks must learn to produce as output, through error-correcting feedback, both types of name. The results of these experiments show that the networks learn the three tasks successfully: each shape is categorized correctly and the errors in the naming tasks are very low.

In these experiments, the most important data with regard to our purpose come from the analyses of the internal organization of the networks. At the level of the hidden units, the networks build categorical representations which sort each input visual stimulus into its own category. To achieve such a capability, the networks must reduce the category confusability by decreasing the within-category differences and increasing the between-categories differences among the input representations. Indeed, during the prototype learning phase, the elements belonging to the same category move closer to one another and the average between-categories distance increases. Further within-category compression and between-category separation are obtained during the naming phase. After the prototype-based categorisation phase, the distances between the two symmetric shapes and the distances between the two asymmetric ones are smaller than the distances between the symmetric shapes and the asymmetric shapes. This means that the category distinctions can be grounded on the similarity and differences among input data. But the further within-category compressions and between-categories separations during the naming phase also suggest an active role of names: names seem to be able to force the data connected to the same object closer to one another. A further confirmation of the active role of names could derive from the experiments – Cangelosi, Riga, Giolito and Marocco (2004) are developing at the moment – concerning the different roles played by nouns and verbs.

3. Conclusion

These neural network experiments support the psychological plausibility of the previous hypothesis of an active role of proper names in our cognitive processes. In the neural network simulation, the introduction of names modifies the treatment of data the networks are manipulating. If neural networks reproduce the functioning of human mind mechanisms, it seems plausible to suppose that, in the human cognitive processes, the use of a proper name implies some changes in the way we treat information related to that name. Such results also support the general importance of descriptions that we are able to associate to proper names. In the neural network experiments, the category distinctions are grounded in similarities and differences among input data, and names are then associated to such categories. These results seem to suggest that the properties of objects we talk about, and the cognitive content connected to such properties, generally play a role in our language. The names we use are generally connected whit some cognitive content, and this content helps us in differentiating names. The fact that, sometimes, we are not able to associate any description to a name doesn't mean that we are never (nor usually) able to associate to names descriptions concerning some cognitive content related to entities to which names refer. This simply means that names play various roles in language. On the other hand, names are not inert and fixed objects. The psychological nature of this proposal makes them cognitive tools. The connection between names and descriptions is not a rigid relation – for example, a necessary translation *à la* Russell (1905) – but a useful cooperation between flexible cognitive tools.

References

Bonomi A. (1987). *Le immagini dei nomi.* Milano: Garzanti.

Cangelosi A., Greco A., Harnad S. (2000). From robotic toil to symbolic theft: grounding transfer from entry-level to higher-level categories. *Connection Science*, 12(2), 143-162.

Cangelosi A., Riga T., Giolito B., Marocco D. (2004). Language emergence and grounding in sensorimotor agents and robots. In *First International Workshop on Emergence and Evolution of Linguistic Communication*, Kanazawa, Japan.

Kripke S. (1972). Naming and Necessity. In Davidson D. and Harman G. (Eds.), *Semantics of Natural Language.* Dordrecht: Reidel (pp. 253-355).

Russell B. (1905). On Denoting. *Mind*, 14, 479-493.

Searle J. (1958). Proper Names. *Mind*, 67, 166-173.

Wittgenstein L. (1953). *Philosophische Untersuchungen.* Oxford: Blackwell.

Cooperation and Communication

EMERGENCE OF COMMUNICATION IN EMBODIED AGENTS: CO-ADAPTING COMMUNICATIVE AND NON-COMMUNICATIVE BEHAVIOURS

DAVIDE MAROCCO STEFANO NOLFI

Institute of Cognitive Science and Technologies, CNR, Viale Marx 15
Rome, 00137, Italy
[davide.marocco; s.nolfi]@istc.cnr.it

We show how a population of simulated robots developed their communication capabilities in order to solve a collective navigation problem. The self-organized emergent vocabulary includes four different signals that influence both the motor and signalling behaviour of other robots. The analysis of the evolved behaviours also indicates: (a) the emergence of a simple form of communication protocol that allows individuals to switch signalling on and off, (b) the emergence of tightly co-adapted communicative and non-communicative behaviours, and (c) the exploitation of properties resulting from the dynamical interactions between motor and signalling behaviours produced by interacting robots.

1. Introduction

Existing models of emergence of communication in embodied agents often focus on how a shared communication system and/or communication ability might emerge in a population of interacting individuals (Cangelosi, 2001; Cangelosi & Parisi, 2002; Kirby, 2001; Steels, 1999; Steels and Kaplan, 2001). In contrast, this paper focuses on the more general question of how a population of agents that have to solve a given adaptive problem might develop their ability to display complex forms of interactions and communications that enhance their adaptive capability (Quinn, 2001; Quinn et al. 2003; Baldassarre et al., 2003).

The reason for this choice is twofold. The theoretical motivation is that communication and communication systems are adaptive capabilities shaped by their function. What, when and how individuals communicate (and whether individuals do or do not communicate) depends on the adaptive function of communication. Similarly the type of communication system that might self-organize in a population of interacting agents will strongly depend on the type of behaviour individuals display in isolation and on the complementary functions that interactions and communications might have. The underlying assumption is that communication and language can be properly understood by considering their relation with other important behavioural, social and cognitive processes. The practical motivation is that, from an application point of view, the possibility of developing embodied agents able to solve real life problems by exploiting complex forms of interaction and communication might have huge application potentials.

In this perspective, three additional aspects play a crucial role. In fact, we are interested in models that (a) allow the emergence of a communication ability and a shared communication system and (b) that allow the discovery of categories (or coupled internal/external dynamical processes) that are useful from the communication and cognitive point of view but not explicitly or implicitly identified in the experimental set. This claim is based on the theoretical assumption that communication abilities might enhance the cognitive/adaptive abilities of interacting agents for two main reasons: (1) the fact that the indirect adaptive advantages of communication might force the development of useful and compact ways to categorize the continuous flow of sensory-motor information. (2) The fact that the discoveries of useful categories might be easier through social interactions as opposed to being in isolation.

We are interested in models in which individuals, in addition to having signalling and interaction capabilities, also have a reach sensory and motor non-communicative repertoire that might allow them to improve their ability to solve their cognitive/adaptive problems by improving both their individual and their social/communication capabilities. This claim is based on the assumption that only by co-adapting their behavioural non-communicative and communicative abilities, individuals might develop a really useful communication system grounded in the physical and behavioural characteristics of communicating individuals and furthermore, be able to exploit active perceptual capabilities. Moreover, this claim is based on the assumption that one of the key aspects of communication is the possible reliance of the individual on implicit information that does not need to be communicated.

Finally we are interested in models in which forms of communication of different complexity might be used. By forms of communication we refer to the protocol with which individuals interact during communication and to the way with which communication signals are organized. Forms of communication might range from simple continuous broadcasted signalling to complex regulated communication protocols. Furthermore, communication acts might be episodic and asynchronous or, for example, communication protocols might be negotiated on the fly between the two communicating agents and communication acts might consist of a sequence of signals organized according to a grammar. This claim is based on the assumption that more complex forms of communication are not effective in general terms. Therefore, agents should have the liberty, where possible, to select the communication forms that are most useful to them, given their current behavioural/cognitive capabilities.

In the following section we briefly report the preliminary results obtained in a set of experiments in which a team of robots has been selected to solve a collective navigation problem.

2. Evolving a team of robots for the ability to solve a collective navigation problem

A group of four simulated robots live in a square arena of 270x270cm surrounded by walls that contains two circular target areas (see Figure 1, left). The robots have a circular body and are provided with: two motors controlling the two corresponding wheels, one communication actuator capable of sending signals with varying intensities, eight infrared sensors, one ground sensor (i.e. a binary sensor that detects whether the robot is located on a target area), and four communication sensors that detect signals produced by other robots up to a distance of 100cm from four corresponding directions (i.e. frontal [315°-45°], rear [135°-225°], left [225°-315°], right [45°-135°]).

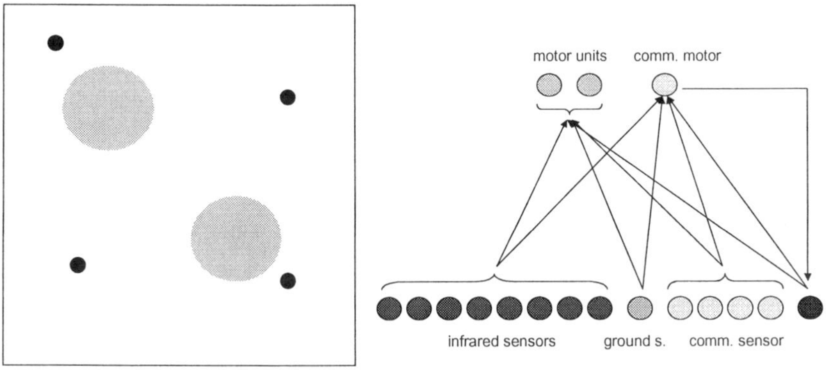

Figure 1. **Left**: The environment and the robots. The square represents the arena surrounded by walls. The two grey circles represent two target areas. The four black circles represent four robots. **Right**: The neural controller of evolving robots.

The robots' neural controllers consist of feed-forward neural networks with 14 sensory neurons (that encode the activation states of the corresponding 13 sensors and the activation state of the communication actuators at times t-1) directly connected to the three motor neurons that control the desired speed of the two wheels and the intensity of the communication signal produced by the robot.

Robots were evolved with the ability to find and remain in the two target areas by subdividing themselves equally between the two areas. Each team of four robots was allowed to "live" for 20 epochs, lasting 1000 time steps of 100 ms each. At the beginning of each epoch the position and the orientation of the four robots was randomly assigned outside the target areas. The fitness of the team of robots consists of the sum of 0.25 scores for each robot located in a target area and a score of -1.25 for each additional robot located in a target area in which there are two others robots already. The total fitness of a team is computed by summing the fitness gathered by the four robots in each time step.

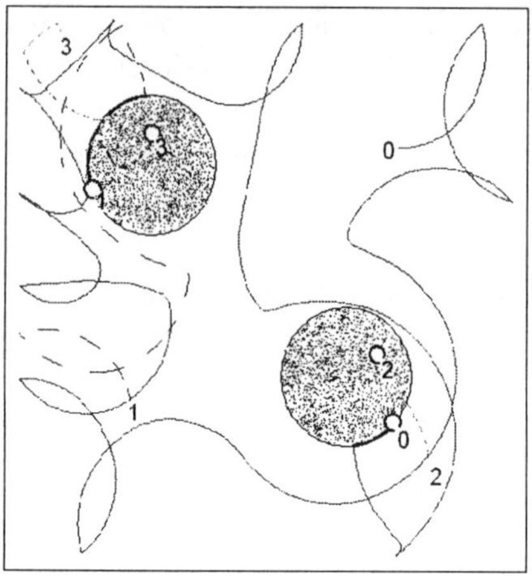

Figure 2. The behaviour of a team of evolved robots of one of the best replications of the experiment. The square and the circles indicate the arena and the target area respectively. Lines inside the arena indicate the trajectory of the four robots during a trial. The numbers indicate the starting and ending position of the corresponding robot (the ending position is marked with a white circle).

The initial population consisted of 100 randomly generated genotypes that encoded the connection weights of 100 corresponding neural controllers (connection weights are encoded with 8 bits each and normalized in the range [–5.0, +5.0]). Each genotype is translated into 4 identical neural controllers that are embodied in the four corresponding robots (i.e. teams consist of four identical robots). The 20 best genotypes of each generation were allowed to reproduce by generating five copies each, with 2% of their bits replaced with **a** new randomly selected value. The evolutionary process lasted 100 generations. The experiment was replicated 10 times.

By analysing the behaviour of evolved robots of one of the best replications of the experiment one can see how they are able to coordinate and to solve the collective navigation problem. In the example shown in Figure 2, robots 2 and 3 quickly reach an empty target area. Later on, robot 1 joins robot 2 in the top-left target area. Finally, robot 0 approaches and then avoids the same target area (that already contains two robots) and later joins the bottom-left target area.

3. Evolved Ontology: grounding and behavioural effects of signals

The analysis of the signals produced by these robots shows how, in order to cooperate, robots developed a non-trivial communication systems. More specifically, the robots described in Figure 2 produce at least four different classes of signals:

(a) a signal *A* with an intensity of about 0.07 produced by robots located outside the target areas not interacting with other robots located inside the target areas;

(b) a signal *B* with an intensity of about 0.45 produced by robots located alone inside a target area;

(c) a signal *C* with an intensity of 0.25 produced by robots located inside a target area that also contains another robot;

(d) a signal *D* with an intensity of about 0.01 produced by robots outside the target areas that are approaching a target area and are interacting with another robot located inside the same target area.

Robots receiving these four types of signals modify their motor and/or signalling behaviour on the basis of the signal received and on other available sensory information in order to coordinate and solve the collective navigation problem. More specifically:

(1) robots located outside the target areas receiving signal *A* tend to modify their motor behaviour to better explore the environment (i.e. they modify their motor trajectory so to explore different areas of the environment);

(2) robots located outside target areas receiving signal *B* tend to modify their motor behaviour (by approaching the robot emitting the signal in order to enter into the corresponding target area) and their signalling behaviour (i.e. by producing signal *D*);

(3) robots located outside the target areas that receive the signal *C* (i.e. the signal produced by two robots located inside the target area) modify their motor behaviour so **as** to move away from the signal source.

4. Coupling and self-organization of primitive forms of communication protocols

The analysis of the evolved individuals also shows that robots are capable of modulating their signalling behaviour on the basis of the signals their receive from other robots. This type of analysis was conducted by testing evolved robots in new environmental conditions that might facilitate the analysis of the interaction between robots and the analysis of the relations between signalling and non-signalling behaviours.

For example, consider a test in which two evolved robots are placed in an arena that includes a single target area (Figure 3). At the beginning the two robots are both outside the target area and produce a signal with an intensity of about 0.07 (signal *A*) that push them apart so that they will have more chance to explore different areas of the environment. Later on individual 0 reaches the target area and starts to produce a signal with an intensity of about 0.45 (signal *B*). Once robot 1 gets close enough to robot 0 (i.e. within a range of 100cm) it modifies its trajectory so as to follow the signal emitted by the other robot and turn off its signalling behaviour (i.e. it starts to produce signal *D*). Later on, when robot 1 joins the area, the two robots start to produce a noisy signal with an intensity of about 0.25 (signal *C*) that will keep other coming robots out the target area.

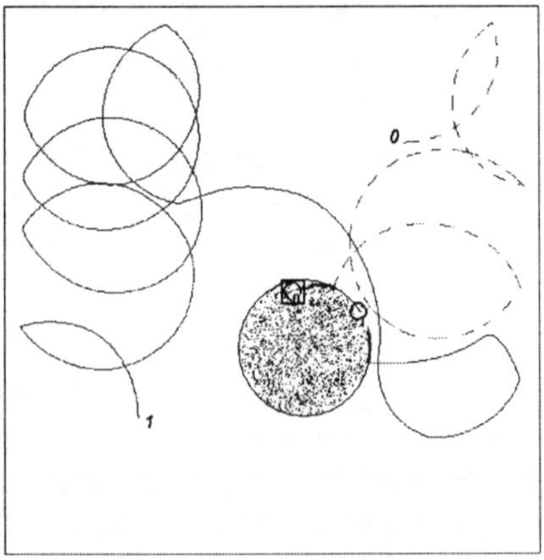

Figure 3. The behaviour of two robots tested in an arena including a single target area. The dashed and full lines represent the trajectory of robot 0 and 1, respectively. The numbers indicate both the starting and ending positions of the corresponding robots.

By analyzing the signals produced by the two robots during the behaviuor shown in Figure 3, we can clearly see two interesting examples in which the signals produced by one robot alter the signals produced by the other robots (see Figure 4).

More specifically, data shown in Figure 4 indicates the emergence of a primitive communication protocol in which signalling is switched on and off according to the needs of the robots. In this experimental setting, the signalling behaviour can be turned off by producing signals with null or close to null intensity. Indeed during the second phase (in which robot 0 is in and robot 1 is out the target area), robot 1 turns its signalling behaviour off as soon as it detect the signal *B* produced by robot 0. The reason that the signalling behaviour is switched off in these circumstances is probably due to the need to not alter the other robot's signalling behaviour. Indeed, by preventing robots from the possibility to switch their signalling behaviour off in these circumstances, a significant decrease in performance can be observed.

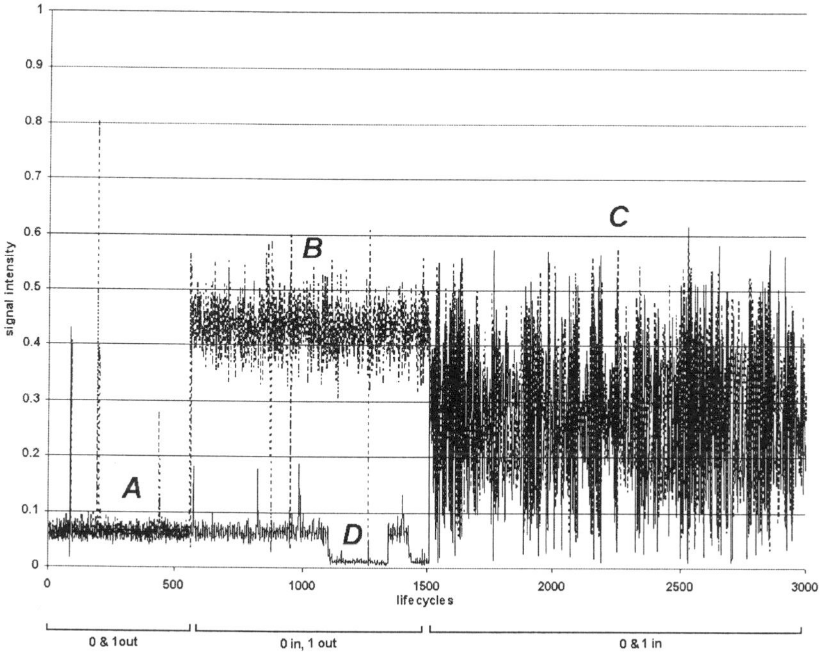

Figure 4. Intensity of the signals produced by the two robots during the behaviour shown in Figure 3. Dashed and full lines indicate the intensity of the signals produced by robot 0 and 1, respectively. Letters (*A*,*B*,*C*, and *D*) indicate the 4 classes of signals produced by the robots. The bottom lines indicate the three phases in which: (1) both robots are out the target area, (2) robot 0 is in and robot 1 is out, and (3) both robots are inside the target area.

More generally, the emergence of the ability to switch signalling behaviours on and off can be explained by considering that communicative actions might have counter-adaptive effects on other robots and on the adaptive ability of the population as a whole. For instance, communication acts might negatively affect the behaviour of other robots thus preventing or delaying the ability of these robots to accomplish their tasks. Signalling might even interfere with the behaviour of the robot producing the signal when, as in the case of the experiments described in this paper, robots perceive their own signals.

A second interesting phase is that in which the two robots located in the target area produce signal *C* that keeps other approaching robots away from the target area. During this phase the bi-directional interaction between the two robots leads to a coupled dynamical process that seems crucial both from the point of view of allowing the two robots to detect the presence of the other robot located inside the area and from the point of view of allowing the two robots to produce signals that push other approaching robots away.

When a single robot is located inside the area, it starts to rotate and to produce signal *B* that attracts other approaching robots. When a second robot joins the same area, it also starts to rotate and to signal. However, as a result of the interaction between the two motor and signalling behaviours, the two robots start to produce signal *C*, a highly varying signal with an average intensity of about 0.25, that rejects other approaching robots.

The fact that the interaction between signalling and not signalling behaviours are crucial for determining the ability of the robots to solve their problem can be demonstrated by running tests in which these forms of behaviours are selectively perturbed. For instance, by preventing one of the two robots to freely rotate at the desired speed when both robots are located inside the target area, we observed that the two robots are no longer able to remain inside the area. Similarly, by preventing one of the two robots from being able to emit signals (while preserving the signalling behaviour of the robot intact), we observed that other approaching robots would erroneously join the area. More generally, we observed that perturbations of the motor behaviour tend to significantly affect signalling behaviour and vice versa and produce a significant degradation of performance.

Overall, the tests performed indicate that signals *B* and *C* are the emergent results of the dynamical interaction between the motor and signalling behaviours of the two robots.

5. Conclusion

In this paper we showed how a population of simulated robots evolved for the ability to solve a collective navigation problem can develop non-trivial communication abilities despite the fact that communication is not explicitly rewarded. The self-organized emergent vocabulary includes four different signals that influence both the motor and signalling behaviours of other robots.

The analysis of the evolved behaviours also indicates: (a) the emergence of a simple form of communication protocol that allows individuals to switch communication on and off according to the current needs of the robots, (b) the emergence of tightly co-adapted communicative and non-communicative behaviours, and (c) the exploitation of properties resulting from the dynamical interactions between communicative and non communicative behaviours produced by interacting robots.

In future research we plan to: (1) study more complex tasks, (2) provide the neural controllers of evolving robots with internal neurons and recurrent connections that might allow them to integrate sensory-motor information over time, and (3) allow evolving individuals to switch the impact of perceived signals on and off.

Acknowledgments

The research has been supported by the ECAGENTS project founded by the Future and Emerging Technologies programme (IST-FET) of the European Community under EU R&D contract IST-2003-1940.

References

Baldassarre G., Nolfi S., Parisi D. (2003). Evolving mobile robots able to display collective behaviuor. *Artificial Life*, 9: 255-267.

Cangelosi A. (2001). Evolution of communication and language using signals, symbols and words. *IEEE Transactions in Evolutionary Computation*, 5(2): 93-101.

Cangelosi A, Parisi D. (Eds.) (2002). *Simulating the Evolution of Language*. London: Springer-Verlag.

Kirby S (2001). Spontaneous evolution of linguistic structure: An iterated learning model of the emergence of regularity and irregularity. *IEEE Transactions on Evolutionary Computation and Cognitive Science*, 5(2): 102-110.

Quinn M. (2001). Evolving communication without dedicated communication channels. In Kelemen, J. and Sosik, P. (Eds.) *Advances in Artificial Life: Sixth European Conference on Artificial Life* (ECAL 2001). Springer Verlag.

Quinn M., Smith L., Mayley G., Husbands P. (2003). Evolving controllers for a homogeneous system of physical robots: Structured cooperation with minimal sensors. *Philosophical Transactions of the Royal Society of London*, Series A: Mathematical, Physical and Engineering Sciences 361, pp. 2321-2344.

Steels L. (1999). *The Talking Heads Experiment*, Antwerpen, Laboratorium. Limited Pre-edition.

Steels L. and Kaplan F. (2001). AIBO's first words: The social learning of language and meaning. *Evolution of Communication*, 4:3-32.

ESTIMATING INTENTION OF OTHERS FOR IMITATION AND COOPERATION

NORIKAZU SUGIMOTO[134] KENJI DOYA[1234] MITSUO KAWATO[13]
norika-s@is.naist.jp *doya@irp.oist.jp* *kawato@atr.jp*
1 *Nara Institute of Science and Technology*
2 *Okinawa Institute of Science and Technology*
3 *ATR Computational Neuroscience Laboratories*
4 *Creating the Brain, CREST, Japan Science and Technology Corporation*

Estimation of others' intention is essential for intelligent social behaviours. However, estimation of intention only from observation of another's actions is in general an ill-posed problem. Here, we propose a reinforcement MOSAIC framework for this difficult problem. We demonstrate the framework in a simulation of a cart-pole swing-up task to show that the proposed estimation scheme can be used for imitation and cooperation.

1. Introduction

Many cognitive skills, such as inference and planning, require the use of internal models that predict dynamics of environments. Furthermore, social interaction requires estimation of the internal states of other agents, such as their intentions or beliefs. However, this estimation is in general an ill-posed problem, and especially difficult when we deal with continuous-valued states, actions, and intentions. In this study, we propose a general framework for estimating another agent's intention by exploiting the modularity and similarity of the control architectures of fellow agents.

Our first assumption is that agents take a modular control scheme based on the MOSAIC architecture (Wolpert & Kawato, 1998; Doya et al., 2002). An agent has multiple forward models to predict environment dynamics and multiple reward models to represent its intention. Our second assumption is that the physical parameters and the possible intentions of the agents of the same species are similar. Under this setup, we derive a method for estimating the intention of another agent, in the sense of estimating the reward model being selected. We will demonstrate how such estimation of intention of a fellow agent enables imitation of its behaviours and cooperative behaviours.

2. Hierarchical Model-based Reinforcement Learning

In this section, we will introduce "hierarchical model-based reinforcement learning (HMRL)" (see Figure 1). It consists of a top layer (TL) which performs discrete reinforcement learning (RL) to select a series of intentions, and a bottom layer (BL) that performs continuous RL to realise the

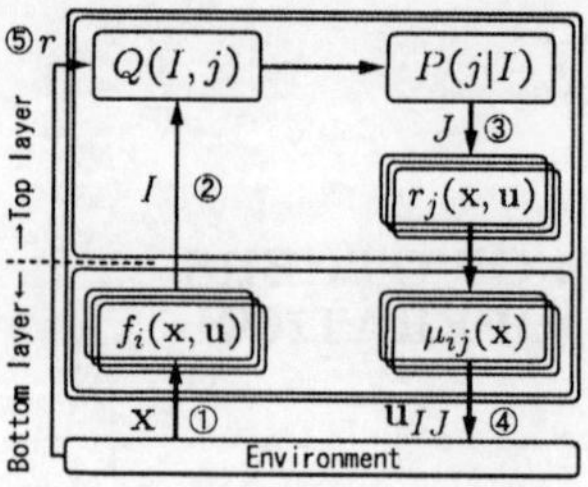

Figure 1. The HMRL architecture.

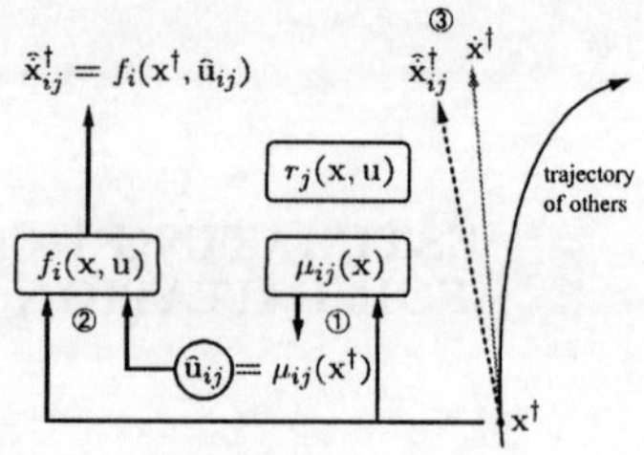

Figure 2. Estimation of another agent's forward and reward models.

intentions through interaction with the environment. The BL has N^f forward models (FM) f_i which locally predict the change of environment, and N^c controllers which generate the actual action output. The TL has N^r internal reward models (RM) r_j corresponding to each possible intention, and a value function $Q(i,j)$ which evaluates how good it is to select the jth RM under ith context.

The HMRL agent optimises a cumulative external reward $r(t)$ from the environment by repeating the following steps. ①The BL observes the environment's state $\mathbf{x}(t)$ and categorises it based on the prediction error of each FM. ②The BL then informs the TL of the context "I", which is the index of the FM that has minimal prediction error. ③Next, the TL assesses a set of value functions Q in terms of context "I" for possible j and selects one RM, denoted as "J", probabilistically in accordance with a selection probability based on values of each RM (e.g. Boltzmann distribution $P(J = j|I) = \frac{\exp[\beta Q(I,j)]}{\sum_{j'} \exp[\beta Q(I,j')]}$). ④The BL generates actual action output to realise the selected RM r_J by using controller derived from the Hamilton-Jacobi-Bellman equation (Fleming & Soner, 1993). In our previous work (Wolpert & Kawato, 1998; Doya et al., 2002), the output of each controller was weighted. In this study, however, the actual output is generated by one selected controller at each time step. ⑤Finally, the agent receives an external reward $r(t)$ from the environment and updates the value function $Q(I, J)$ to maximise the cumulative reward (Sutton & Barto, 1998).

3. Estimating the FM and RM used by another agent

Figure 2 shows the process of estimating the FM and the RM used by another agent. We will denote the variable of the other agent with "†", and the estimated variable with " ^ ". The HMRL agent recognises the other agent's movement by its own internal representations; other words, the agent estimates which module pair is most likely used by the other agent to generate its actions. This is realised by emulating the behaviour of the other agent by all possible combinations of the modules and comparing the predicted outcomes with the observed states.

The way to emulate a certain module pair (f_i, r_j) is as follows. ①The agent first observes the state trajectory $\mathbf{x}^\dagger(t)(0 \leq t \leq T)$ of the other agent and applies it to its own controller μ_{ij} to make the estimated output $\hat{\mathbf{u}}^{\dagger}_{ij} (= \mu_{ij}(\mathbf{x}^{\dagger}))$. ②The agent then inputs the observed state $\mathbf{x}^{\dagger}$ and the estimated output $\hat{\mathbf{u}}^{\dagger}_{ij}$ to the FM f_i to emulate the change of state of the other agent $\hat{\dot{\mathbf{x}}}^{\dagger}_{ij}$ $(= f_i(\mathbf{x}^{\dagger}, \hat{\mathbf{u}}^{\dagger}_{ij}))$. ③Finally, the agent selects the best module pair (f_i, r_j) based on the prediction error $||\hat{\dot{\mathbf{x}}}^{\dagger}_{ij} - \dot{\mathbf{x}}^{\dagger}||^2$, where $\dot{\mathbf{x}}^{\dagger}$ is given by temporal differentiation of $\mathbf{x}^{\dagger}$ and $|| * ||$ is L_2 norm.

3.1. *Model of imitation learning*

In this section, we will formulate an algorithm for imitation learning. The "imitator" agent estimates the RM selection probability of the "demonstrator" agent by observing its behaviour. This means that the HMRL agent can comprehend the other agent's intention.

The imitator estimates the sequence of module pairs $(\hat{I}^{\dagger}(t), \hat{J}^{\dagger}(t))$, $(0 \leq t \leq T)$ from the observed trajectory $\mathbf{x}^{\dagger}(t)$ $(0 \leq t \leq T)$ of the demonstrator. The imitator then estimates the RM selection probability $P^{\dagger}(J^{\dagger}|I^{\dagger})$ of the demonstrator by taking statistics of the sequence of module pairs. The imitator can subsequently use this estimate as the prior information for its own module selection during performance. That is, the RM selection probability of the imitator is determined as follows: $P(J = j|I) = \frac{\hat{P}^{\dagger}(j|I)\exp[\beta Q(I,j)]}{\sum_{j'} \hat{P}^{\dagger}(j'|I)\exp[\beta Q(I,j')]}$. After the observation, the imitator updates the own value function. The imitator can get large reward in the early stage of autonomous learning if the prior is appropriate.

3.2. *Model of cooperation*

When two agents perform a mutually dependent goal-oriented task, each agent must understand the intention of the other agents if efficient cooperation is desired. In this section, we will formulate the cooperation of two agents. We will denote the variable of the two agents with "A" and "B".

The agent estimates the module pair which the co-agent uses and learns the value function of the indices of the estimated module pair. Agent A observes its own state $\mathbf{x}_A$ and estimates the index "I_A" of the FM which has minimal prediction error, and estimates the module pair which the co-agent uses, denoted as $\hat{I}^{\dagger}_B$ and $\hat{J}^{\dagger}_B$. Then the TL of agent A learns a new value function which has the extended state $\{I_A, \hat{I}^{\dagger}_B, \hat{J}^{\dagger}_B\}$, and a new value function $Q(\{I_A, \hat{I}^{\dagger}_B, \hat{J}^{\dagger}_B\}, j_A)$ is updated upon every state transit. Cooperation is recognised by using this value function. Agent B also learns a new value function $Q_B(\{I_B, \hat{I}^{\dagger}_A, \hat{J}^{\dagger}_A\}, j_B)$.

4. Simulations

In this section, we will demonstrate that two cognitive skills, imitation and cooperation can be learned by estimating the intention of the other agent.

4.1. *Imitation task*

First, we demonstrate an example of imitation by estimating the other's intention. In this framework, an agent can perform imitation even with

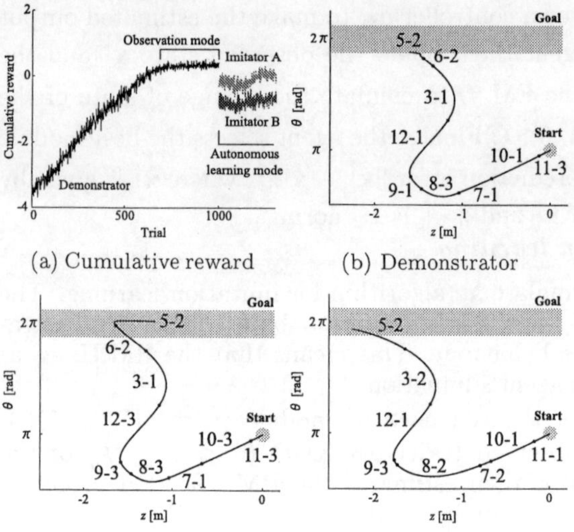

(a) Cumulative reward (b) Demonstrator

(c) Imitator A (d) Imitator B

Figure 3. The trajectory and module pair which each agent used. Each pair of numbers shows the indices of the module pair used in each interval. For example, "2 – 3" means that f_2 and r_3 were used.

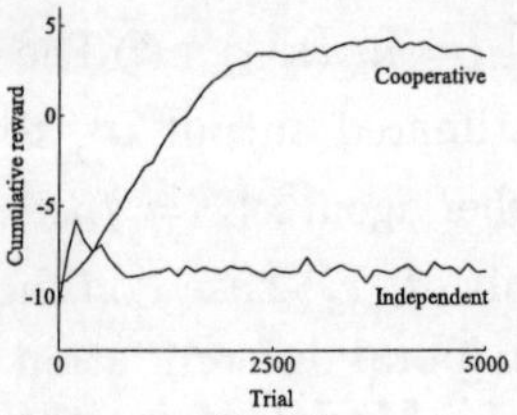

Figure 4. The learning curve of the cooperative and the independent agents. Each line shows the average of 10 simulation runs and 1,000 trials.

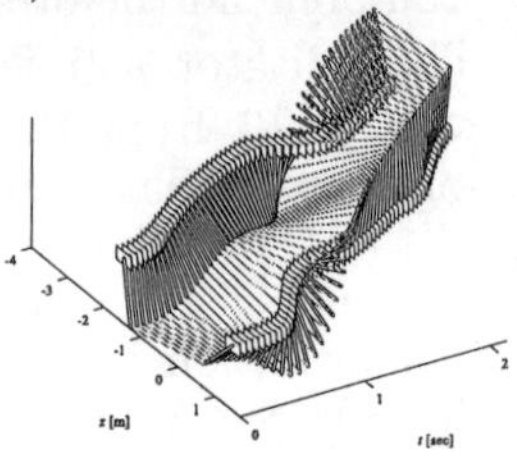

Figure 5. A stroboscopic illustration of the cooperative swinging-up behaviour.

some differences in the physical parameter of body between the demonstrator and imitator. We simulated a cart-pole swing-up task (Doya, 2000) with three kinds of agents. Imitator A performs imitation learning with the same physical parameters as the demonstrator, while imitator B performs imitation learning with a mass 1.2 times heavier than the demonstrator.

The demonstrator learned the task autonomously in advance. Then, the two imitators observed successful trials of the demonstrator and estimated its RM selection probability $P^{\dagger}(J^{\dagger}|I^{\dagger})$. Both imitators then started autonomous learning with the RM selection probability as prior information. The higher the position of the pole tip, the larger was the given reward.

Figure 3(a) illustrates changes of the cumulative reward during each trial, with each line showing the average of 100 simulation runs. The demonstrator needed about 400 trials to swing-up. In contrast, the imitators A and B achieved large rewards from the beginning of the trials. The cumulative reward of imitator B is lower than the imitator A because of its heavier mass. Figures 3(b), (c), and (d) show the trajectory and the sequence of module pairs for each agent. The gray areas are the start and the goal. Clearly, imitator A could imitate the sequence of module pairs of the demonstrator with a small selection error of the RM, while agent B could swing up in a similar trajectory to the demonstrator. This means that the HMRL agent can estimate the other's intentions in order to generate the similar trajectory despite the difference of physical parameters.

4.2. *Cooperation task*

Next, we demonstrate an example of a cooperation task in which two agents control two cart-poles, where their tips are connected by a spring. The agents can't swing up without cooperation due to this physical interaction.

To show the necessity of cooperation, we tested two versions of the simulation. One was a cooperative version in which the agent estimates the intention of a co-agent and acts in accordance with it. The other was an independent version where the agent never estimates the intention of the co-agent. Figure 4 shows the cumulative reward for two simulations at each trial, and it is clear that the cooperative version has better performance than the independent version. Figure 5 shows an example of a successful trial of the cooperative version. In that case, the two agents could swing up in the reverse rotation direction. Other trial types with various initial states are available on the first author's Web site (`http://hawaii.naist.jp/~norika-s/publications/movies/`).

5. Summary

In this study, we introduced a framework to estimate discrete intentions of another agent by observing its continuous state trajectory. Although the intention of others is generally unobservable, our algorithm estimates another's intention using its own hierarchical internal models upon an assumption that a similar architecture is used by the other agent. Furthermore, we demonstrated that two cognitive skills can be realised by our algorithm. In Section 4.1, the imitator A could swing up with the same module sequence and similar trajectory to the demonstrator, while imitator B could understand the demonstrator's intention because the BL could absorb some degree of difference in dynamics. In Section 4.2, we demonstrated that estimating the intention of the co-agent enables cooperative behaviour.

In the real world, the environment is in general partially observable. Furthermore, the number of co-agents isn't necessarily one. In the future, we hope to extend our framework to deal with such situations.

References

Doya, K. (2000). Reinforcement leaning in continuous time and space. *Neural Computation*, *12*, 219-245.

Doya, K., Samejima, K., Katagiri, K., & Kawato, M. (2002). Multiple model-based reinforcement learning. *Neural Computation*, *14*, 1347-1369.

Fleming, W., & Soner, H. (1993). *Controlled Markov Processes and Viscosity Solutions*. NewYork: Springer-Verlag.

Sutton, R., & Barto, A. (1998). *Reinforcement Learning*. MIT Press.

Wolpert, D., & Kawato, M. (1998). Multiple paired forward and inverse models for motor control. *Neural Networks*, *11*, 1317-1329.

Motor Representations and Cognition

MENTAL REPRESENTATIONS AND COGNITIVE BEHAVIOUR – A RECURRENT NEURAL NETWORK APPROACH

SIMONE KÜHN   HOLK CRUSE

Department of Biological Cybernetics, University of Bielefeld
Bielefeld, 33501, Germany
{simone.kuehn, holk.cruse}@uni-bielefeld.de

The ability of building up mental representations of external situations to uncouple the behaviour from direct environmental control can be considered to be a prerequisite for cognitive, adaptive behaviour. The network we present can be used for generation of action but also – due to its attractor characteristics – as a basis for mental representations. In this context, a new learning algorithm is proposed, which leads to a self-organised weight distribution yielding stable states and allows building up several cell assemblies existing simultaneously within a larger network.

1. Introduction

Every living organism has to move around in order to find food to survive and mates to reproduce – thus it faces the problem of sensorimotor control. A solution for this problem requires a suitable control mechanism. To solve this problem we can think of different possibilities: The organism could use the sensory information directly for the control of behaviour as we find it in reactive systems like the so-called Braitenberg-vehicles (Braitenberg, 1984): The problem here is that they only can show appropriate behaviour as long as the external stimulus is present. If it disappears, the behaviour disappears, too. Thus, instead of reacting directly to the input, a solution on a higher level is to use sensory information to construct mental representations of the environment as a basis for subsequent behaviour. Here, the term representation is used in the broad sense of Steels (1995) as being "physical structures (for example electro-chemical states) which have correlations with aspects of the environment".

An advantage of such internal mental models is that the behaviour can be uncoupled from direct environmental control. This enables the organism e.g. to respond to features of the world that are not directly present, to use past experiences to shape present behaviour, to plan ahead, to internally manipulate the content, etc. (Cruse, 2003b). Therefore, we can conclude that these mental representations form an essential prerequisite to explain how organisms can behave in a cognitive adaptive way.

Environmental stimuli can be represented by activation of so-called cell assemblies. Various models to realise such cell assemblies have been proposed, e.g. multi-layer perceptrons (MLPs), Hopfield nets (Hopfield, 1982), Kohonen-maps (Kohonen, 1982), Jordan and Elman nets (Jordan, 1986; Elman, 1990), recurrent networks and recurrent experts (e.g. Wolpert and Kawato, 1998; Tani and Nolfi 1999). Here, we want to refer to a special type of recurrent neural network: the MMC network. These networks can be used on the one hand for the generation of action and on the other hand for building up mental models of the environment due to their attractor characteristics.

2. MMC nets – Generation of action

MMC nets are fully connected recurrent neural networks which were primarily invented to solve geometrical tasks characterised by redundant degrees of freedom like the control of arm movements (Steinkühler and Cruse, 1998), six-legged walking (Kindermann and Cruse, 2002) or landmark navigation (Cruse, 2003a). In all these tasks the nets cope with the problem of sensorimotor integration.

This network exploits the redundancies within the system: A given value is determined simultaneously in several ways as often found in biological systems. By calculating the mean value of these multiple computations (MMC) the final output is obtained. The main feature of this network is that it converges to stable states corresponding to a geometrically correct solution, even when the input does not fully constrain the solution (for detailed information see Cruse at al., 1998). External information is stored within the net by activation of several neurons (units). The activation is maintained because the neurons activate each other via recurrent connections.

Another interesting feature of this network is that the units could not be labelled to be either motor or sensory elements: This is consistent with the view that in biological systems a tight connection between the perception and the motor system exists (e.g. Buccino et al., 2001; „mirror neurons“ Rizzolatti and Arbib, 1998).

3. MMC nets as a basis for mental representations

A step towards a more representational level is to uncouple the motor output from the net: By doing so the nets can serve as internal body models with the attractor state the net reaches at the target point of the movement representing the situation at this moment. Thus, this type of networks can be taken as a neural basis for mental representations as described above.

In all applications mentioned so far the synaptic weights within the MMC net are fixed according to the equations forming the basis of the network, i.e. the body model could be regarded to be „innate“. In contrast, when an organism faces a new situation comprising different objects, the mental model must emerge starting with a “naïve” recurrent network with small random weights before the stimulus appears. Therefore, the goal of this work is to find a way to train the weights to achieve these stable attractors in the end. Furthermore, such a training mechanism may also help to stabilise the weights of an innate networks the synapses of which may change due to physiological effects.

3.1. *The network*

Before explaining the learning algorithm, we want to describe the principles of the network structure applied here in more detail starting with a simple task: A mental model of a situation should be built up that is characterised by two items x_1 and x_2, for example an environmental situation showing two objects. For a first approach we use localist-encoding linguistic units as they code for single linguistic entities, namely words (Cangelosi, 2004). Thus, the corresponding neural system consists of two units each of which represents one object (Fig. 1).

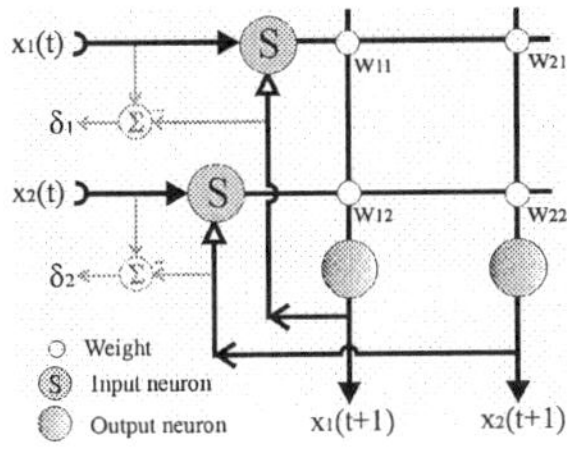

Figure 1: Architecture of a two-unit MMC net: x_i(t) are the inputs, x_i(t+1) the outputs and w_{ij} the weights. Neurons calculating the weighted sum are depicted as shaded circles. The input neurons, which are suppression units symbolised by *S*, are shown as grey circles. Filled arrowheads indicate that the input is suppressed if the alternative channel is activated. The light grey lines show how the δ-error is calculated (see section 3.2.).

The MMC network is a fully connected recurrent neural network which can be described by the following general equation for RNNs: $x(t+1) = W \cdot x(t)$. For a network with two units the system of equations holds:

$$\begin{aligned} x_1(t+1) &= w_{11} \cdot x_1(t) + w_{12} \cdot x_2(t) \\ x_2(t+1) &= w_{21} \cdot x_1(t) + w_{22} \cdot x_2(t) \end{aligned} \qquad (1)$$

The output neurons (shaded grey circles) are “integrate-and-fire” neurons, i.e., they calculate the weighted sum of the incoming activity while the input neurons (depicted as grey circles marked by an S in Fig.1) are suppression units. The external inputs x_i are suppressed by the recurrent connections symbolised by the open arrowheads; this means that these external inputs are only effective during the first time step.

Which distribution of the weights w_{ij} leads to a stable state $x_1(t+1) \neq 0$ and $x_2(t+1) \neq 0$ after the net has been provided with an external input $x_1(t) \neq 0$ and $x_2(t) \neq 0$? A stable solution is achieved if there is a fixed relation between the parameters x_1 and x_2: $x_2 = f(x_1)$. The simplest case is given by a linear equation:

$$a \cdot x_1 - b \cdot x_2 = 0 \tag{2}$$

Using Eq. (2) basic equation for the MMC net leads to:

$$\begin{aligned} x_1 &= w_1 \cdot x_2; \text{ with } w_1 = b/a \\ x_2 &= w_2 \cdot x_1; \text{ with } w_2 = a/b \end{aligned} \tag{3}$$

Thus, Eqs. (3) fulfil the basic Eq. (2) if the weights are constrained by:

$$w_1 \cdot w_2 = 1 \tag{4}$$

Eqs. (3) can be transformed into Eqs. (1) by introducing damping factors $d_i \neq 0$ by which the self-activating connections are weighted:

$$\begin{aligned} x_1(t+1) &= (d_1 \cdot x_1(t) + w_1 \cdot x_2(t))/(1+d_1) \\ x_2(t+1) &= (w_2 \cdot x_1(t) + d_2 \cdot x_2(t))/(1+d_2) \end{aligned} \tag{5}$$

By means of these equations, the four weights in Fig. 1 and equations (1) can be specified in more detail: $w_{11} = d_1/(1+d_1)$, $w_{12} = w_1/(1+d_1)$, $w_{21} = w_2/(1+d_2)$ and $w_{22} = d_2/(1+d_2)$. The damping factors d_i change the dynamics of the net: The higher they are, the slower the net converges to a stable state. Thus, the system obtains low-pass filter properties (Cruse et al., 1998) and oscillations that might occur during the relaxation process can be avoided successfully.

3.2. *Training the weights*

As mentioned above, a mechanism for training the weights is necessary to build up a mental model of a new situation – for example when two objects are presented the net. To this end, the equations forming the MMC net should fulfil Eq. (4) $w_1 \cdot w_2 = 1$.

In general, a matrix of weights is searched for that has one eigenvalue $\lambda_1 = 1$ and a second eigenvalue $\lambda_2 < 1$. The eigenvector $\mathbf{v}_1$ and all linearly dependent vectors $\mathbf{v'}_1$ of $\mathbf{v}_1$ with the eigenvalue $\lambda_1 = 1$ are stable solutions of the net. The target distribution of the weights is obtained by calculating the zeros of the characteristic polynom

$$\det(\mathbf{W} - \lambda \mathbf{I}) = 0 \text{ with } \lambda = 1 \tag{6}$$

Here, **W** is the weight matrix and **I** the identity matrix. Eq. (6) is termed *MMC criterion* in the following as it provides a condition, which yields stability of the net. To evaluate the state of the net during learning, we have to look whether this criterion is fulfilled. As a measure of how much the weights diverge from this criterion a Harmony function H is defined:

$$H = \frac{1}{1+\left(\det(\mathbf{W}-\lambda\mathbf{I})\right)^2} \tag{7}$$

Harmony H approaches the value 1 the more the MMC criterion is fulfilled, i.e., the more the characteristic polynom approaches zero.

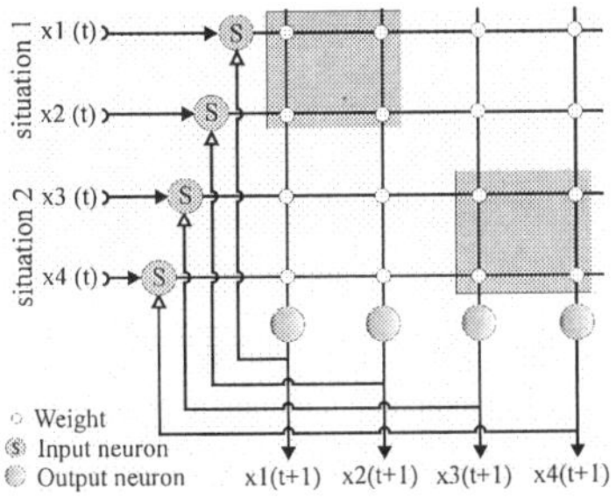

Figure 2: Architecture of a network for an extended task where two situations are presented consisting of two objects respectively. The two grey boxes mark the weights which should be activated when the two situations are presented to the net subsequently.

Now let‘s have a closer look at the task. It should be possible to build up a mental representation of more than one situation simultaneously, i.e., several neural assemblies should be able to coexist (Fig.2). It is the aim to activate only those four synaptic connections, which combine the features belonging to the respective situation (grey boxes in Fig.2). I.e., after having presented situation 1 via an input vector (x_1, x_2) only the weights in the upper grey box should be activated. When later situation 2 is presented (x_3, x_4), the weights in the lower grey box and only these should be activated. Harmony H should go up to the value of 1 which results in stable outputs – the condition for mental representations. How could this be achieved?

If the first new situation is presented, the units coding for the objects of this situation change their activity. This means, because of the recurrent architecture, that there is both a *pre-* and a *post-synaptic change* of activity at those weights that connect the objects of this situation (Fig.2, grey box, upper left). Assume that these weights have already been trained and the activation of the respective neurons has stabilised. If later the net is provided with a second situation, a learning algorithm, which is based on pre- and post-synaptic activity as usual hebbian mechanisms, would not only activate the four weights representing situation 2 (Fig.2, grey box, lower left) but all 16 weights. How could it be avoided to change those eight weights that connect features of different situations and that would produce "crosstalk" between both representations, for example the weight connecting x_2 and x_3? At those weights connecting units of different situations (e.g. the weight connecting x_2 and x_3) there is only a *pre-synaptic change* of activity. The activity of the post-synaptic side does not change anymore, because the unit x_2 has already stabilised after presentation of the first scene.

Consequently we see that the learning algorithm has to meet two requirements: First, only those weights should be activated at which we find both a *pre-* and a *post-synaptic change* of activity. Second, the MMC criterion should be fulfilled in the end to obtain a stable solution. This can be achieved by applying a combination of a hebbian mechanism, that, however, uses the temporal change Δx of the activity instead of the activity x itself, and the δ-error as it is used in the Delta Rule. As a mental model of a current external situation, which is represented by the external input vector $x(1)$, should be built up this input vector can be used as trainer for the net. For this reason the difference δ between the external input vector $x(1)$ and the output vector $x(t+1)$ can be taken as error signal for training the weights (see Fig.1 light grey lines). In doing so, the loop through the environment can be closed. As representations are built up of actually present situations, these internal models are directly grounded in the external world (Cangelosi, 2001).

Therefore, we propose to change the weights by applying the following learning algorithm:

$$\Delta w_{ij} = \eta \cdot \Delta x_i \cdot \Delta x_j \cdot \delta_i$$

with η being the learning rate, $\Delta x_i = x_i(t+1) - x_i(t)$, $\Delta x_j = x_j(t+1) - x_j(t)$ and $\delta_i = x_i(1) - x_i(t)$.

This learning algorithm can be interpreted as a combination of Hebb's rule, however applied to the temporal derivative of pre- and postsynaptic activity, and the Delta Rule. Hence, this algorithm is termed *Dynamic Delta Rule* (DD Rule).

Note that it is not the aim here – as in usual applications of the traditional Delta Rule – to minimise the δ-error, but to minimise the value of the temporal changes Δx_i and Δx_j to obtain stable solutions. So after learning is finished there could still be a δ-error which is however no problem for the concept presented here. Accordingly, the activation of the output units needs not to equate the input activations after learning.

3.3. *Results*

In each simulation the weights w_{ij} are initialised with random values $0 < w_{ij} < 0.02$. Input values are chosen in a range of $1 < x_i < 10$ with $x_i \in \mathbb{N}$. Fig.3 shows the results of two simulations with different input vectors. It can be seen that in both cases the output activation stabilises on values different from zero (first column) and the Harmony function goes up to the value of 1 (last column) which indicates that the MMC criterion is fulfilled. Thus in the end, the external situation is represented within the recurrent connections of the network.

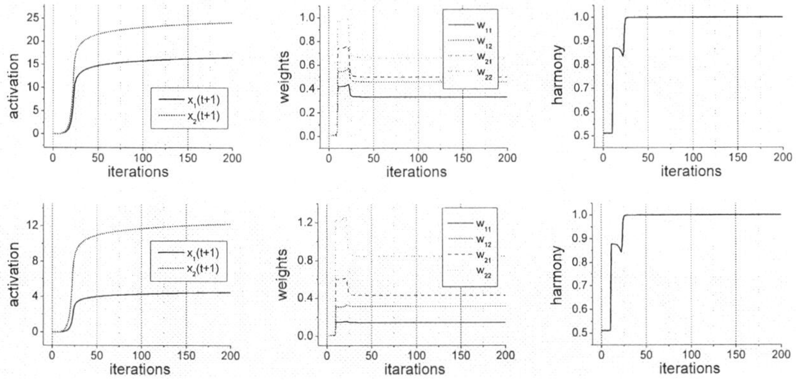

Figure 3: Examples of a net trained using the *Dynamic Delta Rule*: Upper panel: input vector $\mathbf{x}$ = (6, 8), learning rate η=0.002. Lower panel: input vector $\mathbf{x}$ = (2, 4), learning rate η=0.02. The left column shows the output activation of the net, the middle column the changing of the weights and the right column the Harmony function, respectively.

Another fact we can see is that we need very different learning rates to stabilise the system: The results for $\mathbf{x}$ = (6, 8) are obtained by a learning rate of $\eta = 0.002$ and for $\mathbf{x}$ = (2, 4) by a learning rate of $\eta = 0.02$. Here we touch a general problem if the weights are updated online in recurrent neural networks: Online-learning couples two dynamics, the dynamics of the recurrent network as well as the dynamics of the effects of learning (Steil, 1999). A learning rate that is too large can easily cause the net to diverge. On the other hand, a learning rate that is to small results in output activities falling down to zero. Therefore, the learning rate η is a critical parameter. Even though a mathematical proof of convergence appears to be a difficult task due to the nonlinearity of the system, based on many numerical tests we can state that to yield stability of the net the learning rate η has to be chosen the smaller the larger the inputs x_1 and x_2 are.

The task can, however, be simplified, if the weights have upper and lower bounds, which is physiologically plausible. By applying bounds of

$0.001 \le w_{ij} \le 3$, the net is prevented from diverging. If the weights reach the bounds during training, oscillations occur, but – for a broad range of learning rates – the net stabilises after some time. If the dimension of the input vector is increased, it gets more and more difficult to find appropriate learning rates that stabilise the output of the net because of the above-mentioned high non-linearity of the system.

But at least for low dimensional input vectors the DD Rule is able to stabilise the system on values different from zero. This means that a mental model of an environmental situation can successfully be built up within the net even if the situation contains several separate groups of elements.

4. Discussion

The MMC nets, which were primarily used for geometrical tasks like arm-movements or six-legged walking and which are able to cope with the problem of sensorimotor integration, can serve as a basis for internal representations because of their attractor characteristics. By applying the Dynamic Delta Rule the weights could be trained in a way that, in the end the MMC criterion is fulfilled and only sub-populations of a larger net are activated. As the situations have to be presented successively, as a prerequisite a shift of attention from one situation to the next must be assumed (concerning the importance of attention see for example Reynolds and Desimone, 1999).

The capability of the MMC nets of converging to stable states is similar to the attractors of Hopfield type networks (Hopfield, 1982). But in contrast to Hopfield nets, which only have discrete attractors, MMC nets show a continuous attractor space. Thus, looking at the example of the arm movement, every geometrically possible solution can be adopted. Additionally, neither a symmetrical weight distribution nor a bounded activation function is needed.

In the network presented here single units are used to represent single words to describe the situation (Cangelosi, 2004). This appears to be a sensible approach, especially when relating it to the language of young children. When they start speaking in the very beginning, they make use of single-word utterances. These early utterances are called holophrases (De Laguna, 1927) to express that already single words could represent complex ideas by which entire speech acts are intended (Tomasello, 1999). But already in animals representation of higher level information has been observed. Perret et al. (1989) have, for example, shown that neurons exist in the brains of monkeys, which are sensitive to characteristics of interactions between agent and object, e.g. a hand running over a fur. Therefore, it should be taken into consideration that other levels of representation must be possible such as phonemes, morphemes or even more complex phenomena.

The internal models represented within the network can then be used as a basis to produce sequences of behaviour (Kühn & Cruse, in prep). An example to start with is the language of children at the age of about two when they use mainly two- and three word utterances. As the DD rule is still limited to a small number of units a further step will be, to combine such modules to cope with more complex situations.

A general question is still open, which is often called the readout problem (Roskies, 1999): How can the firing of neuronal groups be interpreted by other neurons, in our case how do the subsequent neurons "know" which of the activated neurons belong together? Also within other proposals for solutions of the so-called binding problem (for a review see *Neuron 24*, 1999), this question is not adequately answered yet. One way to potentially avoid this problem is pointed out by Roskies (1999): She states that a possible solution could be "to close the loop between the model and the world, so that the world impinges upon the network, which then acts upon and affects the world, thus generating semantics through action". This loop through the world is – at least in a way – realised in our model by always coupling the activity of the net with sensory input.

Acknowledgments

This work was founded by the DFG grants GK-518.

References

Braitenberg V (1984). *Vehicles: experiments in synthetic psychology*. Cambridge, MA: MIT Press.

Buccino B, Binkofski F, Fink GR, Fadiga L, Fogassi L, Gallese V, Seitz RJ, Zilles K, Rizzolatti G, Freund H-J (2001). Action observation activates premotor and parietal areas in a somatotopic manner: an fMRI study. *Eur J Neurosci*, 13, 400-404

Cangelosi A (2001). Evolution of communication and language using signals, symbols and words. *IEEE Transactions in Evolution Computation,*5, 93-101.

Cangelosi A (2004). The sensorimotor Bases of Linguistic Structure: Experiments with Grounded Adaptive Agents. In Schaal S., Ijspeert A., Billard A., Vijayakumar S.,Hallam J. and Meyer J.-A. (Eds), *From Animals to Animats 8*. Cambridge, MA: MIT Press. (pp 487-496)

Cruse H (2003a). Landmark-based navigation. *Biol Cybern*, 88, 425-437

Cruse H (2003b). The evolution of cognition - a hypothesis. *Cognitive Science*, 27, 135-155

Cruse H, Steinkühler U, Burkamp C (1998). MMC - a recurrent neural network which can be used as manipulable body model. In Pfeifer R., Blumberg B., Meyer J.-A. and Wilson S.W. (Eds.), *From Animal to Animats 5*. Cambridge, MA: MIT Press. (pp. 381-389)

De Laguna G (1927). *Speech: It's Function and Developement*. New Haven: Yale University Press.

Elman JL (1990). Finding Structure in Time. *Cognitive Science*, 14, 179-211

Hopfield JJ (1982). Neural networks and physical systems with emergent collective computational abilities. *Proc Natl Acad Sci USA*, 79, 2554-2558

Jordan MI (1986). Attractor dynamics dynamics and parallelism in a connectionist sequential machine. In *Proc. 8th annual Conf. of the Cognitive Sci. Society*. (pp. 531-546)

Kindermann Th, Cruse H (2002). MMC - a new numerical approach to the kinematics of complex manipulators. *Mechanism and Machine Theory*, 37, 375-394.

Kohonen T (1982). Self-organized formation of topollogically correct feature maps. *Biol Cybern*, 43, 59-69

Kühn S, Cruse H (in prep). Static mental representations in recurrent neural networks for the control of dynamic behavioural sequences

Perret DI, Harries MU, Devan R, Thomas S, Benson PJ, Mistlin AJ, Chitty AJ, Uietanen JK, Ortega JF (1989). Frameworks of analysis for the neural representation of animate objects and actions. *J Exp Biol*, 146, 87-113.

Reynolds JH, Desimone R (1999). The Role of Neural Mechanisms of Attention in Solving the Binding Problem. *Neuron*, 24, 19-29

Rizzolatti G, Arbib MA (1998). Language within our grasp. *Trends Neurosci*, 21, 188-194

Roskies AL (1999). The Binding Problem. *Neuron*, 24, 7-9

Steels L (1995). Intelligence - Dynamics and Representations. In Steels L. (Ed.) *The Biology and Technology of Intelligent Autonomous Agents*. Berlin: Springer.

Steil JJ (1999). *Input-Output Stability of Recurrent Neural Networks*. Göttingen: Cuvillier Verlag.

Steinkühler U, Cruse H (1998). A holistic model for an internal representation to control movement of a manipulator with redundant degrees of freedom. *Biol Cybern*, 79, 457-466

Tomasello M (1999). *The cultural origins of human cognition*. Cambridge, MA, London: Harvard University Press.

Tani J, Nolfi S (1999). Learning to perceive the world articulated: an approach for hierarchical learning in sensory-motor systems. *Neural Netw*, 12: 1131-1141

Wolpert DM, Kawato M (1998). Multiple paired forward and inverse models for motor control. *Neural Netw*, 11, 1317-1329

SPATIAL COGNITION IN ACTION - SCA MODEL: CHILDREN'S GESTURAL IMAGERY IN ACTION

M. PANAYI[A, B] D.M. ROY [A]

[A]*City University Faculty of Social Science, Institute of Health Science Department of Language & Communication Science Northampton Square, London, EC1V OHB, UK*
m.panayi@city.ac.uk

[B]*Zeric Ltd. 97, Gloucester Avenue, London, NW1 8LB.UK,*
d.m.roy@ntlworld.com, m.panayi@ntlworld.com

A preliminary model for the development of *Spatial Cognition as Action (SCA)* is proposed. Two propositions underpin the model, where intentional action is predominately driven by the nature of our understanding of the action in the world, and knowledge is considered in part as body schema in relation to the self, the other and interaction with artefacts. Expressive *gesture-in-interaction* is considered part of an *emergent dynamic non-linear system.* Neural network modelling techniques, video analysis and simulated animation were applied to the recognition of dynamic gesture of people with severe speech and motor impairment. This work re-examines the gestural corpora from an inter-modal interaction perspective. An action based annotation system uses feature sets that represent aspects of child spatial and kinaesthetic cognition. The model is informed by current thinking with regard to the motor theory, neural basis of language and evidence from imaging studies within a framework of narrative theory to understand spatial conceptual integration.

'We know our human spiritual potencies 'ex ipsorum actuum qualitate' through the quality of the actions of which they are the origin'

Umberto Eco in Limits of Interpretation, Contra gentiles 3.46.1984:2.2.4 p221.1994

1. Introduction

1.1. *Spatial Cognition & Gesture*

Research into gesture in human communication has a long history, reviews can be found in Condillac, Gourdian, Cherry, Corballis, Place, Kendon, Mc Neill, reviewed by Roy et al (1996). Recent advances in cognitive science and motor theory, complimented by advances in neuro-imaging and neural-computing have created an exciting place for ideas and methodologies to intersect.

The role and importance of gesture, in the evolution of human language and experience is enjoying a revival of interest and is being revisited by researchers from these domains, as illustrated by a review of recent literature Roy and

Panayi (1998, 2000a,b). Robotic research applications in particular, have exploited a convergence not only at the level of biologically inspired modeling of action, but also at the level of cognitive operations where research into canonical mirror neuron systems are being applied to robotic action (Oztop et al. 2002). Illustrations from the neuroscience arena are give by recent work of V.S. Ramachandran et al 1998, suggest that gesture may play a significant role in the evolution of human cognition, by being crucial to our understanding of the world. In particular work with synaesthetic patients, supports not only the authors propositions of considering interaction as dynamic emergent systems, but also the cross/inter modal nature of our understanding of the world. In addition recent functional analysis of brain imaging work with apraxic populations, suggests that *'one single memory store may be used for to-be-perceived and to-be-produced gestural representations, departing from Rothi et al.'s proposal.'* (Peigneux et al., 2004).

A preliminary model for the development of *Spatial Cognition as Action (SCA)* is proposed. Two propositions underpin the model, where intentional action is predominately driven by the nature of our understanding of the action in the world, and knowledge is considered in part as body schema in relation to the self, the other and interaction with artefacts. Expressive *gesture-in-interaction* is considered part of an *emergent dynamic non-linear system.* Neural network modeling techniques were applied to the recognition of dynamic gesture of people with severe speech and motor impairment (Roy et al. 1994a, 1994b). This work re-examines examples from the gestural corpora from an inter-modal interaction perspective. An action based annotation system (ABAS) uses feature sets that represent aspects of child spatial and kinaesthetic cognition. The model is informed by current thinking with regard to the motor theory, neural basis of language and evidence from imaging studies within a framework of narrative theory to understand spatial conceptual integration (Panayi 1998, 2001). The study of gesture and spatial cognition in children, in particular particularly those with atypical experiences offer a particular insight into the question of spatial cognition and human evolution.

1.2. *Imagery, Action and Neurological Correlates*

Various studies have attempted to dissociate peripheral signals from central command. Illustrative examples, reviewed in Thomson (2003), include the work of Decety on imagining movement, Jørgensen on contracting muscles before and after regional anaesthesia, or the use of neuro-muscular blockade by Secher. The brain regions implicated in these studies and other studies include: SMA, supplementary motor area; PMA, premotor area; DLPFC, dorso-lateral prefrontal cortex; TH, thalamus; SLSMC, superolateral sensorimotor cortex. It should be noted that these protocol involved hypnosis as a way of further separating subjects from the environment. However they do not discuss the potential of imagery to re-created the environmental context.

In comparison, in the work of the authors, narrative context is used with an aim to enhance the responses, in particular those of atypical communicators. In these studies the protocol design uses imagery c.f. imitation. Gestural data is used to understand neurological correlates substrates work in motor control imagery and how narrative conceptual structure enable children with cerebral palsy to effectively dissociate gestural responses from typical central motor control command responses. A critical motivation being the implication of the basal ganglia, motor cortex and cerebellum in the involvement of conceptual integration and motor control, which in the case of these atypical communicators is severely compromised (Roy & Panayi, 2000b).

2 Neural network recognition of Dynamic Movement

2.1 *Emergent Systems - Dynamic Action - Generative Interaction*

Neural network modeling techniques, video analysis and simulation animation were applied successfully to the recognition of dynamic gesture of people with severe speech and motor impairment (Roy et al., 1996, 1994a,b). The research questions in these studies focus on examining the nature of gestural repertoires of individuals and small groups of children. Elicitation is in response to observed/execution condition using words (charades) later extending to visual stimuli with and without sound tracks and non-speech sounds. The nature of the interaction and elicitation protocols, create conditions where, the repertoires of intentional movements are typically *not rehearsed or imitative*. The work of the authors departs from previous studies where imitative protocols are used. An adaptive generative interaction design was developed in these studies to examine aspects of interaction that could best allow the ***'emergence'*** of gestural interaction in the first instance of 'atypical' and later typical communicators (Panayi et al., 2000).

2.2 *Gesture Recognition*

The 'Flock of Birds' six degree of freedom magnetic tracker (Ascension Technology Corporation' 93) was used to calculate 3D position and orientation data (x,y,z). The strategy was developed for gesture recognition, in response to the factors affecting design and determined by the nature of the gesture set. These included: - idiosyncratic gesture patterns; co-articulated pattern (produced in a connected sequence); low-signal-to-noise ratio data; small number of gesture classes; gestures set comprises gestures that are 'static, signal or periodic movements'. System functioning was verified by inspection of the data stream using MATLAB graphics and by comparing the Silicon Graphics computer animation of tracker movement with video. For technical reasons and limitation of the systems, manually segmented gestural data were divided into data sets for training and testing. Feed forward neural networks arranged in time-delay scheme were chosen, 160ms to 1120 ms time windows were compared using filtering and re-sampled x,y,z position data as input features. The nature of cerebral palsy 'jerky' movement and some aspects of sensor noise meant signal conditioning and data reduction was applied through low-pass filtering of the data stream. The best FFNN recognized 12/26 gestures with an average recognition rate of 90% (Roy et al., 1994a,b; Roy, 1996). More recent work has extended the analysis of unit of interest to the "Gesture Action Entity' (Panayi, 1998). The empirical data that forms the current gestural corpus contain exemplars for both atypical and typical communicators. Exemplar video data exist for the various narrative conditions for children with severe speech and motor impairment (SSMI).

3 Gesture-As-Action

3.1 *Action Based Annotation System*

Analysis of 'Gesture-As-Action' (GAA), goes beyond conventional transcriptions of gesture that are based on restrictive topological classifications that arose out of ethnographic studies, dance notation and linguistically driven conversational analysis methodologies. An action based annotation system, (ABAS) has been developed using feature sets that represent aspects of child spatial and kinaesthetic cognition. Feature analysis of the gestural corpora contributed to the development of the (ABAS). The work has been influenced by the work of Washburn (motor theory of imagery, 1916), Jeannerod (motor action), Shultz (action theory), aspects of ecological interaction and development by James & Eleanor Gibson, Mandler (body schema) and by Gallese, Galati and Varlea in terms of conceptualisation of space. Features found in examples and exemplars of the corpora include- *spontaneity, abstraction, congruency, virtual object manipulation, 'synthesis on the fly', consistency, synergy of interaction in the real (veridical) or imaginary world.* Analysis in (ABAS) is driven in the first instance by *action representation and action manipulation.* Four major levels are identified and described as follows:

- ***Abstraction of Action [AOA]*** and divided into trace of path, action representation schema and location
- ***Manipulation of Action [MOA]*** deals with transformations
- ***The Virtual Object Representation [VOR]*** provides the resource to consider object representation in terms of:- self, as art of self, in relation to self and the quality of the object
- ***The Interaction Description [ID]*** affords the research tools to annotate details aspects of interaction including:- volume/dimension, boundary, quality of action e.g. weight, creation of surface, time, space, emotion

Empirical data and findings contributed to the action-based analysis for 'Gesture-As-Action' ' in the *production* aspect of the model (Panayi & Roy, 1998).

4 Spatial Cognition in Action - SCA Model

4.1 *Descriptive Model - Propositions*

This preliminary descriptive model is proposed as a tool for exploring spatial cognition in children and as a challenging framework for approaching developmental neuro-computing models of children's interaction with the world. Typical and atypical (alternate) communicators are considered to be very similar and co-existing in terms of underlying structure with variance in terms of the dynamics of the developmental pathways and cognitive strategies.

The model is informed by narrative theory structure, recent neuro-scientific imaging advances and cognitive and motor theory that begin to enable us to re-examine spatial/kinaesthetic development. These research data contribute to our understandings of the neural basis of language and motor theory in terms of spatial representation. Two propositions underpin the SCA model. Firstly that:

'Intentional actions as gestures are predominately driven by the nature of our understanding of the ***action in the world'*** and secondly that:

'Knowledge of the world can be represented and explored in terms of **body schema** in relation to the **self** the **other** and interaction with **artefacts'** (Panayi, 1998). Implications for theories of action and evolution of language and human-machine-interaction (HMI), including the potential for technological/therapeutic applications have been discussed elsewhere Panayi & Roy (1998, 1999).

4.2 *Gesture: Conceptual Integration & Narrativity*

This work re-examines the Roy-Panayi Gestural Corpora [RPGC] in terms of context and complexity. In particular consideration is given to the role of conceptual integration, space and motor control. The later aspects, until very recently has received reduced emphasis, particularly in neuroscience and neuro-computing research, where complexity and context are inaccessible or offer significant challenge to the experimental methodologies. Gestural features of inter-modal interaction provide an opportunity to investigate *'conceptual integration through play'*. The generative interaction methodologies used the concept of 'Ludic Interaction' where physical and cognitive frameworks are offered to the children in a variety of narrative contexts. ϑ_1, ϑ_2, ϑ_3, examine the conceptual access and control of body action schema in rclation to the self and other, how cross-modality is use to code for action in multi-sensory space. ϑ_3, describe children's abilities to execute dynamic transitions between physical and conceptual worlds.

ϑ_1 Conception,access and control of body action schema, self and other

ϑ_2 Use of cross-modality to code for action in multi-sensory space.

ϑ_3 Ability to execute dynamic transitions between the physical and conceptual world

The overall description of the SCA model is presented in schematic form in Figure 1, illustrates the topology and flow of aspects of spatial cognition in children. This preliminary model proposes that spatial cognition in children can be informed by the macro and micro analysis of gestural interaction. The model argues that gesture is inter-modal and interactive across several cognitive domains. The responses are corporal in nature and embodied in interaction, within communicative function in ludic play environments. Children's interactivity and engagement is often driven or may result from exploration and experimentation with their environment. GAA is offered as a construct based on empirical data elicited in various narrative ecologies.

The children's canvas for exploration, experimentation and play is considered in terms of kinaesthetic and spatial knowledge described as features of space, time and action in both veridical and imaginary space. The early work of Washburn on movement and mental imagery put the case for a motor theory of imagery. Pavio, Arbib and others have re-examined aspects of cognitive functions these include: imagery motor imagery, imaginary tasks. All these tasks require conscious, detailed representation of movement. Pavio, Logie, Richardson, Icobini et al have looked at motor imagery used for visual hand shape discrimination, which seem to activate pre-motor and posterior parietal cortex. Within the SCA model, veridical and non-veridical spatial cognition is schematized as a layer of interactivity where: 'reflection' bind and release imagery through 'rehearsal' to 'embodied' action. Through meta-cognitive process we are able to evaluate aspects of our interaction with the world in terms of e.g. relevance, salience coupled with perceptions and visualization of the world through imagery and action. Similarly binding and release is effected through our physical and tangible interactivity with the world, our sensory information and the common ground and situatedness of our experiences.

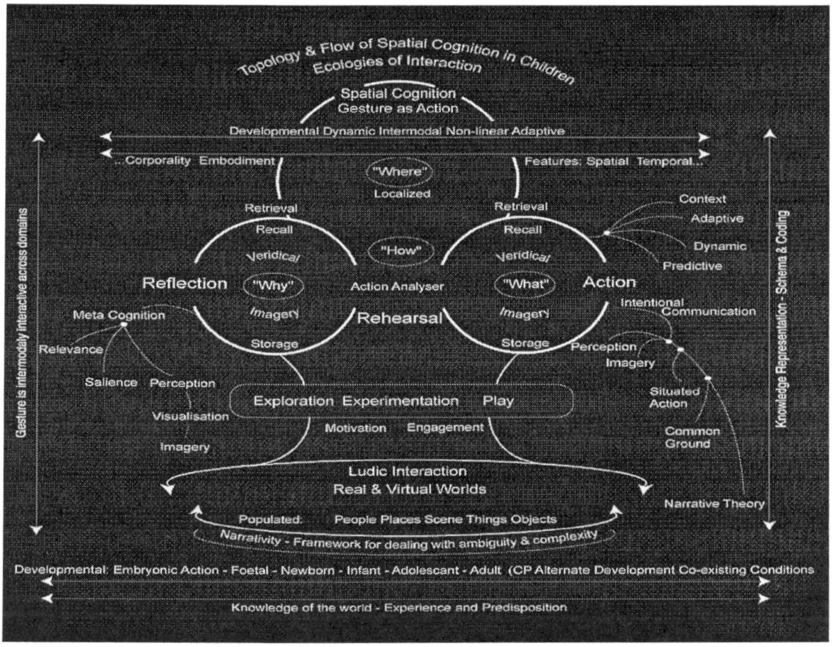

Figure 1.The SCA Model I - Topology and flow of Spatial Cognition in Children

Topology and flow of Spatial Cognition in Children* 4.5 *Spatial Cognition in Action- Implication for Neural-computing

The descriptive model could aid the development of future neuro-computational models that attempt to examine inter-modal interactivity in more detail. Knowledge representation of gesture is in part achieved by schema and coding, however these scheme are still approximations of highly complex dynamical systems. The action based scheme proposed, enables researchers to go beyond conventional gestural categorization as seen in a number of domains including anthropology (e.g. Birdwhistell's), psychology (e.g. Efron, Ekman & Friesen, Mc Neill, Cole et al), psychotherapy (Freedman & Hoffman), psycholinguistic (e.g. Butterworth & Beatie, Butterworth & Hadar). This also applies to research in the arenas of human computer interaction, in early work including e.g. Bolt, Wexelblat, Roy et al (1996) and more recently in establishing standards in language engineering (Roy et al 2000a,b). The SCA model extends an ecological modeling approach, based on the authors previous work and inspired by the Gibsons and the theoretical concepts of 'Physicality & Tangibility of Interaction' developed from early simulation and machine recognition work by Roy et al. It is envisioned that intelligent machine systems of the future would be able to synthesize and/or analysis any given gestural repertoire.

Neural-computing strategies presented more recently begin to incorporate consideration of several more complex biological features and systems. These approaches have potential for modeling the phenomenon of *developmental spatial cognition in action* as described by the SCA model and relate closely to current modeling of motor action and perception. Specific examples included the simulation work modeling vision as oscillators-resonance systems, which operate within the constraints of inhibition, coupled with a theory of epineuronal memory (Borisyuk et al., 2004). Similarly, early work on motor phyletic memory (Fuster, 2004), and recent pseudo-pattern models of infant perceptual categorization proposed and simulated by French.

The importance of content sensitivity is particularly highlighted in recent work, where fMRI and dynamic causal modeling are combined to investigate category-selective regions in ventral occipito-temporal cortex that respond preferentially to faces and other objects. Interestingly for the SCA model, during *'visual perception, category-selective patterns of activation in extrastriate cortex are mediated by content-sensitive forward connections from early visual areas. In contrast, during visual imagery, category-selective activation is mediated by content-sensitive backward connections from prefrontal cortex and content-unrelated connectivity between parietal cortex and the category-selective regions, during both perception and imagery.'* (Mechelli et al., 2004).

The corpora re-examined in this study offer unique perspective on the understanding of spatial cognition in children, particularly those with developmental disorders. The SCA model begins to illustrate the potential role played by imagery in constructing our world knowledge of self in relation to our actions, action in the world and our interaction with artifacts and how this can be exploited in the development of spatial cognition. Re-examination of *enactment* gesture of both typical and atypical communicators, indicates that there is a strong link between the role of imagery and motor production of the concepts that underlie these spatial acts (Panayi 2001, 2003, 2004). The recent developments in cognitive neuroscience and evidence emerging from brain imaging studies including the work of Arbib, Dectey, Gallese, Rizzolati, Jeannerod, and Fogassi are being interpreted in these studies with respect to the role of imagery and motor schema in intentional gestural action. The empirical work, theoretical reviews and analysis presented contribute towards the model of 'Spatial Cognition in Action' - SCA. Feature analysis and interpretations of the corpora proposes that '*gestures are motor acts not necessary synthesized through linguistic pathways, not necessarily dependent on physical interactivity, but accessed via a more distributed, dynamic multi-sensory brain mechanism,'* (Panayi & Roy, 1998, 2001). The forthcoming descriptive stage of the SCA model [II] incorporates data from neuro-imaging and neuro-computing to inform levels of interactivity of 'Gesture-as Action'.

Acknowledgements

Special thanks to all the student co-participants/Kidsearchers, therapists, teachers, parents, carers. Widener Memorial, HMS and Tuke School, NCB. Clinicians, academicians and researchers F. Miller, M. Alexander, T. Sicoli, C. Sinha, G.Strong, H.Van Balkom, A.Baker, B. Woll. Ev Sekkides for graphics support. Previous research by D.M. Roy, financially supported by EPSRC, U.K., A.I. Dupont Institute, Prentke Romich and ISAAC, USA, University of Southern Denmark. Recent research supported by Zeric Ltd., U.K. Presentation of recent work facilitated by Language Acquisition Network in Scandinavia, University of Southern Denmark and supported by BBSRC, EPSRC, City University in the UK.

References

Borisyuk et al. (2004). In H. Bowman and C. Labiouse (Eds.). *Proceedings of the eighth neural computation and psychology workshop: Connectionist models of cognition and perception II*. Singapore: World Scientific.

Fuster M.J. (2004). Upper processing stages of the perception-action cycle. *Trends in Cognitive Sciences*, 8(4).

Mechelli A., Price C.J., Friston K.J., Ishai A. (2004). Where bottom-up meets top-down: Neuronal interactions during perception and imagery. *Cerebral Cortex*, 14(11), 1256-1265

Oztop E., Arbib M.A. (2002). Schema design and implementation of the grasp-related mirror neuron system. *Biological Cybernetics* 87(2),116-140

Panayi M. (1998-2004). Spatial Cognition In Children: Gesture & Narrative. Technical reports and extracts, Ph.D. City University, London UK

Panayi M. (2001). Imagery & gesture in children - Towards a model for generation & synthesis of intentional action I, Gesture & neurological correlates II, *International Workshop on Gesture and Sign Language based Human-Computer Interaction*, City University, London.

Panayi M., Roy D.M (1998). BodyTek: Technology enhanced interactive physical theatre for people with cognitive impairment. In Nakatsu R., Altman E. J. and Pinhanez C. (Eds.): *Proceedings of ACM Multimedia* 98 *Workshop on Technologies for Interactive Movies*, 35-38.

Panayi M., Roy D.M (1998). Gestural corpora, analysis schema & neurological correlates. *European Science Foundation INTERSIGN Workshop on Acquisition*, Hamburg.

Panayi M., Roy D.M (2000). Magic of today: Tomorrows technology wearables for kids. *Workshop on Situated Interaction in Ubiquitous Computing,* ACM CHI 2000

Peigneux P., Van der Linden M., Garraux G., Laureys S., Degueldre C., Aerts J., Del Fiore G., Moonen G., Luxen A., Salmon E. (2004). Imaging a cognitive model of apraxia: the neural substrate of gesture-specific cognitive processes. *Hum Brain Mapp.* 21(3), 119-42.

Ramachandran V.S., Blakeslee S. (1998). *Phantoms in the Brain* New York: Wm. Morrow and Co

Roy D.M. (1996). Gestural human-machine interaction using neural networks for people with severe speech and motor impairment due to cerebral palsy, Ph.D. City University, London, UK

Roy D.M., et al (1994). Computer recognition of impressive dynamic arm gestures of people with severe motor impairment. *Proceedings of World Congress on Neural Networks*, San Diego, Vol.2, pp.191-196

Roy D.M., et al (1994). The enhancement of interaction for people with severe speech and physical impairment through the computer recognition of gesture and manipulation. *PRESENCE - Teleoperators and Virtual Environments*, MIT Press, Vol.3, No.3, pp.227-235.

Roy D., Panayi D.M. (1999). Towards paradigms for physical action and tangibility in interaction. In Roy D.M. & Panayi M. (Eds.) *Proceedings of the First International Workshop on Physicality and Tangibility in Interaction: Towards New Paradigms for Interaction Beyond the Desktop*, Sienna, Italy, October 1999. Odense, Denmark: i3Net.

Roy D.M., Panayi M. (2000a). ISLE project. International standards in language engineering, EU/NSF www.mpi.nl/ISLE/, *Gesture, Language and Meaning*, Porto Portugal.

Roy D.M., Panayi M. (2000b). Gestural annotation schema & neurological correlates, *Gesture, Language and Meaning,* International Gesture Society, Porto Portugal.

Thomson N.J.T. (2003). 'Mental Imagery' philosophical issues. In Nadel L. (Ed.), *Encyclopaedia of Cognitive Science*, Volume 2, pp. 1147-1153. London: Nature Publishing/Macmillan.

Washburn M.F. (1916). 'A motor theory of imagery'. See Dunlap (1914) for another version. Washburn, M.F. (1932). Some Recollections. In C. Murchison (Ed.), *A History of Psychology in Autobiography*, Vol.2. Worcester, MA: Clark University Press. (pp. 333-358), Stanford Encyclopedia of Philosophy, cited in Thomson 2003

EVOLVING A SIMULATED ROBOTIC ARM ABLE TO GRASP OBJECTS

GIANLUCA MASSERA[A] STEFANO NOLFI[A] ANGELO CANGELOSI[B]

[A]*Institute of Cognitive Science and Technologies, National Research Council (CNR, Italy)*
[B]*Adaptive Behaviour & Cognition Research Group (University of Plymouth, UK)*
gmassera@istc.cnr.it, nolfi@istc.cnr.it, acangelosi@plymouth.ac.uk

In this paper we will present the result of a set of experiments in which a robotic arm with a two-fingered hand is evolved for the ability to grasp objects. The robot is controlled by a neural network controller. Preliminary results demonstrate that evolutionary robotics techniques might scale up to situations involving robots with several degrees of freedom and problems that require an ability to produce sequential behaviour.

1. Introduction

In this paper we will present the results of a set of experiments in which a robotic arm with a two-fingered hand is evolved for the ability to grasp objects. This is partially based on a preliminary robotic arm model previously developed by Bianco and Nolfi (2004). This work is also part of the general research effort to design controllers for robotic arms through evolutionary techniques (e.g. Moriarty and Mikkulainen, 1996; Skopelitis, 2002; Marocco et al. 2003; Buehrmann and Di Paolo, 2004). Preliminary investigations demonstrate that the problem of grasping objects in varying environmental conditions can be solved by producing rather simple forms of behaviour. These forms of behaviour exploit properties emerging from the interaction between the body of the robot, its control system and the environment. Evolved neural controllers, however, are not able to master cases in which the position of the objects to be grasped vary significantly. In the following sections we briefly describe: the model and the experimental setup, the obtained results, and the future research directions.

2. Model and Experimental Setup

2.1. *The Robotic Arm*

The simulated robot consists of various objects articulated by joints, as illustrated in Figure 1. The arm consists of three segments (the arm, the forearm, and the wrist) that are attached to the previous segments (the shoulder, the arm, and the forearm, respectively) through three joints. Each joint has two degrees of freedom (DOF). The hand consists of two fingers (a thumb and an index finger) attached to the wrist with two corresponding joints with one DOF. The index finger also includes an additional segment attached to the previous segment using a joint with one DOF. Therefore the whole arm consists of six motorized joints and six segments with 9 DOF.

To model the physical dynamics as accurately as possible, we used a library for simulating rigid body dynamics called ODE (Open Dynamics Engine www.ode.org).

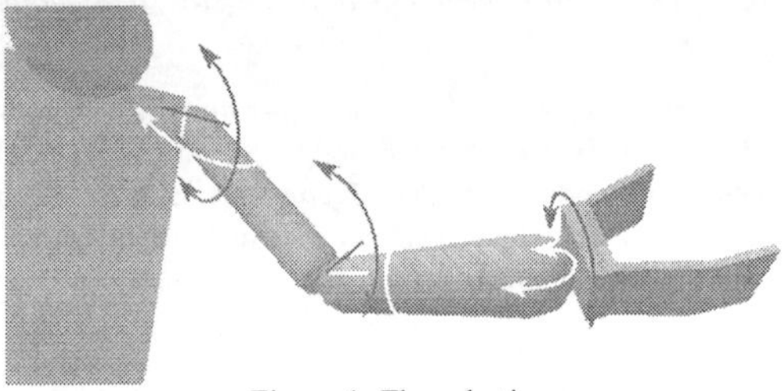

Figure 1. The robotic arm.

2.2. *The Neural Controller*

The neural controller consists of a recurrent neural network with 33 sensory neurons, 2 internal neurons, and 9 motor neurons. Sensory neurons are connected to the internal and motor neurons. Internal neurons are connected to themselves and to the motor neurons. The sensory neurons encode: the current angular position of the joints controlling the 9 DOF (9 neurons), the state of 16 contact sensors placed on the fingers and on the forearm (16 neurons), and the current relative position of the target object with respect to the hand (8 neurons). The 9 motor neurons control the actuators of the 9 corresponding joints. The output of the neurons is normalized within the range of movement of the corresponding joint and is used to set the desired position of the corresponding joint. More precisely, motors are activated so as to reach a speed proportional to the difference between the current and the desired position of the joint.

2.3. *Genetic Algorithm*

The neural controllers have been evolved for the ability to grasp objects placed on a planar surface using the evolutionary robotics technique (Nolfi and Floreano, 2000).

The genotype of evolving individuals encodes the connection weights and the biases of the neural controller. Each parameter is encoded with 8 bits.

Weights and biases are normalized between –5.0 and 5.0. Population size is 100. The 20 best individuals of each generation were allowed to reproduce by generating 5 copies of their genotype, which were mutated by flipping 1% of randomly selected bits.

Each individual of the population was tested for 10 trials, with each trial consisting of 400 steps (each step lasts 15 ms). At the beginning of each trial, the arm and the hand are set in a similar but randomly varied position (i.e. the initial angles of the joints varies within a 2% range). After 260 steps the planer surface is removed to allow un-grasped objects to fall down. Individual are selected on the basis of an ability to grasp the objects. To facilitate the emergence of this ability, however, individuals are also rewarded for the ability to touch the objects with their fingers. More precisely, the fitness is calculated according to this formula: NC + NG*1000, where NC is the numbers of contacts between the fingers and the objects occurring during the first 260 time steps, and NG is the number of time steps in which the object is inside the robot's hand during the successive time steps in which the planar surface has been removed.

3. Results

We ran four sets of experiments. For each experimental set we ran 10 different replications starting from different randomly generated population of genotypes.

In the first set of experiments, evolving individuals were asked to grasp spherical objects located on the right side with respect to the initial position of the arm. By analysing the evolved behaviour at the last generation, we observed that individuals are able to successfully grasp the objects in most of the replications. More specifically, in the case of the best replication, evolved individuals show an ability to successfully grasp objects in 71% of the cases.

One complex aspect of this task is that, due to the low friction with the table, spheres can easily roll away as a result of contact between the arm and the object, thus preventing the possibility to successfully grasp them. Evolved individuals master this complex situation by carefully calibrating their movements and by appropriately orienting their hand while moving toward the object. Indeed, in the best evolved individuals, the grasping behaviour is triggered by internal neurons that detect when the hand is correctly oriented with respect to the object (Fig. 2).

In the second set of experiments, evolving individuals were asked to grasp cubic objects located on the right side with respect to the initial position of the arm. By analysing the evolved behaviour we observed that individuals are able to successfully grasp the objects in most of the replications. More specifically, in the case of the best replication, evolved individuals show an ability to successfully grasp objects in the 87% of the cases.

Figure 2. Four snapshots of a typical evolved behaviour. The robot approaches the object by orienting its hand appropriately and then successfully grasping the sphere.

Cubic objects do not tend to roll away as easily as spheres. As a consequence, evolved robots approach the objects through fast movement and hit them, causing them to start to move. The dynamics of the interaction between the moving object and the moving hand is exploited in order to modify the relative orientation of the object, until an effective grasping is obtained.

In the last two sets of experiments evolving individuals were asked to grasp objects located both on the left and the right side with respect to the initial position of the arm. These two experimental sets differ with respect to the type of objects (spheres or cubes) to be grasped.

The obtained performance was much worse in both sets of experiments and no evolved individual displayed an ability to grasp objects located in both sides. Indeed, evolving individuals tended to specialize on only one of the two sides.

4. Discussion and Future Research Directions

The results obtained in these preliminary simulations are encouraging and demonstrate that evolutionary robotics techniques might scale up to situations involving robots with several degrees of freedom and problems that require an ability to produce sequential behaviors.

However, we also showed that evolved individuals are not capable of successfully grasping objects placed in significantly different positions. This failure might be explained by considering the fact that grasping objects located in significantly different positions requires the ability to display qualitatively different sequential behaviours.

To tackle this problem in future research we plan to provide evolving neural controllers with additional internal units that might be used to categorize different environmental situations. This will also permit the modulation of the behaviour of the basic neural controller appropriately, so that qualitatively different sequential behaviour can be produced in different environmental circumstances.

In particular, we are interested in robots able to develop *dynamical internal categories*, that is internal states that tend to vary while the robot is interacting with the environment by producing different dynamical paths in different environmental situations. Dynamical categories of this form might not only allow robots to deal with situations requiring the ability to produce qualitatively different sequential behaviours, but would also constitute an abstract high-level description (dynamically coupled with the sensory-motor flow) of the sequence of micro-movements required.

Acknowledgments

This research has been supported by the grant FIRB:RBNE01SZB4 of the Italian Ministry of Education, University and Research (MIUR) and by the ECAgents project funded by the Future and Emerging Technologies program (IST-FET) of the European Commission under the EU RD contract IST-1940.

References

Bianco R., Nolfi S. (2004). Evolving the neural controller for a robotic arm able to grasp objects on the basis of tactile sensors. *Adaptive Behavior*, 12(1):37-45.

Buehrmann T. and Di Paolo E.A., (2004). Closing the loop: Evolving a model-free visually-guided robot arm. *Proceedings of the Ninth International Conference on the Simulation and Synthesis of Living Systems, ALIFE'9* Boston, September 12th-15th, 2004. MIT Press

Marocco D., Cangelosi A., Nolfi S. (2003). The emergence of communication is evolutionary robots. *Philosophical Transactions of the Royal Society London – A*, 361: 2397-2421

Moriarty D.E., Mikkulainen R. (1996), *Evolving obstacle avoidance behavior in a robot arm*, In Maes P., Mataric M., Meyer J.-A., Pollack J., and Wilson S.W (eds.), *Proceedings of the Fourth International Conference on Simulation of Adaptive Behaviors*. Cambridge MA: MIT Press.

Nolfi S., Floreano D. (2000). *Evolutionary Robotics: The Biology, Intelligence, and Technology of Self-Organizing Machines*. Cambridge, MA: MIT Press/Bradford Books

Skopelitis C. (2002), *Control System for a Robotic Arm*, Master Thesis, School of Cognitive and Computing Sciences (COGS), University of Sussex, U.K.

References

IS LANGUAGE NECESSARY TO MERGE GEOMETRIC AND NON-GEOMETRIC SPATIAL CUES? THE CASE OF THE "BLUE- WALL TASK"

MICHELA PONTICORVO [A] ORAZIO MIGLINO [B]

[A] *Departmen of Linguisticst, University of Calabria, Via Ponte P.Bucci, Cubo 17b, 87036 Arcavacata di Rende (CS) Italy*
m.ponticorvo@unical.it

[B] *Institute of Cognition Science and Technologies, National Research Council, Viale Marx 15, I-00137 Rome Italy;*
Cognitive Technology Laboratory, Psychology Department, University of Naples II Via Vivaldi 43, 81100 Caserta Italy;
orazio.miglino@unina2.it

Hermer and Spelke (1994, 1996) have claimed that the merging of geometric (e.g the shape of the environment) and non-geometric information (e.g. the colour of landmarks) requires language. To support such an hypothesis, they consider evidences from children that, before language acquisition, can orient in space only using geometric cues. However, it has also been demonstrated that fish (Sovrano et al., 2002) are also able to integrate both kinds of cue in the same experimental setting. Here we propose the hypothesis that the observed stages in the development of spatial orientation abilities may reflect changes in the frequency of exposition to different classes of stimuli. We test this hypothesis using a genetic algorithm to train a population of simulated robots to solve the blue-wall task. By manipulating the frequency with which the robots come into contact with different classes of spatial information we show that, at specific frequencies, the sequence observed in the robots corresponds to the pattern observed in human beings.

1. Introduction

According to Hermer and Spelke (1994, 1996, 1999), language plays an important role in the human interpretation of environmental cues during spatial cognition. According to these authors, language allows the formation of representations that transcend the limits of the core knowledge system (the "geometry module") proposed by Cheng and Gallistel (1984). In particular, they have hypothesised that language makes it possible to merge geometric (i.e. environmental shape) and non-geometric spatial information (e.g. colour of landmarks). A number of experimental observations support this observation. When children who are less than six years old are placed in a rectangular room with a blue-wall, they rely entirely on geometric cues (the shape of the environment) for their orientation and completely ignore other kinds of spatial information (the colour of the walls). By contrast, older children use both kinds of information (Hermer e Spelke, 1994). The fact that six is also the age at which children complete language acquisition seems to further support Hermer and Spelke's hypothesis. However, recent experiments by Sovrano et al. (2002) have shown that *Xenotoca Eiseni* (fresh-water little fish) can replicate adult human behaviour in Hermer and Spelke's experimental setting, exploiting both geometric and non-geometric spatial cues. Fish are not traditionally believed to have strong linguistic capabilities. Sovrano's et al. observations thus contrast with Hermer and Spelke's argument. If language has no role in the merging of geometric and non-geometric spatial information, how can we explain the different stages in spatial cognition observed in human beings? Or, to be more precise, why do children rely on geometrical information in the early stages of their development, and integrate other kinds of information only later. Sovrano's et al. experiments provide no answer to this question.

In this paper, we propose a simple explanation for the different stages of development observed by Hermer and Spelke. Our explanation starts from the consideration that organisms are exposed to specific characteristics of the environment at different frequencies during different stages of their development. For example, new-born babies live in small spaces and have a relatively undeveloped sensory-motor apparatus. As a result, they can access only certain kinds of spatial information. It is plausible that organisms initially learn to respond to the information they encounter most often and which they find easiest to process; the ability to handle rarer (or more complex) classes of stimuli is acquired at a later stage.

Developmental studies of spatial cognition in fish are very difficult. In the investigation reported in this paper, we adopt an alternative approach, based on computer simulation. In our simulations we evolve a population of simulated robots with the ability to solve the blue-wall task. In this setting, we manipulate the rates at which the robots are exposed to different kinds of stimuli. This procedure allows us to identify a set of conditions in which the robots follow a sequence of developmental stages similar to those observed in humans.

2. The Blue-Wall Task

The "blue-wall" task is an experimental task often used in the study of spatial behaviour. At the beginning of the experiment, the subject is placed in a rectangular room, with one blue and three white walls. In a corner, the experimenter has placed a very inviting object (a reward). The subject is allowed to see the reward, which the experimenter then hides. After a disorienting procedure, designed to eliminate the subject's inertial sense of direction, the experimenter asks the subject to find the reward. Subjects who rely exclusively on the shape of the room have no way of distinguishing between rotationally-equivalent corners, and thus search also in the wrong corner. Subjects who take account of the colour of the walls can avoid this mistake. In brief, successful performance of the Blue Wall tasks requires subjects to integrate geometry and colour information.

3. Materials and Method

In our experiments, we used the same experimental setting defined by Sovrano et al. We then used the EvoRobot simulator (Nolfi & Floreano, 2000) to evolve a population of artificial organisms (software robots) with the ability to solve the task.

Artificial Organisms. Each artificial organism consisted of a physically accurate simulation of a round robot, with a diameter of 5.5 cm.. Each robot is equipped with 8 infrared proximity sensors (capable of detecting objects within 3 cm of the sensor) and a black/white linear camera with a receptive field of 270° and 6 meter range. The robots moved using 2 wheels (one on each side of the robot) powered by separate, independently controlled motors. The control system was an Artificial Neural Network (a perceptron) with 16 input neurons (8 units for the infrared sensors, 8 for the camera) and 3 output neurons (2 units for the wheels, 1 "decision unit"). All input units are connected to all output units.

Training environments. In our experiments, we used two distinct "training environments": a "purely geometric arena" with white walls and a "coloured arena" in which one of the long sides was coloured black. Both environments were rectangular in shape (56.8*25.6 cm). In both environments the "target area" (the reward) was located in the top-left corner.

Training procedures. The "robots" were trained using a Genetic Algorithm (Mitchell, 1998). At the beginning of each experiment, we created 100 simulated robots with random connection weights. We then tested each robot's ability to find the target location. The robot was positioned at the centre of the arena facing in a random direction and allowed to move around for 1500 computation cycles (100ms per cycle). Every time the robot reached and "identified" the target (activation of the decision unit greater than 0.5) it received a one point "reward". Each robot was tested 100 times and assigned a final score consisting of the total number of points received during the tests. At the end of this procedure, the 80 robots with the lowest scores were eliminated (truncation selection). The remaining 20 robots were then cloned (asexual reproduction). Each "parent" produced five "offspring". During cloning, 35% of neural connections were incremented by random values uniformly distributed in the interval [-1, +1]. The testing/selection/cloning cycle was iterated for 130 "generations".

To investigate the effects of differing frequencies of exposure to different classes of spatial information, we performed 9 distinct experiments, under different conditions. Each experiment consisted of 20 simulations with identical parameters. Each simulation began with different, randomly generated connection weights. In the first experiment, 10% of the trials were conducted in the "geometric arena" and 90% in the "coloured" arena. In each subsequent experiment we increased the percentage of trials in the "geometric arena" by 10%.

4. Results and Discussion

To compare the results of the experiments we extracted the best robot from the last generation in each simulation. Figure 1 shows the proportion of these robots that were capable of solving the task in both arenas.

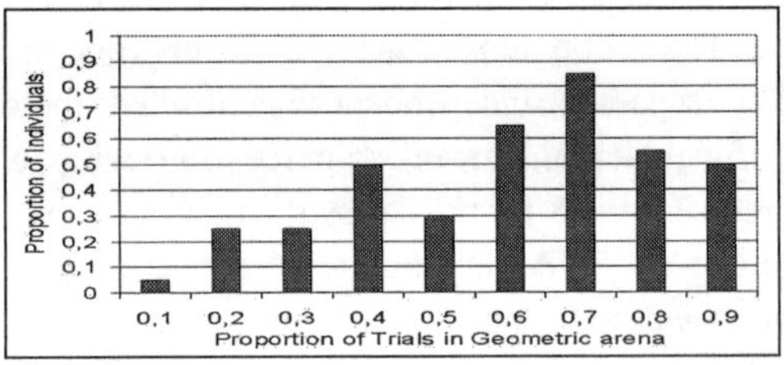

Figure 1 Proportions of robots that behave correctly in both environments.

To see the effects in terms of development, we restricted our analysis to simulations which had produced robots capable of performing correctly in both arenas (figure 2) and examined their performance after 65 "generations": i.e. half way through the evolutionary process.

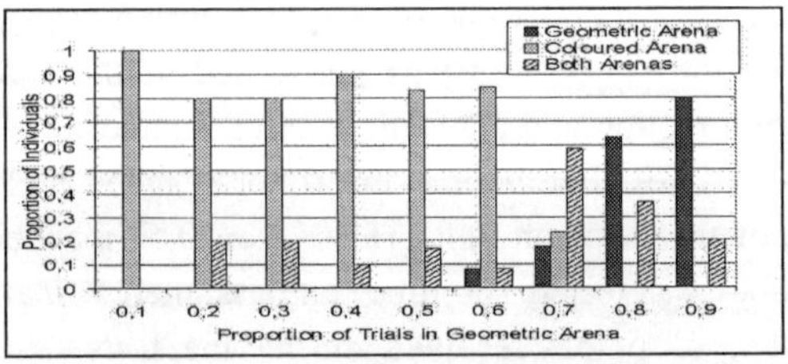

Figure 2. Proportion of individuals behaving correctly after 65 generations. The grey bars represent the proportion of robots that behave correctly in the coloured arena; the dark bar represents the proportion of robots that perform correctly in the geometric arena; the hatched bar represents the proportion of robots which perform correctly in both arenas.

The results show that several development sequences are possible. The developmental pattern that the organisms actually follow depends on the frequency with which they are exposed to different classes of spatial information. When at least 60 or 70% of trials are conducted in the geometric arena, there are robots that replicate the development sequence found in humans. They first learn to behave in a purely "geometric environment" and then acquire the ability to behave correctly in the "coloured environment".

These results do not mean, of course, that human beings do not use language to merge different classes of spatial information. What it does show is that language is not the only possible way of achieving this. In other words, our results are an existence proof: they demonstrate that it is possible to generate the developmental pathways observed in humans without resort to language. This suggests that the way in which natural and artificial organisms acquire spatial orientation capabilities is likely to depend on the precise characteristics of the ecological niche in which they develop. These characteristics may be sufficient to explain their capabilities without resort to specialized cognitive modules or language mediation.

References

Cheng, K., Gallistel, C.R. (1984) Testing the geometric power of an animal's spatial representation. In Roiblat H.L.,.Bever T.G & Terrace H.S. (Eds.), *Animal Cognition*. LEA, Hillsdale, NJ. (pp.409-23)

Hermer, L., Spelke, E. S. (1994). A geometric process for spatial reorientation in young children. *Nature* 370, 57-59.

Hermer, L., Spelke, E. S. (1996). Modularity and development: the case of spatial reorientation. *Cognition*, 61, 195-232.

Hermer-Vasquez, L., Spelke, E. S. (1999). Sources of flexibility in human cognition: dual-task studies of space and language. *Cognitive Psychology*, 39, 3-36.

Mitchell, M. (1996). *An Introduction to Genetic Algorithms*. Cambridge, MA: MIT Press.

Nolfi S., Floreano D. (2000). *Evolutionary Robotics: The Biology, Intelligence, and Technology of Self-Organizing Machines*. Cambridge, MA: MIT Press/Bradford Books

Sovrano, V.A., Bisazza, A., Vallortigara, G. (2002). Modularity and spatial reorientation in a simple mind: Encoding of geometric and nongeometric properties of spatial enviroment by fish. *Cognition*, 85, 51-59.

Attention and Perception

A NEUROCOGNITIVE THEORY OF ATTENTIONAL MODULATION OF SPREAD OF ACTIVATION

EDDY J. DAVELAAR
School of Psychology, Birkbeck, University of London, Malet Street London, WC1E 7HX, United Kingdom
e.davelaar@bbk.ac.uk

In this paper, I will present the initial steps to a neurocognitive theory of semantic priming. This theory is applied to the debate on the automaticity of semantic activation and the modulatory role of attention in the spread of activation. I conclude with an outline of a programme of work that not only addresses the issue of semantic activation, but also relates to the internal structure of the semantic memory.

1. Introduction

In the field of memory, semantic priming is a basic paradigm used to investigate the processes that inter-relate conceptual representations in long-term memory. The basic result is the increase in processing speed of words (i.e., targets), measured in terms of lexical decision times or naming latencies, when they are preceded by a related word (i.e., the prime) relative to control. The most popular explanation of semantic priming is that the prime word activates its long-term memory representation and that activation spreads to other representations in semantic space. Target representations that are semantically related to the prime word will therefore have been pre-activated, leading to the behavioural advantage over unrelated target representations. At its micro-mechanistic level, this standard explanation includes fast, automatic, facilitatory processes and slow, controlled, inhibitory processes, whereas at its macro-mechanistic level, sustained, facilitatory, context-sensitive, expectancy-based processes are operational (see for an extensive review, Neely, 1991).

This paper concerns the debate on the automaticity of semantic facilitation at the micro-mechanistic level. The existence of 'automatic semantic activation' (ASA) has been questioned by findings of context effects at short stimulus onset asynchronies (e.g., Bodner & Masson, 2003; Smith, Besner & Miyoshi, 1994). Although there are strong arguments in favour of ASA (Neely & Kahan, 2001), the findings of attentional modulation of semantic priming in healthy participants, in the elderly and in neuropsychological patients (e.g., Davelaar,

2004; Giffard, et. al., 2001; Moritz, et. al., 2001) begs a closer look at cognitive theories of semantic priming of words.

I will first give an overview of the prime-task effect, the hyperpriming effect and the prime-retention effect, which argue against a strong version of ASA and demonstrate that semantic spread can be modulated by attentional focus. Section 3 outlines a neurocognitive theory of priming. Although this paper does not include detailed simulations, a computational architecture is presented which could handle the three priming phenomena and is compatible with a range of existing working models of semantic priming. The main assumption of the neurocognitive theory is that semantic memory, located in the infero-temporal cortex, has properties that are akin to "receptive fields" and that these "semantic receptive fields" can be tuned by a frontally-mediated attentional system.

2. Semantic Priming and Attention: against ASA

In the empirical literature on short-term memory (and working memory), concepts such as "activation" and attention have been referred to without much consideration of the complex assumptions that are being made. For example, activation is typically used to refer to a state of a memory representation, which can be measured through the amount of priming that this state gives rise to (e.g., Cowan, 1993). An implicit way of reasoning about attention, activation and short-term memory is that when more attention is paid to incoming information, the corresponding representations reach a high level of activation and therefore are more likely to enter the short-term memory system. A competing alternative could be that the more attention is paid the less distracting information interferes with the attended item entering the short-term memory system. How do these alternatives relate to priming effects? According to the first alternative, increased activation would lead to more (and/or further) spread of activation and thus to increased priming effects. However, according to the second alternative, activation of related items is in fact distractor activation and thus more attention to the prime word would lead to *less* priming. As the priming effect is the single determinant of representational activation used in the debate on the existence of ASA, it is critical that the methods used and the measurements taken are unquestionable. How does increased attention to the prime word relate to its activation and to the activation of its associates?

2.1. *The Prime-Task Effect*

The prime-task effect refers to the observation that priming effects are eliminated when a concurrent (non-semantic) task has to be performed on the prime (see for a review, Maxfield, 1997). Examples of concurrent tasks that have been shown to abolish priming effects include: searching for a visual probe in the area surrounding the prime (e.g., Hoffman & MacMillan, 1985), and searching for a letter in the prime word (Henik, Friedrich, Tzelgov & Tramer, 1994; Smith, Theodor & Franklin, 1983).

A substantial body of research on the prime-task effect has used letter search as the prime task. In this task, the participant is presented with the prime word and a string of letters and has to indicate whether the letter in the letter-string is present in the prime word. The prime word is then followed by a target for lexical decision or naming. The absence of priming in this procedure depends on the experimental context. For example, Henik et al (1994) found that priming effects are observed when the number of trials in which the prime and target are related is high, while Stolz and Besner (1996) showed that priming effects are again obtained when the prime-task trials are intermixed with standard control trials.

Smith et al. (1983) used different prime tasks in order to examine the assumption that when the prime task involves 'deep' processing, priming would be obtained whereas if the prime task involves 'shallow' processing, the priming effect would be absent. They found priming effects when the prime task was phonemic analysis, silent reading or semantic analysis, but not when the prime task was visual analysis or letter search. This suggests that attention needs to be allocated at the deeper level of processing in order to obtain priming effects, which is contrary to the stronger position of ASA, in which semantic activation is assumed to be independent of attentional resources.

2.2. *The Hyperpriming Effect*

Hyperpriming refers to a significant reduction in lexical decision times or naming latencies for target words that are preceded by indirectly semantically related prime words (e.g., *lion* and *stripes* are indirectly related via *tiger*). Hyperpriming has been found for thought-disordered schizophrenic patients (e.g., Moritz, et al., 2001), people with Alzheimer's dementia (e.g., Giffard, et al., 2001) and for the elderly (e.g., Myerson, et al., 1997). Spitzer (1997) proposed that semantic activation in schizophrenic patients is enhanced, i.e., activation spreads faster and further in a semantic network. In particular, he suggested that these patients have a decreased signal-to-noise ratio (SNR) in the semantic network, leading to the network to be disinhibited (but see for an alternative view, Condray, et al., 1999). In support of the SNR-view, it was shown that healthy volunteers who received drugs that increase (with L-DOPA;

Kischka, et al., 1996) or decrease (with psilocybin; Spitzer, et al., 1996) the attentional focus exhibited decreased or increased indirect semantic priming, respectively.

2.3. *The Prime-Retention Effect*

Recently, it was found that when the prime word was retained during the lexical decision on the target, priming effects were absent (Davelaar, 2004). Subsequent research revealed that this "prime-retention effect" is particularly evident for targets that are weakly associated to the prime word and when the prime-duration is long (1000ms) followed by a blank prime-target interval (for 250ms) (Davelaar, submitted).

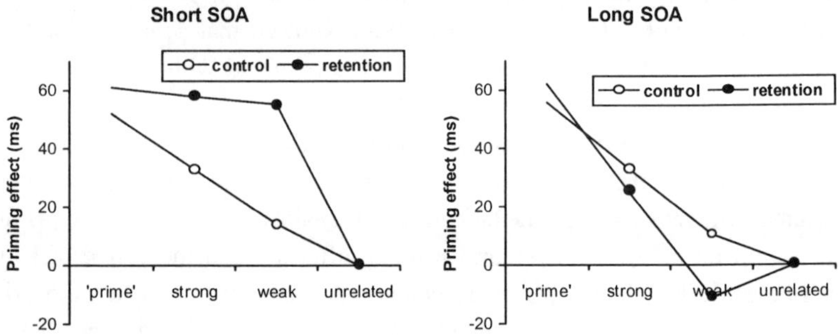

Figure 1. Priming effects as a function of the strength of the prime-target association for short (left) and long (right) stimulus-onset asynchronies (SOA). The values for the 'prime' are linear extrapolations from the values at strong and weak strength.

To illustrate, Figure 1 shows the effect of prime-retention in the short and long SOA conditions. The abscissa represents the associative distance between prime and target, with 'prime' being a linear extrapolation of the datapoints for strong and weak associates. The priming effect, defined as the reaction time for unrelated target minus the reaction time for related targets, is set out on the ordinate. The control condition is a standard lexical decision sequence, with a prime word being followed by a word/nonword target. In the retention condition, the prime word had to be retained while making the lexical decision for later (trial-by-trial) recognition of the prime. The results show that when the prime word is actively retained, i.e., it is in the focus of attention, priming effects are larger in the short SOA condition, whereas in the long SOA condition, priming decreases and in particular for the weakly associated targets.

2.4. *Summary*

The above three findings disprove the strong assumption that semantic activation *always* occurs on presentation of a word, is automatic, attention-independent and can be detected with priming paradigms, such as lexical decision tasks. The prime-task effect suggests strongly that attention needs to be directed to the semantic level in order to obtain reliable semantic priming effects. The hyperpriming effect further suggests that the attention not only needs to be directed to the semantic level, but also that attentional focus modulates the size of the priming effect, with less focus leading to more and further priming. Finally, although still under intensive investigation, the prime-retention effect suggests that too much attention to the prime word limits the spread of activation and therefore leads to less priming, especially for weak associates.

Neely and Kahan (2001) concluded that "unless visual feature integration is impaired through misdirected spatial attention, [semantic activation] is indeed automatic in that it is unaffected by the intention for it to occur and by the amount and quality of the attentional resources allocated to it" (p. 89). This conclusion is valid when only the prime-task effects are considered, but with hyperpriming and prime-retention effects, attentional resources are allocated to the semantic level, but at a defocused or focused manner, respectively. Based on these findings, it seems reasonable to refer to "permitted" spread of activation. This is further explained after the proposal in the next section.

3. An Integrative Neurocognitive Account

The proposal for a neurocognitive theory has largely been informed by the prime-retention effect, but the theory turns out to capture the prime-task and hyperpriming effects. A large number of activation-based priming models exists (e.g., Plaut & Booth, 2000) and only the overall framework and the critical mechanisms will be discussed. It is assumed that these mechanisms are compatible with a wide range of activation-based computational models of semantic memory and semantic priming, which future computational work could explore in sufficient detail.

3.1. *A Neurocognitive Framework*

The proposed framework consists of a posteriorly-localised semantic memory system that contains inhibitory as well as excitatory connections between related concepts. This memory system is interconnected with a frontal cortical system that supports (possibly in conjunction with the superior parietal cortex) focused attention to presented prime words.

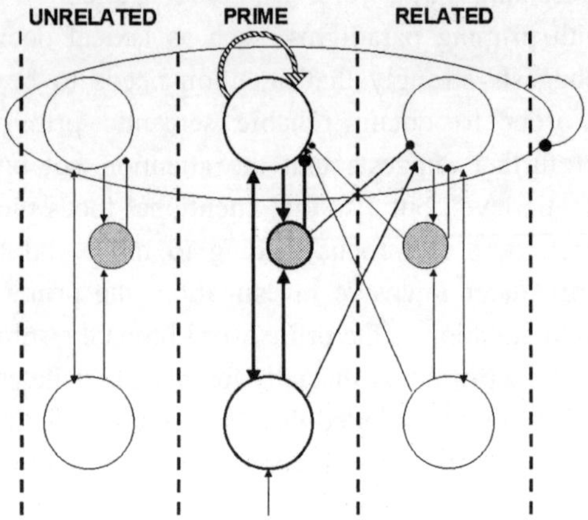

Figure 2. The proposed micro-architecture of semantic memory. External input activates the lower (input) units, whereas attentional focus (denoted by the striped arrow) targets the upper (output) units.

The crucial part of the framework lies in the interconnections *within* the semantic system. Figure 2 shows the proposed architecture of the semantic system in detail. Each representation is presented by a column in which an input unit is connected with an output unit and an inhibitory inter-unit. The inhibitory inter-unit is reciprocally connected with the output unit. This structure has the following dynamics. On presentation of an input, the activation of the output unit increases and activates the inhibitory inter-unit. As activation in the inter-unit takes time to accrue, its effect on the output unit is seen at longer presentation durations as a decrease and a stabilisation of the activation of the output unit.

Each input- and inter-unit of a representation is connected with the output unit of related concepts. This leads to the following dynamics for the output units of concepts related to the concept in the focus of attention. Initially, there is an increase in activation due to the input unit activating the output unit of the related concept. However, as soon as the inter-unit becomes sufficiently active, the activation decreases again and stabilises at an intermediate level. After offset of the input unit of the prime word, the inhibition to the related concepts decreases. Because of the above-threshold intermediate level of activation, a subsequently presented related target word will require less processing time compared to an unrelated target word: the basic semantic priming effect.

This micro-architecture can be categorised as having feedforward activation and lateral inhibition and has properties that are akin to receptive fields in the visual cortex. It is within this microstructure where Dagenbach and Carr's (1994) center-surround hypothesis, hyperpriming and the prime-retention effect come together.

Consider the attentional modulation. The frontally-mediated attentional system is assumed to affect the sensitivity (or gain) of the output unit. In the prime-retention condition, strong attentional modulation keeps the output unit active after offset of the input and thus output units of concepts related to the concept in the focus of attention still receive inhibition from the prime concept's inter-unit. Depending on the strength of the inhibition, a subsequently presented related target word will require less (for strong associates) or equivalent (for weak associates) processing time compared to an unrelated target word: the prime-retention effect. In the case of hyperpriming, inability to focus on the prime word due to reduced attentional resources (Alzheimer's, schizophrenia, old age) or drugs (psilocybin; Spitzer, et al., 1996) leads to less activation of the inhibitory inter-units. This decreased inhibition (*within* the semantic system) leads to weak (or indirect) associates reaching a level of activation at which priming effects can be observed; hyperpriming.

3.2. *Neural Re-description of ASA Debate*

The neurocognitive theory proposes that normal priming effects are due to an interaction between excitatory and inhibitory mechanisms within the semantic system and that attention modulates the balance between these opposing mechanisms. The microstructure that leads to such behaviour implements a form of attention-sensitive semantic receptive fields, where related concepts have excitatory *and* inhibitory connections. Therefore priming effects depend on the *presence* and *focus* of attention on the semantic level. Within the vocabulary of activation-spread, attention permits spread of activation, with focused attention disallowing and relaxed attention allowing activation to spread.

4. Future Work

The theoretical proposal presented in this paper is at present tentative and requires a combination of research efforts. Some major aspects need to be addressed before a true integrative account of attentional modulation of semantic spread of activation can be proposed. Computational work could not only focus on the plausible choice of parameters that make the system behave as described, but also addresses the development of the semantic receptive fields and therefore of semantic memory itself. Empirical work could investigate direct, indirect, semantic and associative priming in healthy young and elderly participants and neuropsychological patients, using the prime-task and the prime-retention paradigms. Although this seems redundant given the vast literature on semantic priming of words, a closer look at the literature already highlights important methodological differences that make comparisons difficult if not impossible. It has been shown that the construction of a block of trials determines whether the prime-task effect (Stolz & Besner, 1996) or even SOA-effects (Smith, Besner & Miyoshi, 1994) will be found. In addition, there exists an increasing body of research reports highlighting the importance of addressing the relation (pure associative, pure semantic) between the prime and the target (e.g., Shelton & Martin, 1992). This distinction has been deliberately ignored in this paper, but becomes crucial when considering the structure of semantic memory.

5. Conclusion

In this paper, I have proposed an integrative account of priming effects that is relevant to the debate on the automaticity of semantic activation. This account was formulated within a neural system and is compatible with existing computational activation-based models of priming. The overall conclusion related to the automaticity of the semantic activation is that at the neural level of description, the cognitive (behavioural) definitions of automatic and controlled processing, and facilitation and inhibition are not suitable. At this level of description "permitted" semantic activation captures some of the internal semantic dynamics. Just as tilting a bathtub makes the water flow to the lowest part, so does attention change the semantic space where activation can spread.

Acknowledgments

This work was supported by a grant from the Economic and Social Research Council (RES-000-22-0655).

References

Bodner, G. E., & Masson, M. E. J. (2003). Beyond spreading activation: an influence of relatedness proportion on masked semantic priming. *Psychonomic Bulletin and Review, 10*, 645-652.

Condray, R., Steinhauer, S. R., Cohen, J. D., van Kammen, D. P., & Kasparek, A. (1999). Modulation of language processing in schizophrenia: effects of context and haloperidol on the event-related potential. *Biological Psychiatry, 45*, 1336-1355.

Cowan, N. (1993). Activation, attention, and short-term memory. *Memory and Cognition, 21*, 162-167.

Dagenbach, D., & Carr, T. H. (1994). Inhibitory processes in perceptiual recognition: evidence for a center-surround attentional mechanism. In D. Dagenbach & T. H. Carr (Eds.), *Inhibitory processes in attention, memory and language*. San Diego, CA: Academic Press.

Davelaar, E. J. (2004). Semantic inhibition due to short-term retention of prime words: the prime-retention effect and a controlled center-surround hypothesis. *Proceedings of the 26th Annual Meeting of the Cognitive Science Society.*

Davelaar, E. J. (submitted). *Semantic inhibition due to short-term retention of prime words: the prime-retention effect.*

Friedrich, F. J., Henik, A., & Tzelgov, J. (1991). Automatic processes in lexical access and spreading activation. *Journal of Experimental Psychology: Human Perception and Performance, 17*, 792-806.

Giffard, B., Desgranges, B., Kerrouche, N., Piolino, P., & Eustache, F. (2003). The hyperpriming phenomenon in normal aging: a consequence of cognitive slowing? *Neuropsychology, 17*, 594-601.

Giffard, B., Desgranges, B., Nore-Mary, F., Lalevee, C., de la Sayette, V., Pasquier, F., & Eustache, F. (2001). The nature of semantic memory deficits in Alzheimer's disease. New insights from hyperpriming effects. *Brain, 124*, 1522-1532.

Henik, A., Friedrich, F. J., Tzelgov, J., & Tramer, S. (1994). Capacity demands of automatic processes in semantic priming. *Memory and Cognition, 22*, 157-168.

Hoffman, J., & MacMillan, F. (1985). Is semantic priming automatic? In M. I. Posner & O. S. Marin (Eds.), *Attention and Performance vol. 11* (pp. 585-600). Hillsdale, NJ: Erlbaum.

Kischka, U., Kammer, T., Maier, S., Weisbrod, M., Thimm, M., & Spitzer, M. (1996). Dopaminergic modulation of semantic network activation. *Neuropsychologia, 34*, 1107-1113.

Maxfield, L. (1997). Attention and semantic priming: a review of prime task effects. *Consciousness and Cognition, 6*, 204-218.

Moritz, S., Mersmann, K., Kloss, M., Jacobsen, D., Wilke, U., Andresen, B., Naber, D., & Pawlik, K. (2001). 'Hyper-priming' in thought-disordered schizophrenic patients. *Psychological Medicine, 31*, 221-229.

Myerson, J., Hale, S., Chen, J., & Lawrence, B. (1997). General lexical slowing and the semantic priming effect: the roles of age ability. *Acta Psychologica, 96*, 83-101.

Neely, J. H. (1991). Semantic priming effects in visual word recognition: a selective review of current findings and theories. In D. Besner & G. W. Humphreys (Eds.), *Basic processes in reading* (pp. 264-336). Hillsdale, NJ: Erlbaum.

Neely, J. H., & Kahan, T. A. (2001). Is semantic activation automatic? A critical re-evaluation. In H. L. Roediger, J. S. Nairne, I. Neath & A. M. Suprenant (Eds.), *The nature of remembering: essays in honor of R. G. Crowder* (pp. 69-93). Washington, DC: American Psychological Association.

Plaut, D. C., & Booth, J. R. (2000). Individual and developmental differences in semantic priming: empirical and computational support for a single-mechanism account of lexical processing. *Psychological Review, 107*, 786-823.

Shelton, J. R., & Martin, R. C. (1992). How semantic is semantic priming? *Journal of Experimental Psychology: Learning, Memory, and Cognition, 18*, 1191-1210.

Smith, M. C., Besner, D., & Miyoshi, H. (1994). New limits to automaticity: context modulates semantic priming. *Journal of Experimental Psychology: Learning, Memory, and Cognition, 20*, 104-115.

Smith, M. C., Theodor, L., & Franklin, P. E. (1983). The relationship between contextual facilitation and depth of processing. *Journal of Experimental Psychology: Learning, Memory, and Cognition, 9*, 697-712.

Spitzer, M. (1997). A cognitive neuroscience view of schizophrenic thought disorder. *Schizophrenia Bulletin, 23*, 29-50.

Spitzer, M., Thimm, M., Hermle, L., Holzmann, P., Kovar, K., Heimann, H., Gouzoulis-Mayfrank, E., Kischka, U., & Schneider, F. (1996). Increased activation of indirect semantic associations under psilocybin. *Biological Psychiatry, 39*, 1055-1057.

Stolz, J. A., & Besner, D. (1996). Role of set in visual word recognition: activation and activation blocking as nonautomatic processes. *Journal of Experimental Psychology: Human Perception and Performance, 22*, 1166-1177.

MODELLING THE ATTENTIONAL BLINK

HOWARD BOWMAN BRADLEY WYBLE
Computing Laboratory, University of Kent at Canterbury
Canterbury, Kent CT2 7NF, UK
hb5@kent.ac.uk, bw5@kent.ac.uk

This paper describes a computational model of the Attentional Blink constructed using the dual-stage model proposed by Chun and Potter (1995) and also incorporating a token based account of working memory Kanwisher (1987). This model reproduces data from a number of blink paradigms and makes predictions that lag-1 sparing is temporal and not sequential in origin. A further prediction is that enhancing the distinctiveness of T2 can impair T1 performance and also provoke order inversions of T1 and T2. Experiments from our lab examined the validity of these predictions. Implications and results are discussed.

1. Introduction

Many studies of temporal attention have employed Rapid Serial Visual Presentation (RSVP), in which stimuli are presented one after the other at the same spatial location. At presentation rates of 10 items per second, targets in the stream are frequently missed. The Attentional Blink (AB) (Raymond and Shapiro 1992; Chun and Potter 1995) paradigm uses such an RSVP stream, with two targets (denoted T1 and T2). The basic finding, depicted in figure 3b, is that there is a period of approximately 500 msec during which processing of T1 seems to impair the ability to detect and report T2. This suggests that the deployment of attention to processing T1 has a temporal window of a little over half a second. This interpretation is complicated by lag 1 sparing, which is the robust finding of almost unimpaired performance on T2 when it immediately follows T1. Lag 1 sparing raises the central question of why attentional resources are too limited to process both T1 and T2 at lags of 100-600 ms, but are sufficient if T2 is at shorter intervals, providing a clear challenge to computational modeling of the AB.

This paper responds to these issues by proposing a computationally detailed model that reproduces a spectrum of experimental data on what can be argued to be a canonical AB task, in which the subject must report the two letter targets that appear in a stream of digit distracters Chun and Potter (1995). Specifically, in the context of this task, our model reproduces a core set of findings, that is (1) a basic blink effect with lag 1 sparing; (2) an impairment in T1 performance at lag 1; (3) an increase in temporal order confusion at lag 1, and to a lesser extent at lag 2; and (4) effects of unmasking T1 and T2, i.e. attenuation of the blink if one or more blanks are inserted after T1 and / or T2 in the RSVP stream.

Two key principles underlie the model. Firstly, in accordance with Seiffert and Di Lollo (1997) and Maki, Bussard, Lopez and Digby (2003), we subscribe to the position that a major source of competition between RSVP items is at the level of visual features, i.e. backwards masking. In our model, masking determines the strength of the activation trace of a target. Thus, weak traces arise from masked targets (e.g. as occur in the basic blink condition), while strong traces arise when targets are unmasked (e.g. when they are followed by blanks). Critically, in our model, "bottom-up" activation strength generated by a stimulus determines how easily the representation of that stimulus can be consolidated into working memory and also, the length of time that attentional resources are occupied with this process. As a result, the severity of the blink bottleneck is regulated by bottom-up trace strength, which is in turn determined by the level of masking to which a target is subjected.

The second key principle that underlies our model is the types – tokens distinction (Kanwisher 1987). Types in our model are detailed representations of the identity of an item, including semantic and perceptual features. However, types are impoverished in their representation of how and when an item occurred. Consequently, it is also necessary to represent token information, which records instance specific details of the occurrence of an item. In particular, in our model, tokens are compact working memory encodings, which record both how and when the item occurred (e.g. its temporal position relative to other items) and which enable type information to be regenerated during retrieval.

An earlier version of this model was presented in the proceedings of the 2003 NCPW conference (Bowman, Wyble and Barnard, 2004). This model improves on that work through the implementation of a transient attentional resource that creates a temporal window of enhanced salience, which allows it to address other types of data. Also, we now separate semantic and task-related processing into different layers with different time courses of activation, an essential step in allowing targets and distracters to prime one another.

2. Methods

2.1. *Overview*

This model implements a representation of the visual processing of an RSVP stream containing 16 distracters and 2 targets presented at a simulated rate of

100 msec per item (Stimulus Onset Asynchrony: SOA). Targets arc separated by 0-6 distracters. The model simulates the neural dynamics associated with the encoding of the targets into tokens that support later retrieval.

2.2. *Neural Elements*

This model uses neural elements designed to broadly capture the generalized excitatory and inhibitory dynamics of cortical areas. Activation functions are a combination of bias, excitation, inhibition and leak currents. Connections between elements are excitatory or inhibitory and are not modifiable, with the exception of binding links described below. The activation function for the membrane potential of all neural elements is show in equation 1,

$$MP_{(i,j,t)} = MP_{(i,j,t-1)} + DT\_VM_{(j)} * [(Bias_{(j)} + Excite_{(i,j,t-1)}) * (EE_{(j)} - MP_{(i,j,t-1)}) + Inhib_{(i,j,t-1)} * (EI_{(j)} - MP_{(i,j,t-1)}) + Leak_{(j)} * (EL_{(j)} - MP_{(i,j,t-1)})] \quad (1)$$

where $MP_{(i,j,t)}$ is the membrane potential for neuron i of layer j at time t, affected by *Bias*, *Excite*, *Inhib* and *Leak* with a time constant *DT_VM*. *EE*, *EI* and *EL* represent reversal potentials. This membrane potential is used in the output function (2) which simulates a sigmoid function that bounds a neuron's output to the range [0,1), where θ and γ represent threshold and scaling parameters.

$$Out_{(i,j,t)} = \frac{[MP_{(i,j,t)} - \theta_{(j)}]_+ * \gamma_{(j)}}{[MP_{(i,j,t)} - \theta_{(j)}]_+ * \gamma_{(j)} + 1} \quad \text{where } [x]_+ = \begin{matrix} x & if \quad x > 0 \\ 0 & otherwise \end{matrix} \quad (2)$$

Apart from the recall and binding link mechanisms, described in more detail below, all elements of this model are created entirely through a network of these simple elements. This paper focuses on describing and analysing the functional characteristics of this architecture, which can be obtained over a broad range of parameter values.

2.3. *Structure*

The structure of this model can be viewed as a two stage architecture similar to that described in Chun and Potter (1995) in which the first stage is capable of operating in parallel, while the second is serial in nature. The second stage implements a tokenization process that binds tokens to types, taking inspiration from Kanwisher (1987).

The first stage comprises the bottom 4 layers depicted in Figure 1. Lateral and feedforward inhibition in the first two layers creates a masking effect that reduces the duration and amplitude of masked relative to unmasked items. Traces also differ systematically in strength based on randomised variance applied at the input layer, to simulate varying degrees of featural masking by digit/letter combinations. All representations are localist, with one neuron per item in each layer of the first stage.

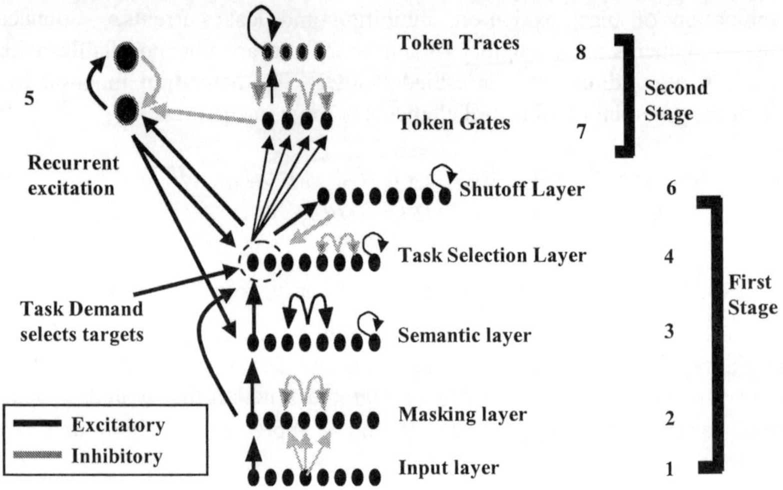

Figure 1. Structure of the model, note that connection patterns between layers are only depicted for some neurons, but apply for the entire layer.

Representations in the masking layer (refer to 2 in Figure 1) are passed directly to both of the next two layers (3 and 4). The first of these layers maintains activation traces for several hundred milliseconds, intended to represent extended semantic processing that is largely immune to conditions of task demand.

The second of these layers (the Task Selection Layer: TSL) invokes a representation at which cognitive control can select a specific category of items for further processing. Task demand input, selectively excites neurons corresponding to targets while suppressing neurons corresponding to distracters. The TSL neurons share weak mutually inhibitory connections, providing interference between co-active representations.

TSL neurons are capable of exciting an attentional resource (5), which uniformly excites all neurons in the semantic and task selection layer with a 50 msec pulse of activity. In applying this pulse across entire layers, our model creates a temporal window of attention, not specific to particular items. The contribution of this enhancement is critical in strengthening weak traces to a level at which they can activate a token and be encoded (see Figure 2). Justification for the inclusion of this excitatory recurrent pulse is provided in the discussion.

This excitatory mechanism is strongly suppressed during a token binding process (as described below). These two methods of interference (lateral inhibition in the TSL and suppression of the excitatory pulse) make it difficult for weak items to become activated in the TSL while an existing representation is undergoing tokenization (see the Lag 3 traces in Figure 2). The TSL is also connected in a 1 to 1 fashion with a shutoff layer (6) that contains neurons capable of selectively inhibiting each task selection neuron after a sufficient amount of activity. This mechanism is critical for preventing strong traces with long durations from erroneously encoding multiple tokens, but can also cause repetition blindness for repeated elements Kanwisher (1987). The second stage of this model implements a process of binding working memory tokens to the item "types" in the TSL. During the presentation of the stream, a token may be activated by items in the TSL, which are above threshold. The activation of a token initiates the construction of binding links between that token and any items that are currently active. The processes of activating the token and building these links take significant time (hundreds of msec), and occur faster with stronger traces than weaker ones. Tokenization is implemented through a dual-layer system consisting of gate (7) and trace neurons (8), arranged in a series of pairs. Together one gate neuron and one trace neuron comprise a token. The gate neurons receive a small bias and are mutually inhibitory such that only one of them is active at any time, ensuring that only one token is available. The trace neurons are strongly self excitatory and switch on once given sufficient input, after which time they self-sustain and inhibit their respective gate neuron, making the bound token unavailable for further binding, and releasing the remaining tokens from inhibition.

A token in the process of binding has a highly active gate neuron and a trace neuron with a steadily accruing level of activation. In this state, binding links are incrementally created from gate neurons to individual items in the TSL. The rate of increase of these links is proportional to the strength of the individual traces in the TSL. When the trace neuron crosses threshold, it becomes strongly self excitatory and shuts down the gate neuron, remaining in this state for the remainder of the trial.

The sustained activation of these trace neurons are used later to determine which tokens were bound. Binding links from their corresponding gate neurons to the TSL determine what item or items were encoded by those tokens. These binding links have no effect on the dynamics of the model during presentation and encoding of the RSVP stream.

This token system can represent multiple instances of the same item. Although not a factor in AB studies that forbid repetition of targets, this feature provides an important facility to a working memory implementation. Second, a single token may point to multiple items. This is a case that may occur when items are presented in close temporal proximity (< 150 msec). Third, the tokens are made available sequentially as a result of the bias currents applied to them, a feature of the model which implicitly encodes order information.

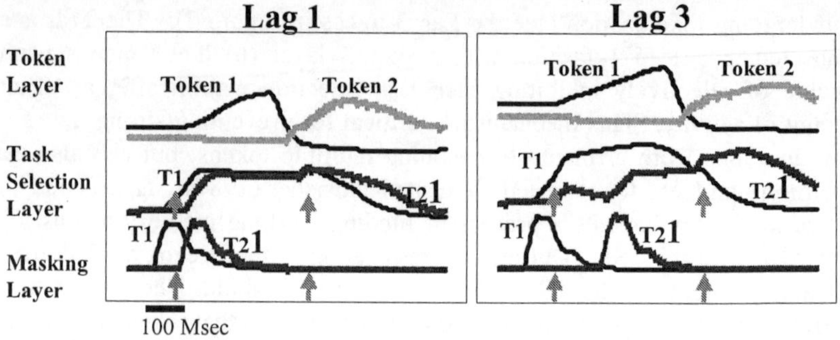

Figure 2. A depiction of the time course of events in the model in two layers of stage 1 and in the token gate neurons. Grey arrowheads indicate activation of the transient attentional resource. At lag 1, T2 is bound into token 1 alongside T1. At lag 3, binding of T2 to token 2 is delayed until the completion of token 1, but by this time T2 strength has decayed to the point at which the binding process fails (binding links not shown)

In this way, the model preserves the order of item encoding when items are presented slowly enough that a second token is available to encode the second item (ie at >300 msec lag).

At the conclusion of an RSVP stream, the model simulates a recall phase to disambiguate the erroneous bindings caused by a presentation rate that is too rapid for the token system to accommodate. This simulation is not performed by further neural dynamics within the network, as it is only our intent to model encoding processes.

For successful recall of a target at the end of a trial, a binding link of sufficient strength must exist from any token to that target, and that token must have an active trace neuron If both targets were successfully bound to either one or two tokens, the recall phase probabilistically determines in which order they were perceived by comparing the relative strength of binding links from each token to each of the two targets. In determining order, we are assuming that a retrieval process makes reasonably efficient use of all available information, including bindings that are below threshold and inactive trace neurons (i.e incomplete tokens). The sum of the token binding links indicating that T1 and T2 are in the correct, or inverted order are compared to determine a probability that a perceived order inversion occurred on a given trial. This approach simulates a robust disambiguation process that makes sense of "sloppy" token bindings.

3. Results

Figure 3 displays model data alongside human data. This model is able to successfully reproduce all of the qualitative features of the standard blink, T1 + 1 Blank, and T2 at the end of the stream conditions, and also creates some predictions for data that have not yet been collected.

The mechanism by which the model generates a blink is largely a result of suppression of the excitatory mechanism that occurs during tokenization, as well as lateral inhibition within the TSL. The combination of these effects prevents a T2 from reaching a level of activation sufficient for tokenization until token 2 is available, at which point T2 may have decayed sufficiently that it cannot be bound.

Lag-1 sparing results from the T2 being close enough in time to T1 to take advantage of the excitatory pulse, allowing it to be encoded alongside T1 into the first token. Via lateral inhibition this dual-encoding of T2 and T1 comes at the partial expense of the T1 binding strength, which is evident in the loss of T1 performance at lag 1. Furthermore, because T1 and T2 are bound to the same token, with only partial information spilling over to a second token, a massive increase in order inversions results at lag 1 (Figure 3c). Figure 2 presents a demonstration of the time course of activation traces in three layers of the model for cases of lag-1 sparing and lag-3 blinking.

Blanks in the T1+1 slot attenuate the blink by producing a stronger T1 trace, which is bound rapidly. Consequently, token 2 is available sooner and T2 is more often able to be bound to it. When the T2 is unmasked, it creates a particularly strong trace, which can either break through or outlast the window of inhibition caused by the T1 token binding process.

4. Discussion

This AB model demonstrates the emergence of a blink from the dynamics of a neural network that embraces a number of pre-existing theories (Chun and Potter 1995; Chun 1997). Further, the contour of this blink is manipulated by modifications of the RSVP stream in a way that is very similar to that shown in human data and makes predictions for what might be observed in variations of the AB paradigm.

One important tenet of this model is the way in which the recurrent excitatory pulse affects all items. This implementation was chosen because it fits available data by Chua and Goh (2001), demonstrating that a distracter in the T1+1 slot has an ability to prime a T2, while a distracter in the T1+2 slot does not. Our implementation of an excitatory pulse applied to all items achieves this increase in salience restricted to a time slot immediately following a task-relevant item. In a simulation to test the influence of primes in T1+1 and T1+2 slots, the former was capable of strongly priming T2 while the latter had little effect (63% vs 37% accuracy at lag 3 for T1+1 priming).

The real heart of this model is in the dynamics of token binding. We assume this process to be promiscuous in that a token is incapable of binding selectively to individual features. Rather, a token binds to all strongly active representations in the TSL. This method of token binding is simple to implement in a neural architecture such as this, but the caveat is that this binding process must be protected from interfering information. We propose that the AB is the means by which this protection is implemented.

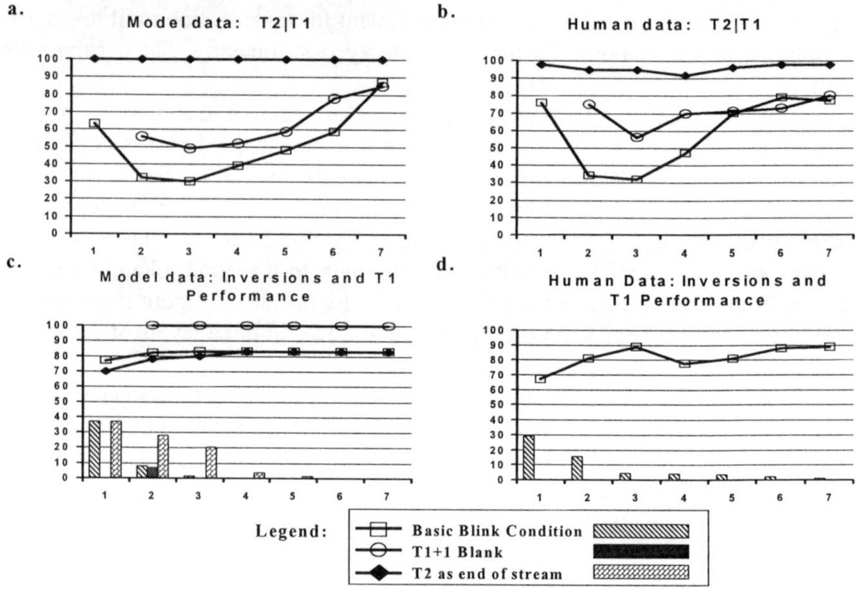

Figure 3a-d: These 4 graphs demonstrate the model's performance (a, c) compared to human data (b, d). The T2 performance data (a, b) represents the accuracy in reporting T2 on trials in which T1 was reported. In c and d, T1 accuracy is shown by the lines at the top of the graph, while the bars at the bottom indicate the percent chance for the reported order of T1 and T2 to be inverted. Human data are from Chun and Potter (1997) except the T2 end of stream data which is from Giesbrecht and Di Lollo (1998). Horizontal axis represents the number of intervening distracters, while the vertical is accuracy. Simulated results in panel c includes data not yet found in the literature.

Our view of token binding as a promiscuous process is in line with other models of AB and working memory (Dehaene and Sergent and Changeux 2003). This model extends the concept of the global workspace by incorporating several key features of the blink lacking from their work, such as lag-1 sparing, binding of multiple targets into a temporal context, and preservation of deeper levels of processing for "blinked" items (i.e. N400 waveforms are evoked by blinked items (Vogel, Luck and Shapiro, 1998).

This theoretical position has the further implication that many of the blink effects are temporal in origin, and not the result of the sequential adjacency of T1 and T1+1 slots. Multiple traces can coexist in the TSL if presented rapidly enough to take advantage of the initial excitatory pulse, and it is their mutual interference that causes order inversions and T1 impairment at lag 1. Any T2 that follows T1 within the appropriate interval (approx 100-150 msec) can be joined into the binding process of the tokenization initiated by T1. Consequently our model makes the further predictions that at SOA's of 50 msec, items presented at lag 2 should exhibit both sparing as well as order inversions. Preliminary data from our lab for RSVP rates of 10 and 20 items/sec (100 and 50 msec SOA) demonstrates that this is correct. Comparison of the fast and slow curves reveal that the blink curve is a function of time, and not the sequential nature of the presentation. Strong lag-2 sparing was obtained for the 50msec SOA and the maximal depth of the blink was at lag 6 (Figure 4).

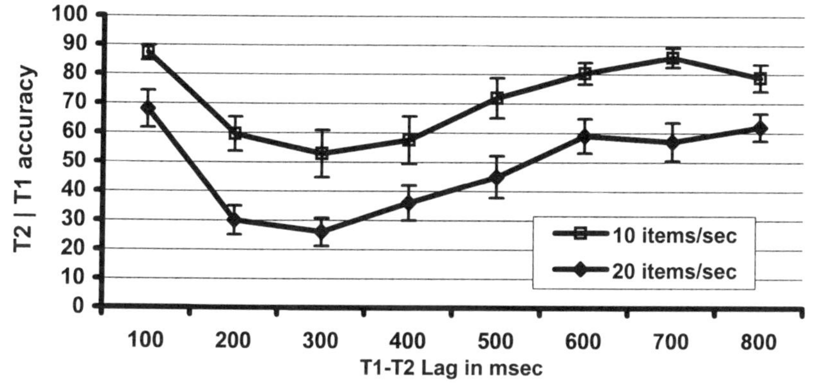

Figure 4. Comparison of blink curves for fast and slow presentation rates. Note that for the fast presentation rate, lag 2 is 100 msec, lag 4 is 200 msec, etc.

Another major prediction of this model is that enhancing the strength of T2 increases interference with T1 at lag 1. Through analysis of our data we have confirmed this prediction for different classes of letters grouped by featural difficulty. A letter by letter analysis of single target RSVP streams reveals that some have consistently higher accuracy rates, presumably because of features that are more easy to recognize through digit masks. Easy letters were: A H N Y T and L. Difficult letters were B C E P J and V.

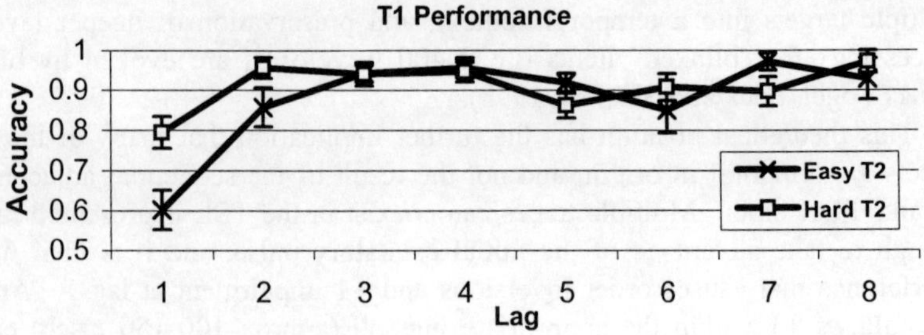

Figure 5. Comparison of T1 performance for easy and hard T2's.

In an AB experiment (Figure 5), 14 subjects saw random pairings of letters, at lags 1-8, with 92 msec SOA's. Examination of all 8 lags discovered that in the comparison of trials which used easy or hard letters for T2, the easy T2's selectively impaired T1 performance only at lag 1, (lag-1 two-tailed T-test, $p < .006$, ns at other lags). The fact that T2's with presumably strong feature traces interfere with T1 more than T2's with weak traces, and only at lag-1, constitutes indirect empirical validation of our prediction that lag-1 sparing is the result of a single combined token being created for targets presented at near 100 msec SOA.

5. Conclusion

This model succeeds in integrating a token based theory of working memory with a dual stage model of the AB, and in doing so exposes important issues that need to be explored to further our understanding of this phenomenon. It is our position that much can be learned about the nature of working memory by exploring further how T2's within the sparing window are capable of interfering with T1. Further, is it the case that T1 and T2 are bound to the same token at lag-1? Finally, do T1+1 blanks, attenuate the blink by reducing the time to bind a token? Experiments designed to address these issues will not only improve our understanding of the Attentional Blink phenomenon, but will also shed light on the structure of working memory.

Acknowledgements

The work presented here was partially supported by EPSRC Grant Number GR/S15075/01.

References

Bowman, H., Wyble, B. P., & Barnard, P. J. (2003, 2004). *Towards a Neural Network of the Attentional Blink.* Paper presented at the Eigth Neural Computational and Psychology Workshop, Canterbury, Kent.

Chua, F. K., Goh, J., & Hon, N. (2001). Nature of codes extracted during the attentional blink. *J Exp Psychol Hum Percept Perform, 27*(5), 1229-1242.

Chun, M. M., & Potter, M. C. (1995). A two-stage model for multiple target detection in rapid serial visual presentation. *J Exp Psychol Hum Percept Perform, 21*(1), 109-127.

Chun, M. M. (1997). Types and tokens in visual processing: a double dissociation between the attentional blink and repetition blindness. *J Exp Psychol Hum Percept Perform, 23*(3), 738-755.

Dehaene, S., Sergent, C., & Changeux, J. P. (2003). A neuronal network model linking subjective reports and objective physiological data during conscious perception. *Proc Natl Acad Sci U S A, 100*(14), 8520-8525.

Giesbrecht, B., & Di Lollo, V. (1998). Beyond the attentional blink: visual masking by object substitution. *J Exp Psychol Hum Percept Perform, 24*(5), 1454-1466.

Kanwisher, N. G. (1987). Repetition blindness: type recognition without token individuation. *Cognition, 27*(2), 117-143.

Maki, W. S., Bussard, G., Lopez, K., & Digby, B. (2003). Sources of interference in the attentional blink: target-distractor similarity revisited. *Percept Psychophys, 65*(2), 188-201.

Raymond, J. E., Shapiro, K. L., & Arnell, K. M. (1992). Temporary suppression of visual processing in an RSVP task: an attentional blink? *J Exp Psychol Hum Percept Perform, 18*(3), 849-860.

Seiffert, A. E., & Di Lollo, V. (1997). Low-Level Masking in the Attentional Blink. *Journal of Experimental Psychology: Hum Perc & Perf, 23*(4), 1061-1073.

Shapiro, K. L., Driver, J., Ward, R., & Sorensen, R. E. (1997). Priming from the attentional blink: A failure to extract visual tokens but not visual types. *Psychological Science, 8*, 95-100.

Vogel, E. K., Luck, S. J., & Shapiro, K. L. (1998). Electrophysiological evidence for a postperceptual locus of suppression during the attentional blink. *J Exp Psychol Hum Percept Perform, 24*(6), 1656-1674.

A MODEL OF SPATIAL AND OBJECT-BASED ATTENTION FOR ACTIVE VISUAL SEARCH

LINDA LANYON SUSAN DENHAM

Centre for Theoretical & Computational Neuroscience, University of Plymouth, Drakes Circus, Plymouth, PL4 8AA, U.K.
linda.lanyon@plymouth.ac.uk

The *biased competition hypothesis* is of particular interest in the visual attention literature currently. We describe the first computational model to apply biased competition in active vision. At the cellular level, the model simulates both spatial and object-based attentional effects over time courses seen in single cell recordings in ventral stream areas. Such effects at the cellular level lead to systems level behaviour that replicates that observed during active visual search for orientation and colour conjunction targets, where the scan path is guided to target coloured locations in preference to locations containing the target orientation or blank areas.

1. Introduction

This chapter discusses a computational model of visual attention that has been previously described formally, with results presented at the systems (Lanyon & Denham, 2004b) and cellular (Lanyon & Denham, 2005) levels. Here, we review the contribution of the model at both levels.

Traditional views of visual attention included the simple spatial "spotlight", which is a serial spatial process that enhances responses at a particular location(s). A spatial enhancement is corroborated by cellular recordings in monkeys (Connor, Callant, Preddie & Van Essen, 1996, Luck, Chelazzi, Hillyard & Desimone, 1997). However, many single cell recordings (e.g. Chelazzi, Miller, Duncan & Desimone, 1998; 2001) as well as studies in psychophysics (e.g. Blaser, Pylyshyn & Holcombe, 2000), functional magnetic resonance imaging (e.g. O'Craven, Downing & Kanwisher 1999) and event-related potential recordings (e.g. Valdes-Sosa, Bobes, Rodriguez & Pinilla, 1998) have provided convincing evidence that attention is also able to operate in a more complex object-based manner. Object-based modulation of the responses of single cells leads to the response of the cell being determined by which object within its receptive field is attended. This effect is known as the 'target effect' and results in the cell's receptive field appearing to contract around the object that is the target for visual search when more than one object is present in the cell's receptive field (Chelazzi et al., 1998, 2001). This effect is explained by the *biased competition hypothesis* (Desimone & Duncan, 1995), which is currently very influential in the study of visual attention. Biased competition suggests that responses of neurons encoding features of visual objects are determined by competitive interactions. It is suggested that there is competition at each level of the ventral "what" stream (Ungerleider & Mishkin, 1982) hierarchy. This competition is subject to a number of biases, such as "bottom-up" stimulus information and "top-down" cognitive requirements, for example a working memory template of the target object during visual search. Early small-scale models proved the feasibility of the hypothesis (e.g. Usher & Niebur, 1996) and recent larger-scale models (e.g. Deco, 2001; Hamker, 2004) have been successful in replicating a range of experimental data. Lanyon and Denham (2004b) presented, for the first time, a systems level biased competition model for active visual search, where retinal inputs change as the focus of attention is shifted, as happens during normal everyday search.

Event-related potential (Hillyard & Anllo-Vento, 1998) and single cell recordings suggest that spatial attention may be able to modulate responses earlier in the stimulus-related response than object-based attention. In ventral pathway area V4, which encodes form and colour features, spatial attention affects the earliest stimulus-invoked response at 60ms post-stimulus and modulates baseline responses before this (Luck et al., 1997). However, object-based effects in V4 appear to take longer to be resolved and have been recorded from approximately 150ms post-stimulus (Chelazzi et al., 2001; Motter, 1994). In inferior temporal cortex (IT), baseline responses during a delay period following a cue and the initial sensory response invoked by the search array are slightly modulated by the target object, but the most significant target effect begins from ~150-200ms post-onset (Chelazzi et al., 1998). Lanyon and Denham (2005) replicated the development of these attentional effects in IT and V4, over the same time course as those observed by Chelazzi et al. (1998, 2001) and Motter (1994). Such effects at the neuronal level allow this model to produce systems level behaviour where scan paths are attracted to behaviourally relevant locations, in particular replicating search behaviour observed in humans (Williams & Reingold, 2001) and monkeys (Motter & Belky, 1998b) where attention was preferentially attracted to locations containing target colour.

2. The Model

Figure 1 shows that the model consists of a retina and V1, which detect form and colour features in a retinotopic manner, and three dynamic modules representing areas V4, IT and the lateral intraparietal area (LIP).

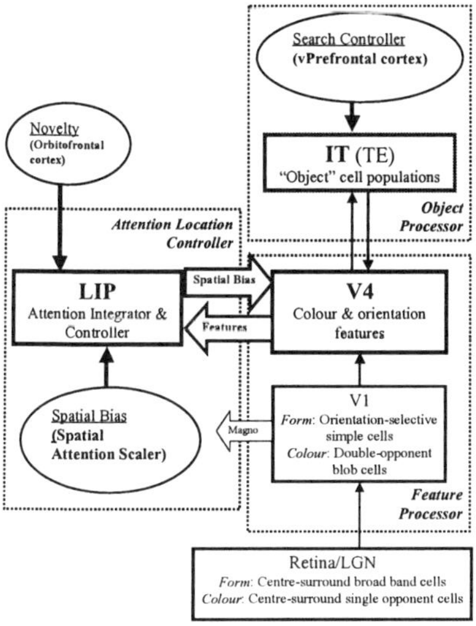

Figure 1. Overview of the model

The model adopts a similar framework to that described by Deco (2001). However, object and spatial biases operate concurrently here to influence the system dynamics and produce attentional effects at the cellular level over time courses observed in single cell recordings. Also in contrast to Deco's model, the retina here moves around the scene in an active vision manner, although a covert mode is also available. The dynamic modules are modelled using a mean field approach where the level of representation is a population of cells with similar properties, known as a cell assembly. Currently, V1 is not included in the dynamic portion of the system and we do not replicate attentional effects therein but, instead, focus on intermediate stages of visual processing, such as area V4. V1 acts as a feature detector, with a neuron for each feature at every pixel location, and features are fed forward to V4 in a convergent manner over the V4 receptive field area. V4 is retinotopic with biologically plausible (Wallis & Rolls, 1997) larger receptive fields than found in V1. Stimuli are coloured oriented bars, chosen to match the feature conjunctions used by Motter & Belky (1998b), in order to reproduce the effects seen in that psychophysical experiment. V4 neurons are functionally segregated (Ghose & Ts'O, 1997) and, like other similar models (e.g. Hamker, 2004), the V4 module encodes colour and orientation in separate feature layers. In contrast to Deco (2001), competition is local, mediated by inhibitory interneurons, and operates between different features of the same type at the same location. Features within the same receptive field compete, as has been found in single cell recordings (e.g. Chelazzi et al., 1998, 2001; Luck et al., 1997; Moran & Desimone, 1985). V4 featural information feeds forward to activate non-retinotopic invariant object populations in IT. Thus, the model's ventral pathway (V1, V4 and IT) operates as a feature/object processing hierarchy with receptive field sizes increasing towards IT, where receptive fields span the entire retina. Anterior areas of IT, such as area TE, are not retinotopic but have large receptive fields and encode objects in an invariant manner (Wallis & Rolls, 1997).

Dorsal pathway module LIP contains a cell assembly for every retinotopic location in V4 and competition is spatial so that LIP is used to control the spatial focus of attention. Each LIP assembly is reciprocally connected to the assemblies in every feature layer in V4 at that location. During fixation, a spatial bias, linked to the eye movement and possibly originating in the frontal eye field, creates an early spatial attention window (AW). The AW is scaled according to local stimulus density (Lanyon & Denham, 2004a) on the basis of low-resolution orientation information assumed to be conveyed rapidly through the magnocellular pathway (Cheng, Eysel & Vidyasagar, 2004). This reflects the scaling of the area around fixation within which targets can be detected (Motter & Belky, 1998a). This spatial bias to LIP creates an anticipatory spatial attention effect, as has been found in single cell recordings in this area (Colby, Duhamel & Goldberg, 1996). LIP provides a spatial bias to V4 and this creates an early spatial attention effect therein, modulating baseline firing before the onset of the stimulus evoked response and, subsequently, modulating this response from its onset at ~60ms post-stimulus, as found by Luck et al. (1997).

Within V4, the early spatial focus of attention becomes object-based over time. V4 is subject to two concurrent biases: A spatial bias from LIP and an object-related bias from IT. Object-based attention evolves in the model's ventral pathway due to its object-related feedback biases. The development of object-based attention in V4 is dependent on the resolution of competition between objects in IT. In addition to the feed forward inputs from V4, competition between objects in IT is biased by a feedback current from prefrontal cortex, which is assumed to hold a working memory template of the target object. IT feeds back an inhibitory bias to V4 that suppresses features not related to the object. Over time, the target object is able to win the competition in IT, and then target features become enhanced whilst non-target features are suppressed across V4, as has been found in single cell recordings in monkeys (Motter, 1994) over the same time course. Once a significant object-based effect is established in IT (determined simply by the activity of the strongest IT assembly being at least twice as active as the second strongest assembly), a saccade is initiated 70ms later, to reflect motor preparation, and in accordance with the relationship between object effects and saccade onset recorded IT and V4 (Chelazzi et al., 1998, 2001).

Following this object-based "highlighting" of target features in parallel across V4, the featural inputs from V4 to LIP provide a bias that results in locations containing target-related features becoming most active in LIP. Thus, LIP is able to represent the locations of salient, behaviourally relevant stimuli (Colby et al., 1996; Gottlieb, Kusunoki & Goldberg, 1998). The connection from V4 to LIP is set such that colour features provide a marginally stronger bias than orientation features. Thus, LIP tends to represent locations containing target coloured stimuli more strongly than locations containing the target orientation but non-target colour. Therefore, colour is given a priority in guiding the scan path, as found by Motter & Belky (1998b).

Inhibition of return (IOR) in the scan path cannot be implemented by inhibition of previously active cortical locations in retinotopic coordinates because the retina moves around the scene. A map containing the relative novelty of locations in the scene provides a bias to LIP and, therefore LIP mediates IOR. Parietal damage is linked to an inability to retain a spatial working memory of searched locations across saccades so that locations are repeatedly re-fixated (Husain, Mannan, Hodgson, Wojciulik, Driver, J. & Kennard, 2001). The source of such a bias could be a frontal area, such as orbitofrontal cortex, which is linked with reward association and, when damaged, affects IOR (Hodgson, Mort, Chamberlain, Hutton, O'Neill & Kennard, 2002).

3. Results

3.1. *Cellular Level Results*

At the cellular level, attentional effects within our model's V4 evolve from being purely spatial to being object-based, over similar times to those seen in monkey cell recordings (Chelazzi et al., 2001; Luck et al., 1997; Motter, 1994). The enhancement of target features across V4, which is dependant upon the development of the target object effect in IT, facilitates the guidance of the scan path. Object-based attention in our model's IT similarly develops over time (Chelazzi et al., 1998) and Figure 2 shows the development of these effects in V4 and IT. Replication of data from older and more highly trained monkeys

(Chelazzi et al., 2001) compared to (Chelazzi et al., 1998) suggested that IT feedback to V4 might be tuned by learning (Lanyon & Denham, 2005).

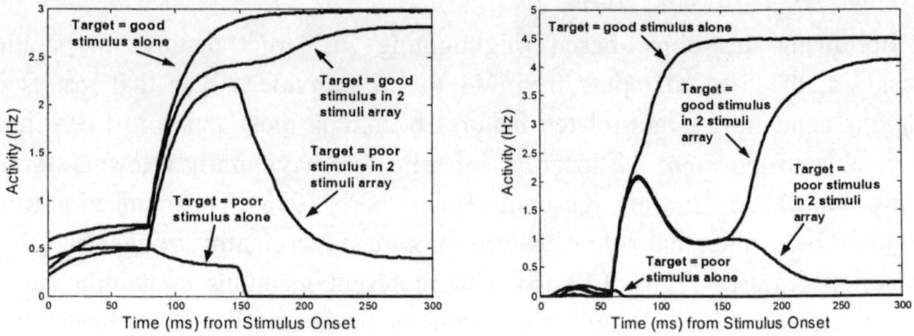

Figure 2. Left: Replicates Chelazzi et al. (1998, *figure 12a*) IT recording. Right: Replicates Chelazzi et al. (2001, *figure 5*) V4 recording. Shows the responses of assemblies in IT and V4 from array onset until the time of the saccade under 4 conditions in separate simulations. When a good stimulus for the cell (i.e. one that causes a high response) is presented alone in the receptive field a high response is produced: Top line. When a poor stimulus is presented alone, the cell's response is suppressed: Bottom line. The most interesting situation occurs when both stimuli are simultaneously presented within the receptive field: Middle two lines. The initial high response normally invoked by the good stimulus is reduced due to the competing presence of the poor stimulus. However, once object-based effects become significant from ~150ms, the response is determined by which of the two stimuli is the target, with activity tending towards that of the single stimulus case for the target object. When the target is the poor stimulus the response is suppressed despite the presence of the good stimulus in the receptive field. This effect enables V4 to represent target stimuli most strongly and bias LIP to guide the scan path to these locations. Here, saccades have been suppressed but, for example, for the two stimuli case in V4, would have occurred at ~235-240ms post-stimulus, which is similar to the time found by Chelazzi et al. (2001).

3.2. *Systems Level Results*

Due to the weighted connections from V4 to LIP, scan paths are attracted to target coloured locations, as was found by Motter & Belky (1998b). Saccades tend to be shorter in dense scenes compared to sparse scenes because the AW is scaled according to local stimulus density. This attentional focus zooms in and out as it moves around a scene of mixed density, resulting in smaller saccades in the dense areas and larger saccades in sparse areas. Updates to the novelty map are also based on the size of the AW and this means that a smaller region is inhibited for return in areas of dense stimuli. Thus, the scan path is able to investigate the dense, and potentially more interesting, areas of a scene with a series of shorter amplitude saccades, as shown in Figure 3. Increasing the weight (i.e. importance) of the novelty bias to LIP slightly reduces the likelihood of fixating target coloured stimuli and, in sparse scenes, increases the number of fixations landing in blank areas of the display (Lanyon & Denham, 2004b).

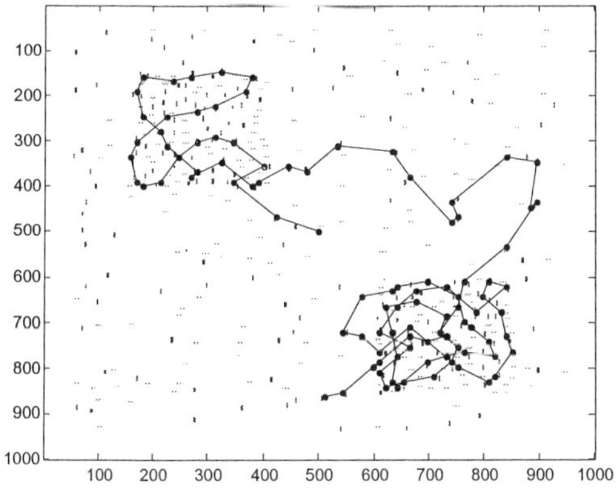

Figure 3. Scan path through a mixed density scene where the target colour is red: Average amplitude of saccades in the dense areas is 4.5°, compared to 7.5° in the slightly more sparse areas. (N.B. In grayscale, the red bars appear as a darker gray than the green bars.)

Weaker object-related feedback within the ventral pathway (prefrontal to IT; IT to V4) reduces the object-based effect in IT and V4, resulting in saccade onset being delayed and more non-target coloured stimuli being able to capture attention. Replication of single cell data suggested that IT feedback to V4 could be tuned by learning and experience. Therefore, search within familiar objects or tasks may be faster and the scan path would be expected to fixate more target coloured locations than when the task or objects were unfamiliar.

4. Conclusion

This model shows that attentional effects modeled at the cellular level, in order to replicate data from monkeys, are able to give rise to systems level search behaviour similar to that seen under certain conditions in humans and monkeys. At the neuronal level, the model shows attentional effects within V4 evolving from being purely spatial to being object-based, over time courses found in single cell recordings from monkeys (Chelazzi et al., 2001; Luck et al., 1997). Object-based effects in IT also accurately reproduce those recorded in monkey IT during visual search (Chelazzi ct al., 1998). Saccade onset times, linked to the development of a significant effect in IT, are very similar to those found during the money cell recordings in IT and V4 (Chelazzi et al., 1998, 2001). Replication of the monkey single cell data suggested a possible influence of learning upon the development of object-based attention and, hence, saccade onset and the behavioural relevance of locations visited by the scan path (Lanyon and Denham, 2005).

Object-based effects at the cellular level were used to guide the search scan path to behaviourally relevant locations. Prefrontal feedback provides a top-down cognitive bias towards the target object. This leads to the target object in IT winning the competition and, consequently, target features being enhanced across V4. The V4 bias to posterior parietal area LIP causes it to represent locations containing target features most strongly and allow attention to be attracted to these behaviourally salient locations. Thus, similar to Wolfe's (1994) *activation map*, LIP acts as a retinotopic "saliency map" (Gottlieb et al., 1998) combining information from the ventral stream, which has resolved top-down and bottom-up influences. Parietal cortex, the frontal eye field (FEF), the pulvinar nucleus of the thalamus and superior colliculus (SC) have all been implicated in providing saliency map functions and may cooperate in this. Shipp (2004) reviewed the neural architecture controlling the spatial focus of attention in relation to current models of attention. His Real Neural Architecture model suggests a distributed saliency system in which the pulvinar acts as the central hub saliency map providing the attentional spotlight and communicating with SC, ventral stream areas and, indirectly, with LIP and FEF. In the context of the model presented here, the LIP module can be considered part of the distributed saliency system and the pulvinar could be a possible source of the spatial bias to LIP and V4.

Representations in the model LIP are retinotopic and based on the current fixation point. The current model does not fully address issues of saccadic re-mapping of the scene and associated saliency. Cortical transformations and compression of space around the time of a saccade are very complex issues and are discussed in relation to this model in Lanyon and Denham (2004b). A scene-based novelty bias, possibly from orbitofrontal cortex, to LIP provided inhibition of return in the scan path. This suggests that parietal areas may be involved in transforming a memory trace of locations visited in the scene into retinotopic coordinates. The connections from V4 to LIP suggest one possible mechanism by which different features could gain priority in attracting attention. We allowed colour to have priority in order to replicate its strong influence on attentional capture found by Motter and Belky (1998) using the same stimuli. Our stimuli were equally salient in terms of bottom-up contribution. However, a feature type having strong bottom-up salience (represented by high activity in V1 and V4) could be capable of overriding the in-built bias from the V4 to LIP connection, since the difference in connection strength is marginal. In future, the model could be extended to allow LIP to represent bottom-up saliency, for example luminance or motion, more strongly than currently, possibly by allowing V1 to feedforward to LIP directly. The saliency map could be further enhanced by additional modules, representing the pulvinar and SC, participating in the system dynamics. In summary, the biased competition approach presented in this model provides an extensible framework that, for example, allows further endogenous and exogenous factors to compete for the capture of attention in LIP.

References

Blaser E., Pylyshyn Z.W., Holcombe, A.O. (2000) Tracking an object through feature space. *Nature*, 408(6809), 196-199

Cheng A., Eysel U.T., Vidyasagar T.R. (2004) The role of the magnocellular pathway in serial deployment of visual attention. *European Journal of Neuroscience*, 20, 2188-2192

Chelazzi L., Duncan J., Miller E.K. (1998) Responses of neurons in inferior temporal cortex during memory-guided visual search. *J. Neurophysiology*, 80, 2918-2940

Chelazzi L., Miller E.K., Duncan J., Desimone R. (2001) Responses of neurons in macaque area V4 during memory-guided visual search. *Cerebral Cortex*, 11, 761-772

Colby C.L., Duhamel J.R., Goldberg M.E. (1996) Visual, presaccadic, and cognitive activation of single neurons in monkey lateral intraparietal area. *J. Neurophysiology*, 76(5), 2841-2852

Connor C.E., Callant J.L., Preddie D.C., Van Essen D.C. (1996) Responses in area V4 depend on the spatial relationship between stimulus and attention. *J. Neurophysiology*, 75, 1306-1308

Deco G. (2001) Biased competition mechanisms for visual attention in a multimodular neurodynamical system. In: Wermter, S, Austin, J. and Willshaw, D. (Eds), *Emergent Neural Computational Architectures Based on Neuroscience: Towards Neuroscience-Inspired Computing*. Heidelberg: Springer-Verlag. (pp. 114-126)

Desimone R., Duncan J. (1995) Neural mechanisms of selective visual attention. *Ann. Rev. Neuroscience*, 18, 193-222

Ghose G.M., Ts'O D.Y. (1997) Form processing modules in primate area V4. *J. Neurophysiology*, 77(4), 2191-2196

Gottlieb J.P., Kusunoki M., Goldberg M.E. (1998) The representation of visual salience in monkey parietal cortex. *Nature*, 391, 481-484

Hamker F.H. (2004) A dynamic model of how feature cues guide spatial attention. *Vision Research*, 44, 501-521

Hillyard S.A., Anllo-Vento L. (1998) Event-related brain potentials in the study of visual selective attention. *Proc. Nat. Acad, Sci. USA*, 95, 781-787

Hodgson T.L., Mort D., Chamberlain M.M., Hutton S.B., O'Neill K.S., Kennard C. (2002) Orbitofrontal cortex mediates inhibition of return. *Neuropsychologia*, 431, 1-11

Husain, M., Mannan, S., Hodgson, T., Wojciulik, E., Driver, J., Kennard, C. Impaired spatial working memory across saccades contributes to abnormal search in parietal neglect. Brain. Vol. 124 (2001) 941-952

Lanyon L.J., Denham S.L. (2004a) A biased competition computational model of spatial and object-based attention mediating active visual search. *Neurocomputing*, 58-60C, 655-662

Lanyon L.J., Denham, S.L. (2004b) A Model of Active Visual Search with Object-Based Attention Guiding Scan Paths. *Neural Networks*, 17, 873-897

Lanyon L.J., Denham S.L. (2005) A biased competition computational model of spatial and object-based attention. *Submitted*

Luck S.J., Chelazzi L., Hillyard S.A., Desimone R. (1997) Neural mechanisms of spatial attention in areas V1, V2 and V4 of macaque visual cortex. *J. Neurophysiology*, 77, 24-42

Moran J., Desimone R. (1985) Selective attention gates visual processing in the extrastriate cortex. *Science*, 229, 782-784

Motter B.C. (1994) Neural correlates of attentive selection for color or luminance in extrastriate area V4. *J. Neuroscience*, 14(4), 2178-2189

Motter B.C., Belky E.J. (1998a) The zone of focal attention during active visual search, *Vision Research*, 38(7), 1007-1022

Motter, B.C., Belky, E.J. (1998b) The guidance of eye movements during active visual search. *Vision Research*. Vol. 38(12) (1998) 1805-1815

O'Craven K.M., Downing P.E., Kanwisher N. (1999) fMRI evidence for objects as the units of attentional selection. *Nature*, 401, 584-587

Shipp, S (2004) The brain circuitry of attention. *Trends Cog. Sci.*, 8(5), 223-230

Ungerleider L.G., Mishkin M. (1982) Two cortical visual systems. In Ingle D.J., Goodale M.A., Mansfield R.W.J. (Eds.), *Analysis of Visual Behaviour*. MA: MIT Press. (pp. 549-586)

Usher M., Niebur E. (1996) Modeling the temporal dynamics of IT neurons in visual search: A mechanism for top-down selective attention. *J. Cognitive Neuroscience*, 8(4), 311-327

Valdes-Sosa M., Bobes M.A., Rodriguez V., Pinilla T. (1998) Switching attention without shifting the spotlight: Object-based attentional modulation of brain potentials. *J. Cognitive Neuroscience*, 10(1), 137-151

Wallis G., Rolls E.T. (1997) Invariant face and object recognition in the visual system. *Prog. Neurobiology*, 51, 167-194

Williams D.E., Reingold E.M. (2001) Preattentive guidance of eye movements during triple conjunction search tasks: The effects of feature discriminability and saccadic amplitude. *Psychonomic Bulletin & Review*, 8(3), 476-488

Wolfe, J.M. (1994) Guided Search 2.0: A revised model of visual search. *Psychonomic Bulletin & Review*, 1, 202-238

HEMISPHERIC ASYMMETRIES IN THE NEGLECT SYNDROME: A COMPUTATIONAL STUDY

ANDREA DI FERDINANDO[A] MARCO CASAROTTI[A]
GIUSEPPE VALLAR[B] MARZO ZORZI[A]

[A] *Dipartimento di Psicologia Generale, Università di Padova*
via Venezia 8 - 35131 Padova (Italy)
andrea.diferdinando@unipd.it marco.casarotti@unipd.it marco.zorzi@unipd.it

[B] *Dipartimento di Psicologia, Università di Milano-Bicocca*
Piazza dell'Ateneo Nuovo 1 - 20126 Milano (Italy)
giuseppe.vallar@unimib.it

One of the most controversial aspects of hemispatial neglect is its asymmetric nature. According to most theories of neglect, this asymmetry would depend on asymmetrical representations of space in the left and right hemisphere, although there is little agreement on their exact nature and their functional meaning. In the present paper, we show a series of neural networks simulations which implement different theories of neglect to verify their neuropsychological plausibility. The results show that asymmetries are best explained by a right hemispheric dominance for spatial representations, expressed in terms of a higher number of neurons involved in spatial tasks.

1. Introduction

Unilateral spatial neglect is a neuropsychological syndrome characterized by a significant inability to perceive and explore the contralesional side of the space (see Vallar, 1998, for review). Essentially, neglect patients behave as if one half of the external world (typically the left) has ceased to exist. For instance, they may eat only from half of their plate, bump into obstacles, or dress only one side of the body. This behaviour, however, can not be accounted for in sensory or motor terms, because neglect symptoms may be present without sensorimotor deficits. Instead, neglect appears to be a higher-order cognitive deficit, involving both spatial representation and attention, although there is no agreement on their relative contribution.

Hemispatial neglect is typically associated with damage to the posterior parietal cortex (PPC), although in many patients lesions are more extensive, involving both the posterior parietal and the premotor areas. PPC is thought to play a crucial role in the computation of sensorimotor transformations, that is in linking sensation to action. PPC neurons integrate sensory information from different modalities with posture signals in order to represent object locations in a format which facilitates motor planning. Pouget and Sejnowski (1997) suggested that parietal neurons compute *basis functions* of the input variables. In particular, the response of a parietal neuron can be approximated by the product of a Gaussian function of retinal location and a Sigmoid function of eye position. The resulting representation allows to perform sensorimotor transformations using single linear projections, facilitating both computation and learning.

Pouget and Sejnowski (2001) developed a computational model of sensorimotor transformations based on the basis function approach, showing that lesions to the model produce a deficit similar to that of neglect patients. However, their model is unable to explain one of the most controversial aspects of hemispatial neglect, that is its asymmetric nature. Indeed, neglect is more frequent, long lasting and severe after right sided lesions than after equivalent lesions in the left hemisphere (Vallar, 1993). According to most theories of neglect, this asymmetry would depend on asymmetrical representations of space in the left and right hemisphere, although there is little agreement on their exact nature and their functional meaning.

Our work builds on the basis functions approach to develop a computational model of sensorimotor transformations that can explain the asymmetric nature of neglect. Rather than presenting a single computational model, we performed a series of simulations, based on the same computational architecture, implementing different theories of neglect to verify their neuropsychological plausibility.

2. The Computational Framework

The model is implemented with a neural network, consisting of three layers of units, with an additional inhibition map (Figure 1a).

a) Input maps. The input of the model consists in a two-dimensional visual map of 21 x 21 topographically organized units with Gaussian receptive fields, and a postural map of 2 x 7 units with sigmoid tuning curves encoding the position of the eye in the orbit in head-centered coordinates. The properties of the visual map are similar to those of the early visual areas. Each neuron V_i responds in proportion to the distance d_{ri} between the center of its retinal receptive field (r_{xi}, r_{yi}) and the retinal coordinates of the visual target (r_x, r_y). In particular:

$$V_i = e^{-\frac{d_{ri}^2}{2\sigma^2}} \tag{1}$$

where $d_{ri}^2 = (r_x - r_{xi})^2 + (r_y - r_{yi})^2$ and σ is the width of the Gaussian, that was set to 5^0. Both the horizontal and vertical coordinate of the units' receptive field center were spread uniformly between -40° and +40° with increments of 4°, for a total of 21 x 21 units.

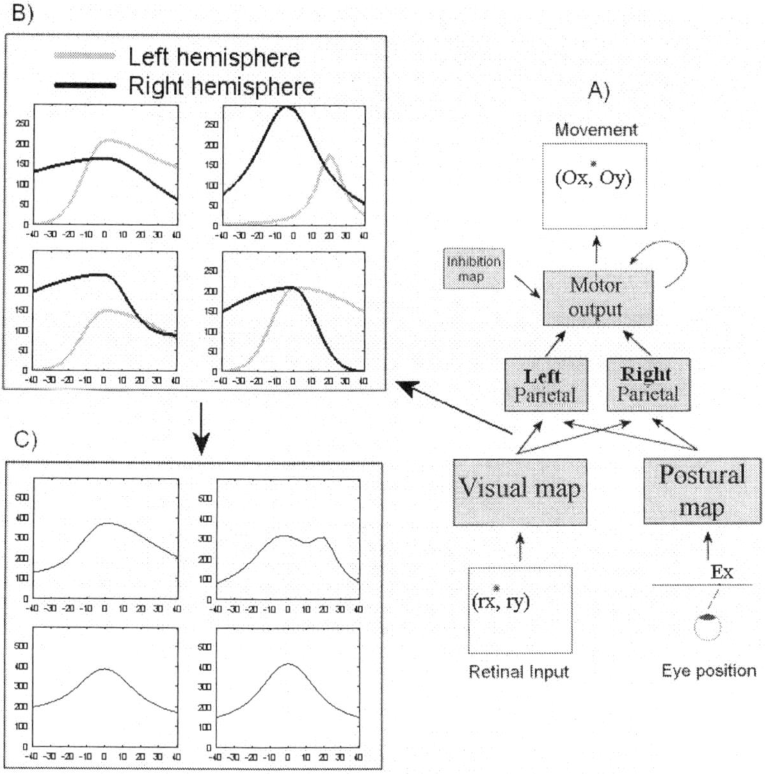

Figure 1. Network's architecture (A) and connectivity patterns used in the simulations (B). Each graph shows the number of neurons in the two parietal maps, as a function of the visual horizontal position. The corresponding global representation of the space is showed in (C).

The 14 postural units (E_i) respond with a sigmoid function of eye position:

$$E_i = \frac{1}{1+e^{-sd_{ei}}} \tag{2}$$

where $d_{ei} = |e_x - e_{xi}|$ is the distance between the abscissa of the inflection point of the sigmoid (e_{xi}) and the horizontal position of the eye in the orbit (e_x), and s is the slope of the sigmoid. We assigned to half the postural units a negative slope of -1/8 and to the remaining half a positive slope of +1/8, as both positive and negative slopes have been found in the PPC. The inflection point of the sigmoid was spread uniformly between -24° and +24° with increments of 8°, for a total of 7 units for each slope. To reduce the complexity of the task we considered only the horizontal component of the eye position.

b) Parietal maps. Both input maps are connected to two modules of basis function units simulating some of the properties of the PPC. Each parietal neuron P_{ij} is connected with the i-th unit in the visual map and the j-th unit in the postural map. These two input are combined multiplicatively: $P_{ij} = V_i \times E_j$.

c) Motor map. The parietal maps are connected to a motor output map, that encodes the movement to be performed in head-centered coordinates. Head-centered maps have been described in several cortical areas (see, for example, Galletti et al., 1993; Graziano & Gross, 1997). The synaptic weights between parietal and output maps are learned using the *delta* rule. They are the only learnable weights in the networks.

The output map is composed of 21 x 21 units. Each neuron O_i receives inputs from all the parietal neurons, from one unit in the inhibition map and from all the other output units. These different input are combined in the following way:

$$O_i = \begin{cases} \dfrac{2}{1+e^{-NetInput_i}} - 1 - Inib_i & \text{if NetInput}_i > 0 \\ 0 & \text{otherwise} \end{cases} \tag{3}$$

$$\text{Net Input}_i = \sum_j W_{ij} P_j (1 - Inib_i) + \sum_j W_{ij} O_j \tag{4}$$

where $NetInput_i$ is the input arriving to the unit and $Inib_i$ is the activation value of the i-th unit in the inhibition map. The left-hand side of equation (4) computes the external input (from visual and postural map, via the parietal maps), modulated by the corresponding inhibition unit $Inib_i$, where P_j is the j-th unit in the parietal maps, and W_{ij} is the weight of its connection with the output unit. The right-hand side of equation (4) computes the internal input, that is the input coming from the other output units. The output map contains symmetric lateral connections with fixed-value inhibitory weights that depend on the distance between neurons. When more than one stimulus is presented to the network, the competition between neurons belonging to different populations resolves over time in favor of one of the stimuli.

d) Inhibition map. The motor map receives input from an inhibition map of the same size (21 x 21): each neuron in the motor map is connected with a corresponding neuron in the inhibition map. The inhibition map has been implemented to simulate the sequential selection of stimuli. When multiple stimuli compete for selection in the motor map, the winning population of neurons activates the inhibition map, which in turn inhibits the corresponding motor units. This ensures that the same population will not be selected again during the sequential selection process.

d) Output decoding. Given that an output unit responds to a broad range of movements (depending on the width of its motor field), a specific motor plan is represented through a population code. To determine the location of the planned movement (O_x, O_y) this population code is decoded through a simple vector method (Salinas & Abbott, 1995).

e) Connectivity patterns. Hemineglect is more frequent and severe after damage to the right hemisphere of right-handed subjects. This asymmetry may be explained assuming that the representation of ipsilateral and contralateral sides of space may be different in the two hemispheres. Our model allows us to investigate the differential contribution of the two hemispheres to spatial representation by modifying the pattern of connectivity from visual to parietal maps. Parietal units do not reflect all possible combinations of visual and postural units. Rather, each combination has a given probability to belong to each of the two parietal maps. In particular, the left parietal map is more connected with visual units whose receptive fields are centered on the right hemiretina, whereas the right parietal map is more connected with units whose receptive fields are centered on the left hemiretina. Moreover, the central part of the visual field is overrepresented with respect to the peripheral part, that is there are more neurons in the parietal maps connected with foveal units. This general gradient of connectivity is consistent with neurophysiological evidence (Hamed, Duhamel, Bremmer & Graf, 2001).

Given this general arrangement, we implemented different gradients of connectivity, simulating different theoretical explanations of the asymmetries found in the neglect syndrome. In the present paper, we show the results of four networks (see Figure 1b):

Network 1. The right hemisphere attends to space bilaterally, whereas the left hemisphere primarily to the right hemispace (Heilman et al., 1987). The total contribution of the two hemispheres to the overall neural representation is equal, as the number of units in the left and right parietal maps is almost the same (2686 vs. 2809). However, the left hemisphere has a steeper gradient of connectivity with respect to the right hemisphere.

Network 2. The model proposed by Anderson (1996; 1999) is another implementation of the Heilman's hypothesis. However, differently from the previous network, the total contribution of the two hemispheres to the overall neural representation is not equal, because the left parietal map has fewer units than the right parietal map (975 vs 3413).

Network 3. The right hemisphere is the major contributor to the overall representation of space, in terms of a different number of neurons in the left and right parietal maps (1842 vs. 3822). However, differently from the two previous models, the gradient of connectivity of the two hemispheres is roughly the same. Moreover, while in the two previous models the right hemispace is slightly overrepresented with respect to the left hemispace, in the third model the opposite is true.

Network 4. The right hemisphere has larger receptive fields than the left hemisphere, while the connectivity patterns of the two hemispheres are completely symmetric. This hypothesis was first modeled by Monaghan and Shillcock (2004).

3. Tasks

A variety of clinical tests have been used to diagnose the neglect syndrome, among them the most widely employed are the "line bisection" task and Albert's test (or "cancellation task"). In the line bisection task, patients are simply asked to mark the midpoint of a horizontal line, but they typically displace it toward the ipsilesional side of space; in the cancellation task, patients are asked to cross out a set of line segments uniformly spread over a sheet of paper, but they typically omit contralesional targets.

4. Results and Discussion

The four networks were trained to solve the two clinical tests and their performance was compared with that of normal subjects. Networks were then damaged to simulate parietal lesions. The lesion consisted in destroying all the units of one parietal map (either left or right). Lesioned units had their activation fixed to a low constant value, independent from the contents of the visual field (Monaghan & Shillcock, 1998).

All the data presented are the average of 10 replications of the simulations, obtained by varying the random conditions of simulations (starting connection weights and presentation order of the training patterns).

4.1. *Bisection Task*

Normal subjects bisect a line slightly toward the left. This phenomenon has been called "pseudoneglect". After training, the four networks accurately performed the bisection task. When a line is presented in the visual field, the motor map produces an activation pattern that, after decoding, corresponds approximately to the middle of the line. However, the different gradients of connectivity produce slightly different behaviours. Figure 2 shows the performance of the four "normal" networks in the bisection task, for lines of different length.

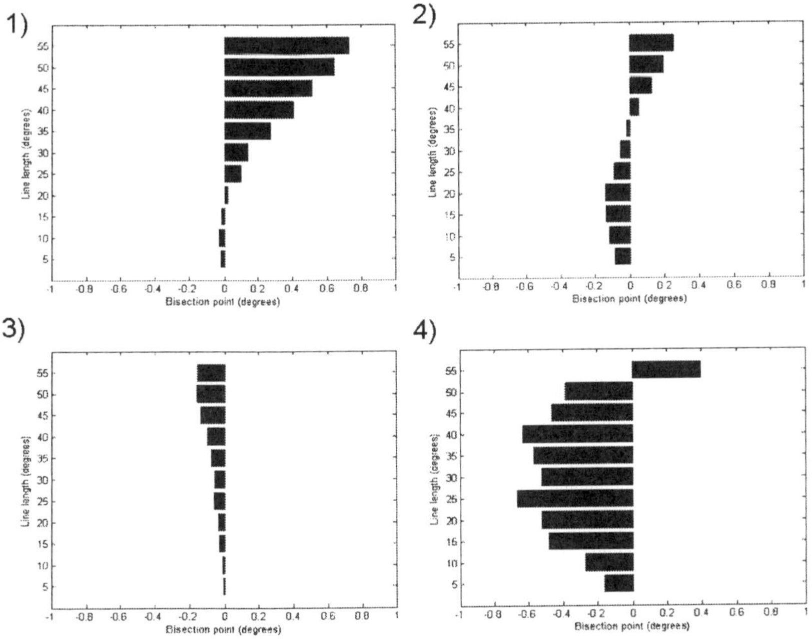

Figure 2. Bisection error as a function of line length, for "normal" networks. Notice the very small range of errors (less than one degree).

All networks showed, to some extent, the pseudoneglect phenomenon, but only in Network 3 was it associated with lines of all lengths. This result seems to be related to the fact that in Network 3 left hemispace is slightly overrepresented with respect to right hemispace.

After a lesion, all networks behave similarly to neglect patients, placing the bisection mark toward the ipsilesional side of space. Moreover, the four networks, as neglect patients, tend to produce larger errors for longer lines. However, while Network 1 shows similar results for left and right lesions, the other networks perform differently depending on the side of the lesion (Figure 3). Nevertheless, the behaviour of Network 2 and Network 4 appears to be less consistent with the performance of right neglect patients.

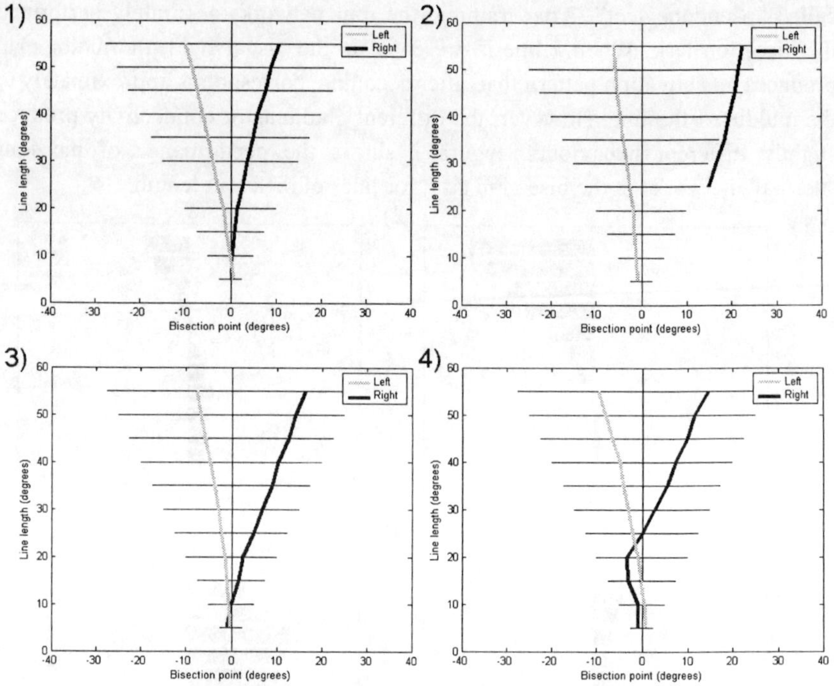

Figure 3. Bisection error as a function of line length, after a left or right lesion.

It is important to notice the different performance of Network 1 and Network 2: given that both networks are implementations of Heilman's hypothesis, the asymmetry observed in the latter seems to be produced by the different number of neurons in the two hemispheres, as in Network 3. Moreover, Network 3 shows for a right lesion the so called "cross-over effect": for very short lines neglect patients tend to make errors toward the contralesional side of the space.

4.2. *Cancellation Task*

The cancellation task has been implemented by presenting the model with 16 stimuli, divided into 4 columns. Lateral connections in the motor map and the addition of an inhibition map simulate a selection mechanism that allows to select only one stimulus at a time.

Like neglect patients, the four networks tend to omit the stimuli in the contralesional side of space after a parietal lesion. Figure 4 shows the number of stimuli cancelled by the four networks depending on their position in the visual field (i.e what column they belong to). While Network 1 shows roughly symmetric results for left and right lesions, the other three networks show a more asymmetric result, especially Network 2 and Network 3. Again, the similar performance of these two networks seems to be related to the different number of neurons in the two hemispheres.

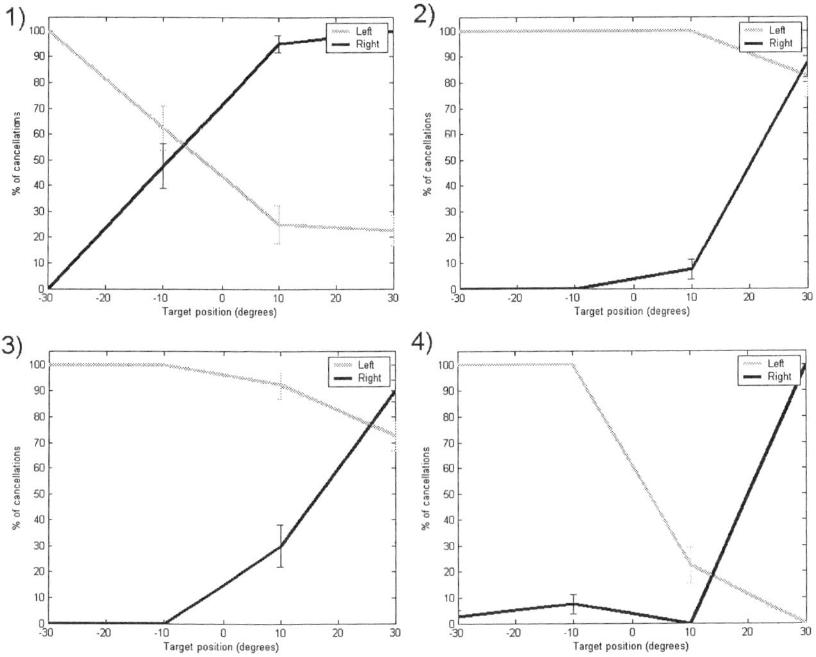

Figure 4. Percentage of cancellations as a function of the position of stimuli in the visual field, after a left or right lesion.

5. Conclusions

This work aims to lead the way to a deeper understanding of the neural mechanisms involved in the hemineglect syndrome. Using the computational model presented here, we obtained a behavior similar to that of neglect patients. Furthermore, manipulating the pattern of connectivity between visual and parietal maps we could investigate the role of hemispheric spatial representations in explaining the asymmetric nature of neglect. This is an important point: neural representations are difficult to study in humans, while the hemispheric asymmetry so apparent in humans is not evident in monkeys. Computational modelling can help to overcome this problem and test theoretical explanations of neglect. From this point of view, the versatility of our model represents a strength with respect to previous computational models.

Simulation results suggest in particular that asymmetries in the neglect syndrome are best explained by a right hemispheric dominance for spatial representations, expressed in terms of a higher number of neurons involved in spatial tasks rather than by a different distribution of the two hemispheres. Moreover, the performance of normal subjects in the bisection task (pseudoneglect) can be accounted for by a slightly overrepresentation of the left hemispace.

References

Anderson B. (1996). A mathematical model of line bisection behavior in neglect. *Brain*, 119, 841-850.

Anderson B. (1999). A computational model of neglect dyslexia. *Cortex*, 35, 201–218.

Graziano M.S.A., Hu X.T., Gross, C.G. (1997). Visuospatial Properties of Ventral Premotor Cortex. *J Neurophysiol*,77, 2268-2292.

Galletti C., Battaglini P.P., Fattori P. (1993). Parietal neurons encoding spatial locations in craniotopic coordinates. *Exp. Brain Res*., 96, 221–229.

Hamed S.B., Duhamel J.R., Bremmer F., Graf W. (2001). Representation of the visual field in the lateral intraparietal area of macaque monkeys: a quantitative receptive field analysis. *Exp Brain Res*, 140, 127–144.

Heilman K.M., Bowers D., Valenstein E., Watson R.T. (1987). Hemispace and hemispatial neglect. In Jeannerod M. (Ed.), *Neurophysiological and neuropsychological aspects of spatial neglect*. Amsterdam: North-Holland (pp. 115-150).

Monaghan P., Shillcock R.C (1998). The cross-over effect in unilateral neglect: modelling detailed data in the line-bisection task. *Brain*, 121, 907-921.

Monaghan P., Shillcock R.C. (2004). Hemispheric asymmetries in cognitive modelling: Connectionist modelling of unilateral visual neglect. *Psychological Review*, 111, 283-308.

Pouget A., Sejnowski T.J. (1997). Spatial transformations in the parietal cortex using basis functions. *Journal of Cognitive Neuroscience,* 9 (2), 222-237.

Pouget A., Sejnowski T.J. (2001). Simulating a lesion in a basis function model of spatial representations: comparison with hemineglect. *Psychological review,* 108, 653-673.

Salinas E., Abbott L.F. (1995). Transfer of coded information from sensory to motor networks. *Journal of Neuroscience*, 15(10), 6461-6474.

Vallar G. (1993). The anatomical basis of spatial hemineglect in humans. In Robertson I.H. and Marshall J.C. (Eds.), *Unilateral Neglect: Clinical and Experimental Studies*. Hillsdale, NJ: Erlbaum. (pp. 27–59).

Vallar G. (1998). Spatial hemineglect in humans. *Trends in Cognitive Sciences,* 2, 87-97.

AN OSCILLATORY NEURAL MODEL FOR TRACKING A MOVING OBJECT

Y.B. KAZANOVICH [A] R.M. BORISYUK [B]

[A] *Institute of Mathematical Problems in Biology, Russian Academy of Sciences Pushchino, Moscow Region, 142290, Russia*
yakov_k@impb.psn.ru

[B] *Centre for Theoretical and Computational Neuroscience*
University of Plymouth,
Drake Circus, Plymouth PL4 8AA, UK
rborisyuk@plymouth.ac.uk

We describe an oscillatory neural model of attention that can be used to track a moving target among a set of distractors. The model works with a set of identical visual objects randomly moving on the screen. At the initial stage, the model selects into the focus of attention an arbitrary object that is considered as a target. Other objects are used as distractors. The model aims to preserve the target until one of the following events takes place: If a target moves outside the boundaries of the visual field, another target is selected in the focus of attention. If a target overlaps with a distractor, the focus of attention is spread over the composite object. After separation of the objects only one of them is kept in the attention focus. This scheme of attention focus formation and switching is implemented through a proper interplay of synchronizing and desynchronizing interactions in a neural network with a central element.

1. Introduction

Selective visual attention is a mechanism that allows a living organism to extract from the incoming visual information the part that is most important at a given moment and that should be processed in more detail. This mechanism is necessary due to the limited processing capabilities of the visual system which preclude the rapid analysis of the whole visual scene. Different types of attention are responsible for implementing different strategies of attention focus formation. Traditional theories characterize attention in spatial terms as a spotlight or a "zoom lens" which could move about the visual field focusing on whatever fell within that spatial region (Posner et al., 1980; Eriksen & St. James, 1986). More recent theories of attention state that in some cases the underlying units of selection are discrete objects independently of whatever location these objects occupy. This type of attention is called object-based attention (Scholl, 2001).

The important property of attention is metastability. This means that after being fixed the focus of attention tends to be preserved for some time if the image is not radically changed. Usually slow changes of object parameters or the location of objects in the image do not lead to attention switching. Attention can automatically switch if the attended object suddenly disappears or if a new and important object appears in the scene. There may also be a voluntary change of attention caused by an instruction.

Object-based attention is most evident in experiments with moving objects. An important experimental paradigm of this type is multiple object tracking (MOT). In the canonical MOT experiments (Pylyshin & Storm, 1988), a subject views a display with several identical objects. A certain subset of the objects is briefly flashed to mark them as targets. Others are considered as distractors. Then all objects begin moving independently and unpredictably about the screen without passing too near to each other. It has been shown that subjects can simultaneously track up to five targets.

In the last few years some computational models of attention have appeared that aim to take into account both neurobiological and psychological evidence on attention. These models can be subdivided into two categories. Connectionist network models (Olshousen et al., 1993; Tsotsos et al., 1995; Itti & Koch, 2000) are based on a winner-takes-all procedure and are implemented through a proper modification of the weights of connections in a hierarchical neural network. Such models are difficult to apply to moving objects since the networks have to work in the space of the visual field, therefore for any new position of the objects the weights of the connections should be recomputed. Another type of attention model is represented by oscillatory neural networks (Kazanovich & Borisyuk, 1999; Wang, 1999; Corchs & Deco, 2001). They are more suitable for object-based attention because such networks operate in phase-frequency space, which automatically makes the attention focus invariant to object locations in physical space.

In this paper we present an Attention Model with a Central Oscillator (AMCO) (Kazanovich & Borisyuk, 1999; Kazanovich & Borisyuk, 2002; Kazanovich & Borisyuk, 2003; Borisyuk& Kazanovich, 2004) and describe the results of its application to tracking a single target among several distractors. As far as we know, this is the first publication where oscillatory principles of attention modeling are applied to non-stationary visual scenes. This model should be considered as a preliminary step to MOT modeling. In the discussion we will give a short sketch of what kind of modifications should be made to generalize the model in this direction.

Being able to track only one object, AMCO can do a bit more than is necessary for a traditional MOT experiment. We considered it reasonable for the model to be able to operate also in the case when objects intersect during their movements. The addition of object intersection may increase the number of errors in the recognition of targets but for certain it should not completely disorganize the functioning of the model. Since objects are assumed to move randomly and unpredictably, there is no possibility of identifying a target after it has been occluded by a distractor. In this case the best strategy for the attention system is to keep in the focus of attention either of the two objects that have just been separated. Therefore the model does not allow the attention system to permanently hold more than one isolated object. This is important in order to prevent the multiplication of target objects that otherwise would take place.

2. AMCO: the main principles of design and functioning

AMCO contains a one-layer network of locally coupled oscillators - the so-called Peripheral Oscillators (POs) - whose dynamics are controlled by a special Central Oscillator (CO) through global feedforward and feedback connections (Kryukov, 1991). The POs represent the columns of cortical neurons that respond to specific local features of an image. In particular, a feature can characterize physical properties of light reflected by a single pixel. A CO plays the role of the central executive of the attention system (Baddeley, 1996; Cowan, 1988).

In AMCO, objects are represented by synchronous assemblies of POs and the focus of attention is formed by those POs which work synchronously with the CO. The synchronization with the CO causes the resonance in POs, therefore the attention focus is formed by those POs which demonstrate oscillations with a high (resonant) amplitude (Kazanovich & Borisyuk, 2002; Borisyuk & Kazanovich, 2004).

The architecture of AMCO is shown in Figure 1. The input to the network is an image on the plane grid that contains a number of isolated grayscale objects on a white background. The grid of pixels is of the same size as the network of POs. Each PO receives an external signal from the pixel whose location on the plane is identical to the location of the PO. We consider a simple coding of the image when the external signal to a PO is determined by the contrast between the intensities of the pixel and the background. This signal determines the value of the natural frequency of the PO. The natural frequency is higher if the contrast is higher. The POs corresponding to the pixels of objects are called active, their dynamics are controlled by the equations shown below. The POs corresponding to the pixels of background are called silent. They do not participate in the dynamics of the network.

The POs have synchronizing local connections with their nearest neighbors. These connections are used to bind the POs representing an isolated object into a coherent assembly. This is done in agreement with the synchronization hypothesis in feature binding (Singer, 1999). The CO has desynchronizing feedforward and synchronizing feedback connection to each PO. The synchronizing connections are used to phase-lock the CO by an assembly of POs. The desynchronizing connections are used to segregate different assemblies of POs in frequency space to prevent simultaneous synchronization of the CO with several assemblies of POs.

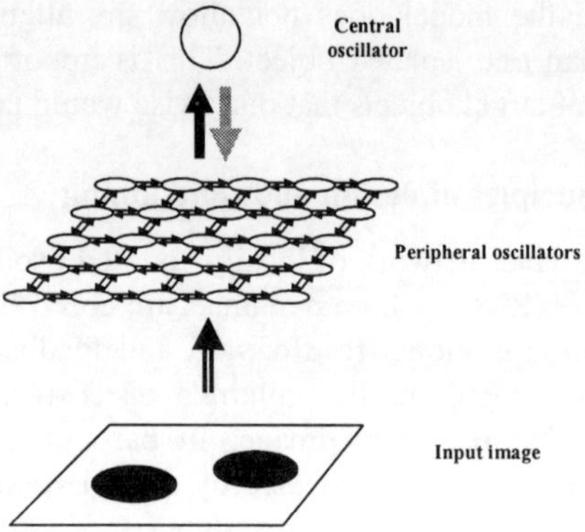

Figure 1. Architecture of AMCO. Hollow arrow shows the assignment of natural frequencies of POs. Black arrows show synchronizing connections that are used (a) to bind the POs coding an object into a synchronous assembly, (b) to synchronize an assembly of POs with the CO. The gray arrow shows desynchronizing connections that are used to avoid simultaneous synchronization of the CO with several assemblies of POs.

Since the CO can be phase-locked by only those POs whose natural frequencies are in some range around the natural frequency of the CO, the natural frequency of the CO is adapted to its current value. Due to this adaptation the CO tends to synchronize with an assembly of POs. The interplay between the synchronizing and desynchronizing connections of the CO results in the competition of different assemblies of POs for synchronization with the CO. Only one assembly of POs can win this competition, therefore in AMCO at each moment only one object can be attended (excluding some short transitory periods). The POs representing this object work with a high (resonant) amplitude. The amplitudes of other oscillators are shut down to a low level.

The oscillators comprising AMCO are described as generalized phase oscillators. The state of such an oscillator is described by three explicitly given variables: the phase of oscillations, the amplitude of oscillations, and the natural frequency of oscillations. The dynamics of AMCO are described by the following equation:

$$\frac{d\theta_0}{dt} = 2\pi\omega_0 + \frac{w_0}{n}\sum_{i=1}^{n} s_i a_i g(\theta_i - \theta_0),$$

$$\frac{d\theta_i}{dt} = 2\pi\omega_i - a_0 w_1 h(\theta_0 - \theta_i) + w_2 \sum_{j \in N_i} a_j p(\theta_j - \theta_i) + \rho,$$

$$\frac{da_i}{dt} = \beta_1(-a_i + \gamma f(\theta_0 - \theta_i))^+ + \beta_2(-a_i + \gamma f(\theta_0 - \theta_i))^-,$$

$$\frac{d\omega_0}{dt} = -\alpha(2\pi\omega_0 - \frac{d\theta_0}{dt}).$$

In these equations, θ_0 is the phase of the CO, θ_i $(i = 1,\ldots,n)$ are the phases of POs, $\frac{d\theta_0}{dt}$ and $\frac{d\theta_i}{dt}$ are the current frequencies of oscillators, ω_0 is the natural frequency of the CO, ω_i are the natural frequencies of POs, a_0 is the amplitude of oscillations of the CO (a constant), a_i are the amplitudes of oscillations of POs, w_0, w_1, w_2, are constant positive parameters that control the strength of interaction between oscillators, s_i is the saliency of a pixel (the set of s_i forms the saliency map; $s_i = 0$ if the *i*th PO belongs to background, otherwise it is positive), *n* is the number of active POs, N_i is the set of active oscillators in the nearest neighborhood of the oscillator *i*, ρ is the Gaussian noise with the mean 0 and standard deviation σ, functions *g, h, p* control the interaction between oscillators, *f* is the function that controls the amplitude of oscillations of POs and their transition to the resonant state, $\alpha, \beta_1, \beta_2, \gamma$ are network parameters (positive constants). The values ω_i are determined by the external input signal; $\theta_0, \theta_i, \omega_0, a_i$ are internal variables that characterize the state of the network. By definition,

$$(x)^+ = \begin{cases} x & if\ x \geq 0 \\ 0 & if\ x < 0 \end{cases}, \quad (x)^- = \begin{cases} x & if\ x \leq 0 \\ 0 & if\ x > 0 \end{cases}.$$

More details can be found in (Kazanovich & Borisyuk, 2002; Borisyuk & Kazanovich, 2004).

It has been shown that under appropriate rules of interaction AMCO can select an object and keep it in the attention focus (Kazanovich & Borisyuk, 2002). The selection is biased by the size and saliency of objects, larger and more salient objects have the advantage in the selection.

Figure 2 gives an example of the selection of a character in an image that displays the printed word HELLO. The chart of the amplitudes of POs is given in the figure for the moment when the character E has been selected into the attention focus. At this moment the CO synchronizes its oscillations with the assembly of POs located in the region E. Due to the resonance with the CO the oscillators in this region sharply increase the amplitudes of their oscillations while the amplitudes of other POs are shut down.

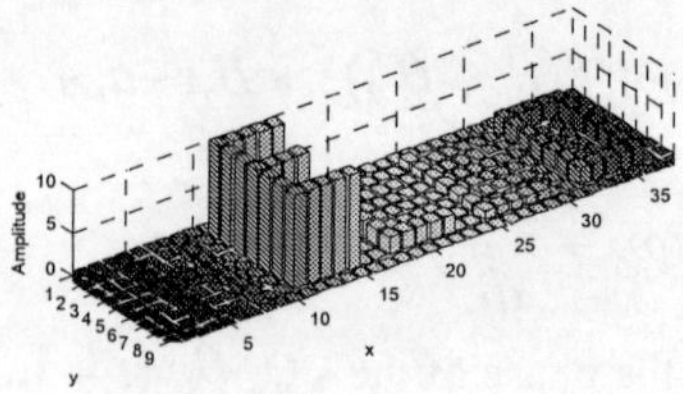

Figure 2. Selection of the character E of the printed word HELLO. The characters of the word represent individual objects any of which can be selected into the focus of attention.

3. Simulation example

Our previous publications on AMCO have dealt with stationary objects only. Images with moving objects are more difficult to process since the attention system should be able to properly react to the changes in object locations and also to the intersection and separation of objects during their movements. We illustrate the functioning of the system in the case of an image of size 25×50 pixels that contains 9 circles of radius 3. Figure 3 displays the snapshots of the dynamics of the image at the moments 0, 1, 2,... The frames are ordered in time from left to right and from top to bottom. The top-left frame shows the initial position of the circles. Later the circles move in any of the four randomly chosen directions (up, down, left, right). A circle can change the direction of its movement at moments which are a multiple of 0.1. With the probability of 0.3, the circle "takes the decision" that a new direction should be chosen. After that any of the four possible directions is selected with the probability 0.25 (the new direction may coincide with the previous one). A moving circle can leave the visual field (then it disappears from the image and becomes invisible to the attention system) and return to it again.

All pixels that belong to the circles receive the same illumination and therefore they have the same contrast relative to the background. We have selected the range of theta frequencies for computations and set the natural frequencies of all active POs equal to $\omega_i = 5$. The saliencies corresponding to active POs are 1, the saliencies for silent POs are 0. The amplitudes of POs vary in the range (0, 11). The threshold for a resonant amplitude is $R = 8.8$. If the amplitude of a PO exceeds *R*, the corresponding pixel is assumed to be included in the focus of attention. In Figure 3 attended pixels are black, non-attended pixels are gray, and pixels of the background are white.

Due to movements, the distribution of the circles in the image becomes random and complex objects appear that represent different combinations of overlapping circles. At the initial moment no object is under attention. After a short lag attention is automatically focused on a randomly chosen circle (second frame in the first row). This object is kept in the attention focus for some time but later (frame 6 in the first row) when it nearly disappears from the visual field attention automatically switches to another object (frame 7 in the first row). which combines two overlapping circles and due to its larger size has had the advantage in attracting attention. Attention is focused on this complex object while it exists but as soon as the circles comprising this object separate (frame 7 in row 2) only one circle is left in the focus of attention.

Frame 6 in row 3 shows a situation when an attended circle (target) intersects with another circle (distractor). In this case attention is spread over the composite object. But again, shortly after the circles separate (frame 7 in row 3) only one of the circles is left in the focus of attention (frame 1 in row 4). Similar examples of attentional selection of one circle from the two separated circles can be found in frame 4 of row 6 and frame 2 of row 8. Such a change of the focus of attention is reminiscent of passing a baton in a relay race with the only difference being that the process has a probabilistic nature.

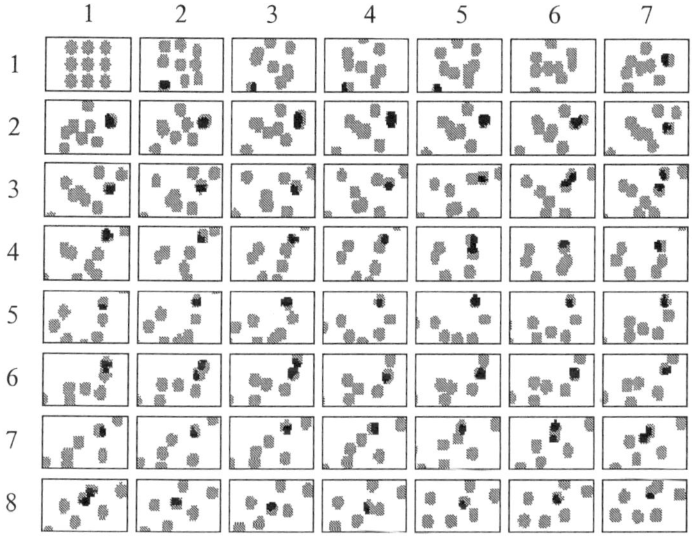

Figure 3. Snapshots of attention focusing and switching in an image that contains 9 randomly moving circles. Attended pixels are black, non-attended pixels of objects are gray, pixels of the background are white.

4. Discussion

Two main ideas are combined in the presented model of attention. Firstly, oscillations and synchronization are used as a key mechanism in attention focus formation. Secondly, it is assumed that attention focusing is controlled by a special neural system that is called a central executive. The evidence that oscillatory and synchronous activity is related to attention has been provided in studies of EEG (Herrmann & Knight, 2001), MEG (Sokolov et al., 1999), and more recently in multiunit and local field potential recordings (Steinmetz et al., 2000, Fries et al., 2001, Fell et al., 2003).

The idea of a central executive of the attention system is attractive from both biological and technical points of view. A biological justification of this idea can be found in (Cowan, 1988; Baddeley, 1996). The question about localization of the central executive in brain structures is still obscure. The most suitable candidates are the prefrontal cortex and the hippocampus whose participation in the work of the attention system is confirmed by many studies (Vinogradova, 2001; Herrmann & Knight, 2001; Duncan, 2001). An important property of a system controlled by a central element is a relatively small number of connections (of the order n, where n is the number of elements in the system) in comparison to the systems with all-to-all connections (the number of connections is of the order n^2). This makes the systems with a central element biologically plausible and technically feasible even in the case of large n.

How to generalise the AMCO to make it suitable for modelling MOT with k targets? The main idea is to use a network that consists of k copies of AMCO as the layers of the network with each copy tracking one particular target. Two types of difficulties arise when implementing this idea. First, one should prevent the situation when the same object is simultaneously taken in the attention focus by several AMCO. Second, the system should be able (as far as possible) to preserve the set of targets in the case when objects intersect during movements. These problems can be solved by a suitable organization of the interactions between the copies of AMCO. The most important element in the MOT model design is the introduction of desynchronizing connections between the central oscillators of different layers. A proper choice of the strength of these connections guarantees that different objects are taken into synchronization by different central elements. A detailed presentation of the MOT model and the results of its simulation is in preparation (Kazanovich & Borisyuk, 2005).

Acknowledgments

This work was supported by the Russian Foundation of Basic Research (Grant 03-04-48482) and by the President Program of the Leading Research Schools (Grant 1872.2003.4).

References

Baddeley A. (1996). Exploring the central executive. *Quarterly J. Experimental Psychol.*, 49A, 5-28.

Borisyuk R., Kazanovich Y. (2004). Oscillatory model of attention-guided object selection and novelty detection. *Neural Netw.*, 17, 899-915.

Corchs S., Deco G. (2001). A neurodynamical model for selective visual attention using oscillators. *Neural Netw.*, 14, 981-990.

Cowan N. (1988). Evolving conceptions of memory storage, selective attention and their mutual constraints within the human information processing system. *Psychol. Bull.,* 104, 163-191.

Duncan J. (2001). An adaptive coding model of neural functions in prefrontal cortex. *Nature Reviews/Neuroscience*, 2, 820-829.

Eriksen C.W., St. James J.D. (1986). Visual attention within and around the field of focal attention: a zoom lens model. *Perception and Psychophysics*, 40, 225-240.

Fell J., Fernandez G., Klaver P., Elger C.E., Fries P. (2003). Is synchronized neuronal gamma activity relevant for selective attention? *Brain Research Reviews*, 42, 265-272.

Fries P., Reynolds J.H., Rorie A.E., Desimone R. (2001). Modulation of oscillatory neuronal synchronization by selective visual attention. *Science*, 291, 1560-1563.

Herrmann C.S., Knight R.T. (2000). Mechanisms of human attention: event related potentials and oscillations. *Neuroscience and Biobehavioral Reviews,* 25, 465-476.

Itti L., Koch C. (2000). A saliency-based search mechanism for overt and covert shifts of visual attention. *Vision Res.,* 40, 1489-1506.

Kazanovich Y.B., Borisyuk R.M. (1999). Dynamics of neural networks with a central element. *Neural Netw.*, 12, 441-454..

Kazanovich Y., Borisyuk R. (2002). Object selection by an oscillatory neural network. *BioSystems*, 67, 103-111.

Kazanovich Y.B., Borisyuk R.M. (2003). Synchronization in oscillator systems with phase shifts. *Progr. Theor. Phys.*, 110, 1047-1058.

Kazanovich Y.B., Borisyuk R.M. (2005). An oscillatory neural model of multiple object tracking (in preparation).

Kryukov V.I. (1991). An attention model based on the principle of dominanta. In Holden A.V. and Kryukov V.I. (Eds.), *Neurocomputers and Attention I. Neurobiology, Synchronization and Chaos*. Manchester: Manchester University Press (pp. 319-352).

Olshausen B.A., Anderson C.H., Van Essen D.C. (1993). A neurobiological model of visual attention and invariant pattern recognition based on dynamic routing of information. *J. Neurosci.*, 13, 4700-4719.

Pylyshin Z. W., Storm R. W. (1988). Tracking multiple independent targets: evidence for a parallel tracking mechanism. *Spatial Vision*, 3, 179-197.

Posner M.I., Snyder C.R.R., Devidson D.J. (1980). Attention and the detection of signals. *J. Exp. Psychol.: General*, 109, 160-174.

Scholl B.J. (2001). Objects and attention: the state of the art. *Cognition*, 80, 1-46.

Singer W. (1999). Neuronal synchrony: A versatile code for the definition of relations. *Neuron*, 24, 49-65.

Sokolov A., Lutzenberger W., Pavlova M., Pressl H., Braun C., Birbauner N. (1999). Gamma-band MEG activity to coherent motion depends on task-driven attention. *Neuroreport*, 10, 1997-2000.

Steinmetz P.N., Roy A., Fitzgerald P., Hsiao S.S., Johnson K.O., Niebur E. (2000). Attention modulates synchronized neuronal firing in primate somatosensory cortex. *Nature*, 404, 187-190.

Tsotsos J.K., Culhane S.M., Wai W.Y.K., Lai Y., Davis N., Nufl, F. (1995). Modeling visual attention via selective tuning. *Artificial Intelligence*, 78, 507-545.

Vinogradova O.S. (2001). Hippocampus as comparator: role of the two input and two output systems of the hippocampus in selection and registration of information. *Hippocampus*, 11, 578-598.

Wang D.L. (1999). Object selection based on oscillatory correlation. *Neural Netw.*, 12, 579-592..

SELECTIVE ATTENTION AND ACTION IN AN ARTIFICIAL, EVOLVED AGENT: REACTIVE INHIBITION

ROBERT WARD[A] RONNIE WARD[B]

[A] Centre for Cognitive Neuroscience, University of Wales, Bangor
Bangor, LL57 2AS, UK
r.ward@bangor.ac.uk

[B] Dept of Computer Science, University of Texas A&M
College Station, TX, 77843-3112, USA
ronniew@pdq.net

We analyzed the mechanisms for selective attention and action in an artificial, evolved agent. Our analysis shows the agent selectively responds to targets through reactive inhibition of nontarget items. Reactive inhibition, in which suppression of nontargets is in proportion to their salience, has been previously shown to be a means of selection in people. Our results suggest that reactive inhibition may be a fundamental process in selective control of action.

1. Control of action

Agents operating in complex environments can have competing demands for action. A simple example is a frog eyeing two flies. If the frog selects one fly for action, it may succeed in catching it. The "intermediate" response, attacking a midpoint between the two flies, is useless. Effective action must therefore be selective -- response mechanisms must be focused on selected targets, and these same mechanisms must be insulated from the possibilities for action offered by nontargets. Theories of attention have proposed different accounts how selection links perception and action. One influential account assumes that perceptual processes lead to the relatively unselective activation of numerous potential actions. Control of action is achieved by inhibitory mechanisms that suppress activation of all but the desired action (e.g. Tipper et al, 1998, Houghton & Tipper, 1996). Another approach uses biased competition (Duncan et al, 1997), and demonstrates how distributed representations of objects can be linked to distributed representations of action, through a competitive selection process (Ward, 1999). This model shows how a stable state of selective network activation, representing one object and its implications for behaviour, can arise from a confluence of what, where, and how information.

In these and other psychological models of attention (e.g., Schneider, 1995), there is a single episode of filtering which defines the target for action. There is a rich literature of computational models of attention, exploring mechanisms for selecting a single target from a static array of nontargets. However, filtering is only part of the problem when considering the control of action. For example, we can return to the case of the frog and the flies to illustrate some of the complexities of selective action. First, the two flies must be prioritized for action, so that a first target is selected. Second, there must be a means for keeping response mechanisms tied to the movements of the first fly, and for insulating responses from movements of the second. In other words the effects of the second fly on action systems must be filtered. However, if possible, sensory processing of the second fly should continue, so that once the first is caught, the frog will be able to rapidly acquire the second. After the first fly is caught, the second fly, once insulated from response mechanisms, must now be allowed to control them. That is, the frog must reallocate selective processing mechanisms. Thus, multiple potential targets for action in a dynamic environment create demand for selective response, a separation of perceptual and response processing, and the reallocation of selective processing from one target to the next.

Current psychological models of attention have generally not addressed this range of processes necessary for selective action in a dynamic environment. Or more accurately, perhaps the theories (e.g. Tipper et al, 1998; Duncan et al 1997) are meant to be applied to such environments, but the computational implementations of these theories (e.g. Houghton & Tipper, 1994; Ward, 1999) have been limited to much simpler worlds. This is a problem, because we don't know how or if psychological mechanisms like distractor inhibition (Tipper et al, 1998) and biased competition (Duncan et al, 1997) work, when brought out of static worlds and into more complex ones.

The aim of this research is to explore the psychological underpinnings of models of perception, attention, and action that operate in dynamic environments, and in which perception, attention, and action are temporally extended processes. We examine these processes in working but tractable artificial agents. Our approach is similar in some ways to comparative psychology. Our aim is to uncover the principles underlying the operation of the agent, and to then compare these principles with models of selective processing in the psychological literature. In comparing the solutions of the artificial agent to the psychological literature, we hope to enrich our understanding of computationally adequate, psychological models of selective action.

Models of embodied, situated, and dynamical (ESD) agents stress what Clark (1998) calls "the unexpected intimacy between the brain, body, and world". The ESD approach emphasizes the contextually-bound nature of solutions to cognitive problems, as opposed to isolated but general-purpose attention, perception, or motor subsystems. Important for our purposes is the "minimally cognitive" form of ESD agent, as developed by Beer (1996). This approach attempts to create simple artificial agents that instantiate interesting cognitive behaviors. Such agents serve as model systems, allowing a tractable analysis of the cognitive processing going on in more complex systems.

An application of the 'minimally cognitive agent' approach to problems of perception, attention, and action was demonstrated recently by Slocum et al (2000) and Goldenberg et al (2004). Slocum et al (2000) used genetic algorithm (GA) methods to evolve a small continuous time recurrent neural network (CTRNN) capable of performing a selective attention task. The "visual agent" (VA) was constrained to run along the bottom of a 2D environment, moving left or right to catch two balls falling from the top of the environment. We extend this work, exploring the operation and abilities of such an minimally cognitive agent in detail. This chapter summarizes some of the findings of Ward and Ward (submitted), in which fuller details and analyses can be found.

2. Methods

2.1. *The agents*

A key part of our research strategy was to consider two types of VAs: a "reflexive" and a "controlled" agent. The reflexive visual agent (RVA) has direct links from the sensory units to the motors with no intervening hidden units. In RVA, perceptual inputs drive motor output directly, and there is no possibility for selective processing of one ball over the other. In contrast, the "controlled" visual agent (CVA) has an intermediate hidden layer, so that potentially at least, CVA can focus processing activity on one target. RVA also used a linear activation function, while CVA used the standard sigmoid activation function. The capabilities of the RVA therefore serve as a baseline by which the effectiveness of selective processing in CVA can be assessed.

For both agents, the input layer consisted of units driven by oriented proximity sensors. The sensors signaled the distance between the VA and a "solid" surface intersecting the sensor ray (as in Slocum et al, 2000). An array of 7 sensors evenly covered 6° of visual angle. The output layer of both agents contained 2 motor units, one for controlling left and one for right movement. In the CVA but not the RVA, the motor units were connected to each other. The hidden layer of the CVA contained 8 fully inter-connected units, each receiving inputs from all sensors and each with reciprocal connections to the motor units. These networks are illustrated in Figure 1.

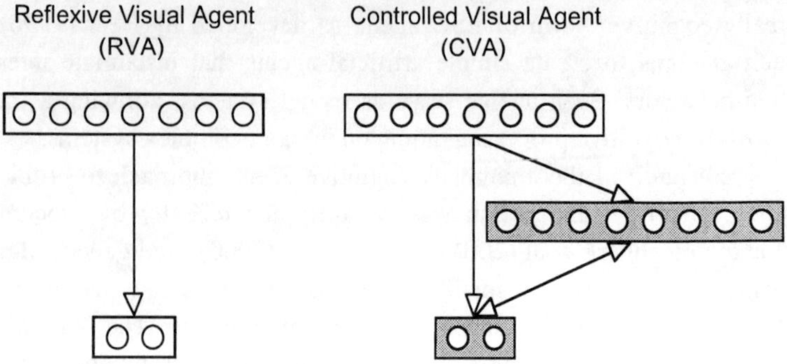

Figure 1. Architectures of the two visual agents. Connections between layers, indicated by arrows, represent bilaterally symmetric connections. Connections within layers are indicated by the shading of the outlining box. Filled boxes indicate that each unit is connected to every other unit within the layer using bilaterally symmetric weights. Unfilled boxes indicate no intralayer connections.

2.2. *Training*

Both VAs were trained to catch targets falling from the top of the simulated environment to the bottom, by moving underneath them before the targets impacted (see Figure 2). There were two types of trials: Divided Attention trials, in which two targets fell, and Focused Attention trials where only a single target fell. First we describe the Divided Attention trials. The two targets, T1 and T2, were identical objects dropped simultaneously at the start of each trial. On a given trial, T1 and T2 fell with a constant vertical velocity and with zero horizontal velocity. However, by definition, T1 traveled faster than T2, and therefore this target needed to be prioritized for response. Once T1 impacted, it was removed from the environment so that it no longer blocked the sensors. The VA always began each trial centered along the bottom of the environment.

The starting position and speed of targets in training were varied factorially. There were four factors describing Divided Attention trials: T1 Side (T1 started 24 units Left or Right of center); T2 Side (T2 was positioned to the Left or Right of T1); Spatial Separation (T1 and T2 were either relatively Near [24 units] or Far [48 units] from each other); and Temporal Separation (again Near or Far; on Temporal Near trials, T1 and T2 fell at close to the same speed [4 and 3] and so impacted at near the same time; on Temporal Far trials there was greater difference in speed [5 and 2] and they impacted further apart in time). This factorial design creates 16 trials types. For greater generalization, these 16 trials were each presented three times each epoch, with a global offset of 8 units to the left, 8 to the right, and with 0 offset.

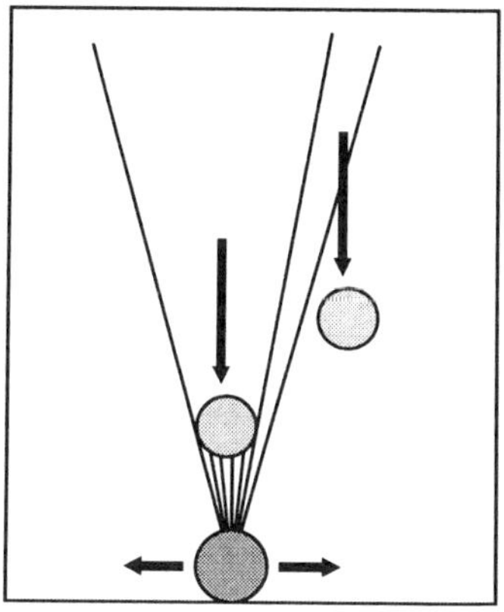

Figure 2. Schematic of catching task with two targets. Targets indicated by the down arrows. VA is located at the bottom of the environment, and can move left and right by activation of its two motor units. VA is here under T1. The rays emitting from VA represent its sensors. The leftmost and two rightmost sen-sors detect nothing, and the others have intersected T1. At this time T2 (on the right) is out of view. Copies of the simulator and associated data files are available for download at http://www.psychology. bangor.ac.uk/ward.

In addition to the Divided Attention trials described above, a training epoch consisted of Focused attention trials, in which a single target appeared. Focused trials were created by simply removing one target from the Divided set, so that there were 48 Focused T1 trials in which T1 was presented without T2, and 48 Focused T2 trials in which T2 appeared without T1. There was then a total of 144 trials in each epoch, requiring the catch of 192 targets.

There are several notable factors about the training set. First, the training set was constructed to ensure T2 catch was possible after T1 catch. The ratio of the final distance separation and the catch time separation were determined so as to be a maximum of about one-half (~55%) of the maximum agent velocity. That is, after catching T1, the agent had to travel at most 55% of its max velocity to catch T2. Second, the 6° spread of the VA sensor array was enough to span both targets at their maximal separation when between 10 and 100% of the sensor range. Finally, the Far Space/Near Time trials were a particular challenge. On these trials, a VA that is centered under T1 at the time of impact will have no sensor contact with T2 (as illustrated in Figure 2). That is, if VA makes an accurate catch of T1 on these trials, it will subsequently be looking at empty space. This leaves VA with two options to avoid failure on T2: initiate a blind search for T2, or remember the last seen location of T2 and move towards that remembered location.

The weights of each CTRNN were selected with a genetic algorithm (GA) procedure. GA search was halted when the best population member did not change after 1000 generations and the agent had evolved at least 5000 generations. Searches were terminated at 15,000 generations if the above conditions were not met. In all cases, the lowest performing population member was nearly equal to the highest performing member, which is indicative of near convergence. The search objective was to minimize the total miss distance divided by the total number of targets processed. Miss distance was calculated as the center-to-center separation between a target and the VA when the target's bottom vertical position equaled or exceeded the top of the VA's vertical position.

2.3. *Test trails*

After training, performance was assessed on a set of 1500 randomly generated test trials, for Divided, Focused T1, and Focused T2 conditions. Five hundred Divided attention trials were constructed as follows. T1 position was randomly sampled in the interval of 169 to 209. T2 was offset a random distance from T1 in the range of 24-48 units left or right of T1. T1 speed was generated as a real number between 4 and 5, and T2 speed in the range of 2 to 3. Focused trials repeated the Divided attention trials, omitting one of the targets.

3. Results

3.1. *Selective attention to T1*

Before investigating possible selective processing in CVA, we had to first verify that CVA was actually capable of selective action. Selectivity was assessed by a simple question: how much did the presence of T2 interfere with the catch of T1? This is illustrated in Figure 3. Looking first at RVA performance, we can see that catch accuracy was almost perfect with single targets in Focused Attention conditions. But relative to Focused Attention, the presence of T2 in Divided Attention conditions produced significant interference on T1. Interference from T2 is best described as a 'capture effect', so that RVA

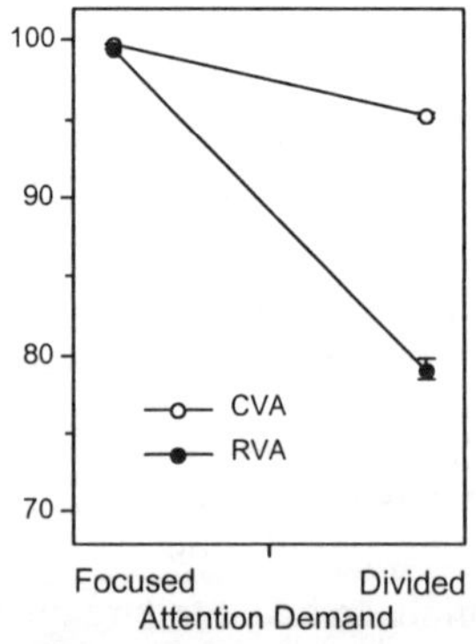

Figure 3. Catch accuracy (% overlap be-tween the agent and the target) on 1500 random test trials. Agents show near-perfect perform-ance with single targets under Focused Atten-tion. When two targets must be caught in Div-ided Attention condi-tions, CVA improves significantly relative to RVA.

would miss T1 by positioning itself between T1 and T2. This tendency to miss towards rather than away from T2 was highly significant, $t(498)= 26.5$, $p<.0005$. For CVA, the interference in Divided Attention conditions was much reduced, as evidenced by the signifcant interaction of Agent Type (CVA or RVA) x Attention Demand (Focused or Divided), $F(1,992)=428$, $p<0.00005$. Relative to the RVA baseline, CVA was clearly capable of selectively responding to T1, and preventing interference from T2.

3.2. *Memory for T2*

Selective processing of T1 created a problem for CVA. When CVA took up a position directly underneath T1, T2 could fall outside of the sensor array (especially Far Space/Near Time trials, see section 2.2). We called these cases "out-of-view" (OOV) trials. On OOV trials, after the T1 catch, CVA would be looking at a completely blank sensor array. Which way should CVA move in such case? To solve the problem of OOV trials, CVA developed a memory for position of the unseen T2. We assessed this memory by determining whether CVA's first move after T1 catch on OOV trials was towards or away from T2. Our training set had the virtue that the position of T1 was uninformative about the direction in which T2 might lie. A random first move would be towards the unseen T2 only 50% of the time. CVA does much better than chance, making the correct first move in 100% of the OOV training trials, on 84% of the OOV trials in our first test set. These levels of performance were significantly better than chance, chi-squared(1) > 78, $p < .000005$. CVA must therefore have developed a memory for T2 position, established earlier in the trial, while T2 was in view.

3.3. *Reactive inhibition: A mechanism for selective action*

We have seen that CVA selectively responds to T1, and insulatse its response mechanisms from the effects of T2. However, after T1 is caught, CVA must reallocate processing to the previously ignored T2. In people, this kind of cognitive reallocation produces robust costs in the form of "negative priming". Negative priming refers to costs in performance when a previously ignored object must be subsequently attended and reported (see Figure 4). The Houghton and Tipper (1994) model of reactive inhibition proposes that the strength of negative priming directly reflect competition for attention. According to the model, a target is selected by inhibiting all nontarget items. The more salient the nontarget -- that is, the more powerfully it tends to attract attention -- the greater the inhibition needed to prevent the object from gaining control of behavior.

Thus, more negative priming is observed for highly salient nontargets than for less salient ones (Grison & Strayer, 2001).

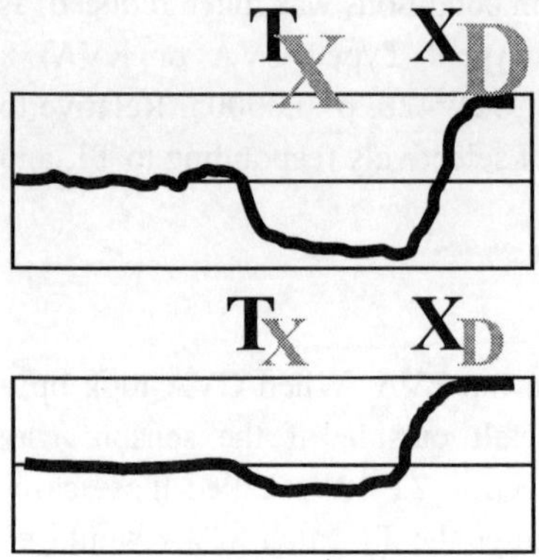

Figure 4. Example of reactive inhibition. X-axis plots time and onset of two displays. Task is to report the black letter and ignore the grey. Y-axis shows activation of "X" repre-sentation. **Upper panel**. X is to be ignored, and so X activation is sup-pressed to allow accurate report of T. However, in next display, X is the target, and now activation must rise from below baseline levels before report. The extra time required for activity to reach baseline reflects negative priming costs. **Lower panel**. Compared to upper panel, a less salient X distractor requires less inhibition to allow accurate report of T. Less inhibition on first display means less cost on second display.

Given that CVA must switch its processing from T1 to the previously ignored T2, we looked for a similar "negative priming" cost. We observed a marked hesitation in CVA movement after the T1 catch, so that, on average, CVA was immobile for roughly 30 timeslices. We measured this period of hesitation when T2 was in-view and OOV after T1 catch. For the intact CVA, the hesitation period was longer for in-view than OOV T2s, $F(1, 498)=95.9$, $p<.0005$.

This may seem like a counter-intuitive result. After all, once T1 is caught and removed from the environment, CVA must now catch T2. And on in-view trials, T2 is setting right in front of CVA --- shouldn't this facilitate T2 catch? This intuition is countered by the predictions of the reactive inhibition model. According to reactive inhibition, a T2 that is visible for longer periods during the trial would be expected to receive greater inhibition, to prevent it from being the target of selective action. Thus, the in-view T2 would be expected to receive more inhibition than an OOV T2. The decay of this inhibition would result in a longer hesitation for in-view T2's.

Reactive inhibition fits the pattern of hesitation we observe, but can we find direct evidence of inhibition? An inhibition model predicts that, during the build-up to T1 catch, CVA is developing an inhibitory signal, acting to keep CVA away from the T2 during the T1 catch. Under normal conditions, once T1 is caught, the external input from T2 would act in opposition to this internal signal. The external input would attract CVA towards T2, while the internal inhibition would keep CVA away from T2. However, as the inhibition of T2 decays, CVA will move gradually in the direction of this external input. We can isolate any internal inhibitory signal by disconnecting sensor inputs right after T1 impact. By doing so, we remove the counteracting external input from T2. CVA is now subject only to the inhibitory signal and should move away from T2. This is what we see in Figure 5. The first move of the intact CVA, following T1 catch, is towards T2, on 417 of 417 trials. With the sensor input of CVA disconnected, this pattern completely reverses. After catching T1, CVA now moves away from T2, on 414 of 417 trials.

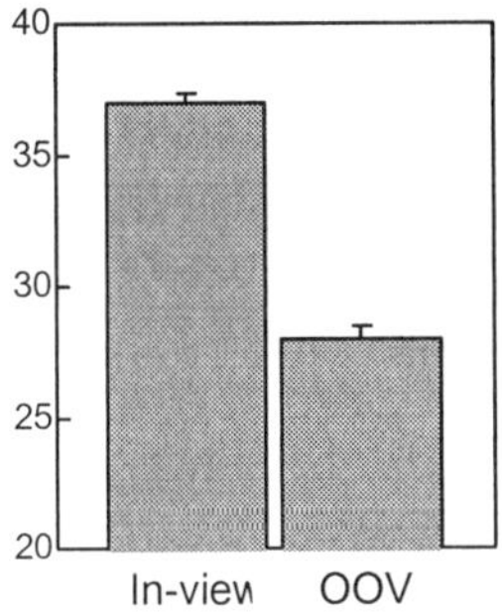

Figure 5. After catching T1, CVA hesitates before moving towards T2. This hes-itation was longer for in-view than OOV T2s. This result may seem surprising but is predicted by reactive inhibition accounts.

4. Summary and conclusions

We analyzed mechanisms for selective action in an artificial evolved agent. We found that our agent, CVA, utilized a reactive inhibition mechanism, strikingly similar to proposals in the psychological literature for control of action (Houghton & Tipper, 1994). In particular, CVA demonstrated costs when reallocating its selective processing, to respond to a previously ignored item. This cost varied significantly with the salience of the ignored item: the more salient the ignored item, the greater the subsequent costs for reallocation. We were also able to directly confirm that selective processing was achieved by inhibition of nontargets. We were able to isolate an inhibitory signal, applied to nontargets, and generated within the internal recurrant connections of CVA. This inhibitory signal allowed CVA to catch the target withoug interference from the nontarget, but did not prevent CVA from perceptual processing of the nontarget, and in particular from forming a memory for nontarget location. Greater detail and analysis on these and other points is available in Ward and Ward (submitted).

The use of reactive inhibition in CVA is notable for several reasons. First, this is the first demonstration of reactive inhibition in an embodied, dynamic agent. The reactive inhibition model appears suitable for selective action involving behaviors extended over time and space. Second, reactive inhibition emerged as a solution for control, despite the simplicity of CVA and its environment. There was nothing in the architecture of CVA to obviously encourage reactive inhibition. This is particularly notable when comparing

CVA's architecture to the relatively complex and sophisticated "wiring" of Houghton and Tipper (1994). Furthermore, the CVA network is about as primitive as one could imagine, in its sensory and motor capabilities, and in its 8-unit hidden layer. The environment in which CVA operates is also about as simple as any environment can be that requires selective action over time. Reactive inhibition may therefore be a fundamental mechanism for selective control, in all sorts of selective agents and dynamic environments.

References

Beer, R. D. (1996). Toward the evolution of dynamical neural networks for minimally cognitive behavior. In P. Maeset al (Eds.), From *Animals to Animats 4: Proceedings of the Fourth International Conference on Simulation of Adaptive Behavior* (pp. 421-429). MIT Press.

Clark, A. (1998). Where brain, body and world collide. *Daedalus: Journal Of The American Academy Of Arts And Sciences), 127*, 257-280.

Duncan, J., Humphreys, G. W., & Ward, R. (1997). Competitive brain activity in visual attention. *Current Opinion in Neurobiology, 7*, 255–261.

Goldenberg, E., Garcowski, J., & Beer, R.D. (2004). May we have your attention: Analysis of a selective attention task. To appear in From *Animals to Animats 8: Proceedings of the Eighth International Conference on the Simulation of Adaptive Behavior*. July 13-17, 2004, Los Angeles, CA.

Grison S., & Strayer , D. L. (2001). Negative priming and perceptual fluency: More than what meets the eye. *Perception & Psychophysics, 63*, 1063-1071.

Houghton, G., & Tipper , S. P. (1994). A model of inhibitory mechanisms in selective attention. In Dagenbach, D., & Carr, T. H. (Eds.), *Inhibitory processes in attention, memory, and language* (pp 53-112). Academic Press.

Schneider, W. X. (1995). VAM A neuro-cognitive model for visual attention control of segmentation, object recognition and space-based motor action. *Visual Cognition, 2*, 331-376.

Slocum, A.C., Downey, D.C. and Beer, R.D. (2000). Further experiments in the evolution of minimally cognitive behavior: From perceiving affordances to selective attention. In J. Meyer, A. et al (Eds.), From *Animals to Animats 6: Proceedings of the Sixth International Conference on Simulation of Adaptive Behavior* (pp. 430-439). MIT Press.

Tipper, S. P., Howard, L. A., & Houghton, G. (1998). Action-based mechanisms of attention. *Philosophical Transactions of the Royal Society,353*,1385-1393

Ward, R. (1999). Interactions between perception and action systems: A model of selective action. In (Humphreys, G.W., eta l., Eds.*) Attention, space, and action: Studies in cognitive neuroscience*. Oxford: OUP.

Ward, R., & Ward, R. (submitted). Selective attention and action in an artificial, evolved agent: Dynamic integration of perception and action.

A PROPOSED MODEL OF REPETITION BLINDNESS

COLM G. CONNOLLY
Dept. of Computer Science,
University College Dublin,
Belfield, Dublin 4, Ireland
E-mail: Colm.Connolly@ucd.ie

RONAN G. REILLY
Dept. of Computer Science,
National University of Ireland, Maynooth
Co. Kildare, Ireland
E-mail: Ronan.Reilly@may.ie

We describe a model of repetition blindness which draws on the dichotomous division of the visual system into two subsystems which process identity and location information. The model is constructed from self-organising networks of spiking neurons which are connected by plastic inhibitory and excitatory synapses. In particular, we describe how these networks are capable of learning translation invariant letter representations and learning to locate a stimulus in the input array using the SOM learning algorithm.

Various psychological phenomena require the establishment of separate episodic representations for each of the objects perceived in the environment. From the point of view of the brain these representations need not be stored in the same place, but may be distributed over many regions. Indeed, there is neurobiological evidence supporting the theory that the episodic context of an object is stored separately from its identity. We describe a model of one such phenomenon which uses separate representations for the identity and location of word information.

1. Repetition Blindness

Repetition blindness (RB) is the failure to detect the repeated occurrence of a stimulus which is presented in a *rapid serial visual presentation* (RSVP) experimental paradigm. If the exposure of the first occurrence of the stimulus, or critical word (C1), is set so that it is just identifiable—about

100 ms—then identification of the second occurrence (C2) is impeded (Kanwisher, 1987).

RB can be observed under a variety of stimulus conditions, including compounds, homographs, homophones and single letters (Kanwisher & Potter, 1990). RB is not confined to the level of whole words, as overlapping fragments of words can be subject to blindness (Harris & Morris, 2000). Furthermore, it appears not be restricted to letter clusters. It has been found between numbers in their verbal (nine) and Arabic (9) forms and between homophones, such as I / eye (Bavelier & Potter, 1992).

1.1. *Types, Tokens and Cortical Streams*

The distinction between types and tokens is well known in language and is critical to the understanding of language and perception in general. Consider the following example: When extracting the different meanings from the following sentences, "Big fish eat little fish" and "Big fish eat little", it is necessary to clearly establish that there are two tokens of the type "fish" (Kanwisher, 1987). Normally, we are proficient at this task. However, in RSVP experiments containing repeated stimuli this process can break down.

Kanwisher (1987) proposed that RB occurs when words are recognised as separate types but not tokenised, hence the description *type activation without token individuation.* Kanwisher also suggested that as each word is presented separate type and token nodes are setup with links established between the two to tokenise the type.

One possible source of the type-token division is the what-where division of the visual system (Ungerleider & Mishkin, 1982; Baylis, Driver, & Rafal, 1993). In this view, the types are processed in the temporal "what" pathway whereas the tokens, or episodic representations, are the domain of the parietal "where" stream. It is assumed that the dorsal pathway is also responsible for the tokens' temporal order. The hypothesis is that the separate spatiotemporal and identity representations are then bound together, hence tokenising the type. Activation of the type representation and creation of the episodic token can happen in parallel, the problem arises when the two are bound together to form a stable percept. In RB, failure to bind the type-token representations causes the disappearance of the second critical stimulus.

One popular proposal is that the brain uses the synchronisation of neural spikes to solve this binding problem (Singer, 1994). In the context of accounting for RB within this framework, one could hypothesise that under normal circumstances there are two ensembles of neurons, one in each of the what and where streams. Over a short period of time these entrain producing the neuronal correlate of a stable percept. Disruption of this entrainment, however, gives rise to RB. In RB experiments, the time scale of stimulus presentation is so fast that sometimes entrainment fails to occur when the type is repeated.

2. A Model of RB

Drawing on the presence of the functional streams in the visual cortex, the model we propose for repetition blindness, shown in figure 1, follows this roughly Y-shaped division. It consists of two networks that share a common input source. One of the networks is specialised for localising where a letter is presented in the input and the other is specialised for identifying the letter regardless of location in the input. The final network in the model plays the role of what is commonly called association cortex. It is bidirectionally connected to both the "what" and the "where" networks and allows for identity and location information to be bound together even though two different networks sub-serve these functions. These three networks are laterally connected self-organising maps (SOM) (Kohonen, 1995) of excitatory and inhibitory spiking neurons.

The input layer roughly models the pattern of orientation sensitive line detectors found in V1 (Hubel & Wiesel, 1965). The layer is fully connected to the excitatory neurons of the what layer while the connections between it and the where layer are arranged in overlapping topographic receptive fields. This allows only one side of the where map to activate in response to a letter presentation on one side of the input. However, some of this activity spreads, via the lateral connections, to the other side of the map.

The final part of the model consists of an "association" cortex layer. Its function is to mediate the binding of letter identity and letter location information into a stable percept. This is accomplished by using a third self-organising layer of laterally connected excitatory and inhibitory neurons. This layer receives input from both the "where" and "what" layers. In addition to this, it also sends an extensive network of reciprocal connections to both the "what" and "where" layers. These connections promote inter-areal synchronisation thus facilitating the binding of the "what" and "where" information (Singer, 1994).

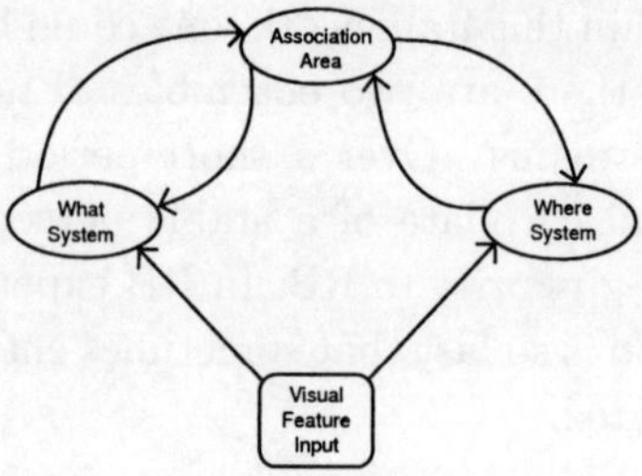

Figure 1. Gross outline of the model of RB.

2.1. *Visual Input*

Input to the model consists of a simplified version of the simple cells described by Hubel and Wiesel (1965). Four different orientations, namely 0, 45, 90 and 135, and 16 possible spatial locations are used. This representation was chosen as it allowed the use of the font described by Rumelhart and Siple (1974). The font together with the oriented line detectors and their locations can be seen in figure 2. Each one of these features corresponds to a neuron capable of generating a single spike and each letter is represented by a spike on a subset of these oriented line detectors.

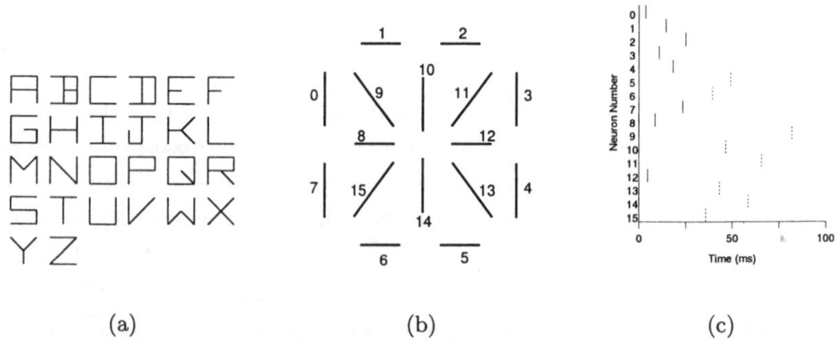

Figure 2. (a): The font used for letter presentation to the model is similar to that used by Rumelhart and Siple (1974). (b): The template showing the location and arrangement of all 16 features. After figure 2, Rumelhart and Siple (1974). (c): The encoding of the letter "A" by spike times. In this case, neurons 0, 1, 2, 3, 4, 7, 8 and 12 signal its presence. These spikes are solid lines. The other neurons, which do not take an active part in encoding the feature have their spikes shown as dashed lines. The neuron numbers correspond to the features numbered in (b).

The visual input to the model consists of two groups of 16 neurons. Each of these groups can represent one letter, consequently the two groups are capable of registering two simultaneously presented letters.

The presence of a particular feature in the input is encoded in the firing time of the appropriate neurons. In this case, the encoding is a delay or latency code (Gerstner & Kistler, 2002). In the training pattern set, those neurons signalling the presence of a particular feature fire early and all other neurons fire much later. This arrangement can be seen in figure 2(c). The inclusion of this positive and negative information is necessitated by the design of the synapses used in the network. Since they only change efficacy in response spike events, some means must be included to indicate that the efficacy must either be increased or decreased. This is accomplished by having spikes encoding the presence of a feature fire early and roughly in synchrony. All the others fire randomly much later than those signalling the presence of a feature. Given an appropriate membrane time constant, τ_m, these early spikes summate and cause the SOM neurons to fire. When the efficacy of the synapses is updated the early spikes will cause an increase in the synaptic efficacy since they, for the most part, lie in the time period before a SOM neuron spike when afferent spikes can potentiate a synapse. The later firing neurons, on the other hand, lie in the time window after a SOM neuron spike which causes a decrease in synaptic efficacy.

This pattern encoding scheme also has the advantage that a large number of patterns can easily be generated for a single letter by addition or deletion of spikes from a canonical letter representation or by shifting one or more spike's time.

The testing pattern set is similar to the training pattern set save for the absence of the late, dashed spikes in figure 2.

Having described how patterns are created for the letters, we now proceed to outline the neural model used in the SOM layers which learn these input patterns.

3. The Neural Model

The spiking model used in the network is the Spike Response Model (SRM) (Gerstner & Kistler, 2002). In order to obtain a computationally efficient function which computes the response of a neuron to afferent spikes, several simplifications described by Connolly, Marian, and Reilly (2004), have to be made to the SRM equations. Firstly, between the arrival of any two spikes at a neuron its membrane potential evolves in a deterministic manner.

Therefore if we know what the membrane potential is at time t_{int} we can compute what it will be at some later time $t_{\text{int}} + \Delta t$. We then need only compute the additional change in the membrane voltage caused by spikes arriving since t_{int}. Secondly, if we assume that the post-synaptic potentials due to the arrival of a pre-synaptic spike contribute instantaneously to the membrane potential of the model neuron then we can eliminate the need to use a biophysically more realistic function to model the change of the post-synaptic potential. In its place we can simply use a Dirac delta function. The modified equation is shown in (1).

$$
\begin{aligned}
V_i(t) = \sum_{t_i^{(f)} \in \mathcal{F}_i} \eta(t - t_i^{(f)}) + \epsilon(t_{\text{int}}) \cdot \exp\left(\frac{-(t - t_{\text{int}})}{\tau_m}\right) + \\
\sum_{j \in \Gamma_i} \sum_{t_j^{(f)} > t_{\text{int}} \in \mathcal{F}_j} w_{ij} \delta(t - t_j^{(f)} - \hat{d}_{ij})
\end{aligned}
\tag{1}
$$

where η describes the response of a neuron to its own spikes, $\epsilon(t_{\text{int}})$ is the post-synaptic potential at time t_{int}, w_{ij} is the efficacy of the synapse between neurons i and j, $t_j^{(f)}$ is the most recent spike time of neuron i, Γ_i is the set of neurons pre-synaptic to neuron i, $\mathcal{F}_j$ is the set of spike times of neuron j, $\hat{d}_{ij}$ is a noisy spike transmission delay between neurons i and j, and δ is the Dirac delta function ($\delta(x) = 1$ iff $x = 0$, 0 otherwise). This simplification of the neural model, though physiologically unrealistic (Gerstner, 2001), has considerable computational benefits since we no longer consider the time course of synaptic input to the neuron (Marian, 2002).

Learning in the model is accomplished by using plastic excitatory and inhibitory spike time-dependent synapses (Abbott & Nelson, 2000). It is thought that in the brain, plasticity processes at synapses mediate long-term learning. At the excitatory synapses a type of Hebbian learning is modelled. If a pre-synaptic spike precedes that of a post-synaptic neuron, potentiate of the synapse can occur. If the reverse occurs then depression of the synapse takes place. Equations (2) and (3), below, govern these two processes.

$$
w_{ij}(t + \Delta t) = w_{ij}(t) + \kappa_E(\hat{t}_j + \hat{d}_{ij} - \hat{t}_i) \tag{2}
$$

$$
\kappa_E(\Delta) = \begin{cases} A_+ \cdot \exp(\Delta/\tau_P) & \text{if } \Delta < 0 \\ B_- \cdot \exp(-\Delta/\tau_D) & \text{if } \Delta > 0 \\ 0 & \text{otherwise} \end{cases} \tag{3}
$$

where $\hat{t}_j$, $\hat{t}_i$ are the most recent spike times of the pre- and post-synaptic neuron, respectively. The amplitude and polarity of the change are determined by the function κ, which implements what is known as the learning window (Gerstner & Kistler, 2002). τ_P and τ_D control the time course over which potentiation and depression occur respectively, and the constants $A_+ > 0$ and $B_- < 0$ control the amplitude of potentiation and depression, respectively. This function differentiates between the case where the pre-synaptic spike, t_{pre}, occurs before the post-synaptic spike, t_{post}, and vice versa, with potentiation occurring in the former instance and depression in the latter.

The similar process takes place at inhibitory synapses (Roberts, 2000). When spiking of the pre-synaptic inhibitory neuron follows that of the post-synaptic excitatory neuron, potentiation can result. Conversely, when pre-synaptic activity precedes post-synaptic activation, depression can be observed. Two similar equations, (4) and (5), can also be defined, with all parameters as previously mentioned.

$$w_{ij}(t + \Delta t) = w_{ij}(t) + \kappa_I(\hat{t}_j - \hat{t}_i) \tag{4}$$

$$\kappa_I(\Delta) = \begin{cases} A_+ \cdot \exp(-\Delta/\tau_P) & \text{if } \Delta > 0 \\ B_- \cdot \exp(\Delta/\tau_D) & \text{if } \Delta < 0 \\ 0 & \text{otherwise} \end{cases} \tag{5}$$

In both cases, a realistic description of synaptic plasticity requires that the efficacy, w_{ij}, be constrained to vary within a certain range of values, namely $[w^{EI}_{min}, w^{EI}_{max}]$. The excitatory (inhibitory) synapse described here must maintain a positive (negative) weight which must not be greater (less) than w^{EI}_{max}. These constraints can be easily realised by setting $B_- = w_{ij}b_-$, $b_- < 0$ and $A_+ = (w^{EI}_{max} - w_{ij})a_+$, $a_+ > 0$ in equations (3) and (5). Thus each positive (negative) term which increments (decrements) the synaptic efficacy is proportional to $w^{EI}_{max} - w_{ij}$ whereas each negative (positive) term which leads to a decrement (increment) of the efficacy is proportional to w_{ij}. When the efficacy reaches its maximum (minimum) value it is no longer incremented (decremented) (Gerstner & Kistler, 2002).

4. Results

Thus far eliciting synchronous oscillation from the network has proved very difficult as the network is very sensitive to small changes in parameter sets.

Seemingly innocuous modifications can change the regular output shown in figure 3 into self-propagating epileptiform spiking patterns where all information about letter identity and location is lost. Balancing excitation and inhibition in a way which maintains the topographic regularity extracted by the SOM algorithm with the need to have regions of the three maps spiking together, and binding what and where information is so doing, is a demanding task. Solving this problem is our current goal.

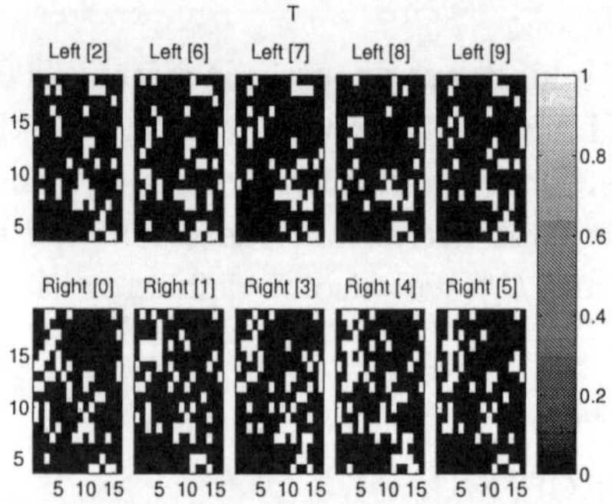

(a) Letter identity.

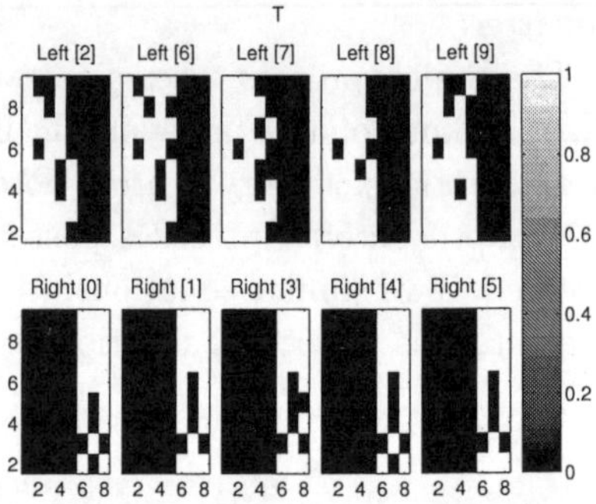

(b) Letter location.

Figure 3. The figures above show the number of spikes generated by 10 presentations of the letter "T" to the model.

Currently the model is capable of correctly learning both letter location and a translation invariant representation of letter identity. This can be seen in figure 3. The letter "T" was presented to a network 10 times. Half of these presentations were to the left side of the input array and the other half to the right. The model learned letter identity, show in figure 3(a), regardless of presentation location. Though the individual spike patterns are not identical, due to noise in spike transmission, they are substantially similar. Likewise in figure 3(b), the model easily located the letter in the input array. This is a consequence of the receptive field connection between the input and where layers.

5. Summary and Conclusions

In this paper we have described a model of orthographic repetition blindness, a disruption in the ability to recognise repeated words, or word fragments, which are rapidly visually presented.

Drawing on the division of the visual system into streams which process object identity and location separately, the model consists of two self-organising maps performing these functions. These two maps are reciprocally connected, via a third SOM, which promotes synchronous firing of the neurons in each map.

We have described a model spiking neuron based on Gerstner's SRM with networks of these neurons connected with plastic excitatory and inhibitory synapses exhibiting spike time-dependant learning.

So far the network has successfully learned both letter and location information. In attempting to get the model to demonstrate synchronous oscillations in different parts of the three maps the model has almost consistently become epileptic, a known problem with these types of networks (Huyck, n.d.).

This may be due to a number of factors. Firstly, the balance between slow and fast excitatory and inhibitory synaptic transmission which can stabilise oscillatory persistent activity (Brunel, 2003) may not be present in the network. Secondly, the simple fire and quickly forget refractory function may decay to swiftly thus depriving the network of its activity history. We are currently investigating whether addressing these two problems will allow the network to enter stable oscillatory persistently active states.

Acknowledgements

We wish to thank Harry Erwin for his useful comments and suggestions for solving the simulated epilepsy demonstrated by the model.

References

Abbott, L. F., & Nelson, S. B. (2000). Synaptic plasticity: taming the beast. *Nature Neuroscience, 3*, 1178–1183.

Bavelier, D., & Potter, M. C. (1992). Visual and phonological codes in repetition blindness. *Journal of Experimental Psychology: Human Perception and Performance, 18*(1), 134–147.

Baylis, G., Driver, J., & Rafal, R. D. (1993). Visual extinction and stimulus repetition. *Journal of Cognitive Neuroscience, 5*, 453–466.

Brunel, N. (2003). Dynamics and plasticity of stimulus-persistent activity in cortical network models. *Cerebral Cortex, 3*(11), 1151–1161.

Connolly, C. G., Marian, I., & Reilly, R. G. (2004). Approaches to efficient simulation with spiking neural networks. In H. Bowman &

C. Labiouse (Eds.), *Connectionist models of cognition and perception II* (Vol. 15, pp. 231–240). London, UK: World Scientific.

Gerstner, W. (2001). What's different with spiking neurons. In H. A. K. Mastebroek & J. E. Vos (Eds.), *Plausible neural networks for biological modelling* (Vol. 13, pp. 23–48). Dordrecht, The Netherlands: Kluwer Academic Publishers.

Gerstner, W., & Kistler, W. M. (2002). *Spiking neuron models: Single neurons, populations, plasticity.* Cambridge, UK: Cambridge University Press.

Harris, C. L., & Morris, A. L. (2000). Orthographic repetition blindness. *The Quarterly Journal of Experimental Psychology, 53A*, 1039–1060.

Hubel, D. H., & Wiesel, T. N. (1965). Receptive fields and functional architecture of two non-striate visual areas (18 and 19) of the cat. *Journal of Neurophysiology, 28*, 229–289.

Huyck, C. (n.d.). *Creating hierarchical categories using cell assemblies.* (Manuscript under review. Retrieved 14th October 2004 from `http://www.cwa.mdx.ac.uk/chris/hebb/hier/hier.ps`)

Kanwisher, N. G. (1987). Repetition blindness: Type recognition without token individuation. *Cognition, 27*, 117–143.

Kanwisher, N. G., & Potter, M. C. (1990). Repetition blindness: Levels of processing. *Journal of Experimental Psychology: Human Perception and Performance, 16*(1), 30–47.

Kohonen, T. (1995). *Self-organizing maps.* Berlin: Springer.

Marian, I. D. (2002). *A biologically inspired model of motor control of direction.* Unpublished master's thesis, National University of Ireland, Dublin, Dublin, Ireland.

Roberts, P. D. (2000). Modeling inhibitory plasticity in the electrosensory system of mormyrid electric fish. *Journal of Neurophysiology, 84*(1), 2035–2047.

Rumelhart, D. E., & Siple, P. (1974). Process of recognizing tachistoscopically presented words. *Psychological Review, 81*(2), 99–118.

Singer, W. (1994). Putative functions of temporal correlations in neocortical processing. In C. Koch & J. L. Davis (Eds.), *Large-scale neuronal theories of the brain* (pp. 210–237). Cambridge, MA: The MIT Press.

Ungerleider, L. G., & Mishkin, M. (1982). Two cortical visual systems. In D. Ingle, M. A. Goodale, & R. J. W. Mansfield (Eds.), *Analysis of visual behavior* (pp. 549–586). Cambridge, MA: The MIT Press.

Cognitive Processing and Control

MODELLING THE SLOW EMOTIONAL STROOP EFFECT: SUPPRESSION OF COGNITIVE CONTROL

BRADLEY WYBLE[A] DINKAR SHARMA[B] HOWARD BOWMAN[A]

[A] *Centre for Cognitive Neuroscience and Cognitive Systems, Computing Laboratory, University of Kent at Canterbury*
Canterbury, Kent CT2 7NP, UK

[B] *Centre for Cognitive Neuroscience and Cognitive Systems, Psychology Department, University of Kent at Canterbury*
Canterbury, Kent CT2 7NP, UK
bw5@kent.ac.uk, D.Sharma@kent.ac.uk, hb5@kent.ac.uk

Our connectionist model provides a theoretical explanation for the existence of slow and fast emotional Stroop effects, and depicts them as independent but interacting phenomena. We build upon previous modelling work by Cohen et al (1990) and Botvinick et al (2001) among others, and incorporate data that suggest a functional division of the Anterior Cingulate Cortex (ACC) into Cognitive and Affective Divisions. This work suggests that slow emotional Stroop effects are caused by activation of the affective portion of the ACC, which inhibits the Cognitive division, reducing top-down cognitive control on the subsequent trial.

1. Introduction

1.1. *Modelling the Stroop Effect*

Stroop interference refers to increases in reaction time to name the ink colour of a word when that word is itself the name of a different colour (Stroop, 1935). There are three types of trials in the Stroop paradigm: congruent, in which the ink and colour word match, neutral, in which the word is not a colour name, and incongruent, in which the ink and colour name do not match. Incongruent trials show interference relative to neutral trials, while congruent trials show facilitation. In this paper we propose a connectionist model of the Stroop effect, built upon previously existing models (Cohen, Dunbar and McClelland 1990; Botvinick, Braver, Barch, Carter and Cohen 2001) that extends their functionality to include fast and slow emotional Stroop effects in a manner that is consistent with recent imaging data about the functionality of the Anterior Cingulate Cortex (ACC).

1.2. *Anterior Cingulate Cortex and Cognitive Control*

The dorsal Anterior Cingulate Cortex (ACC) is thought to assist in the mediation of cognitive tasks, especially those that involve mutually incongruent stimuli pairs (Bush, Whalen, Rosen, Jenike, McInerney and Rauch 1998). Furthermore, it is thought that the ACC is able to detect conflict between the incompatible responses of such tasks and elevates cognitive control from one trial to the next to improve performance in such difficult tasks (Botvinick et al 2001).

There also exists evidence that the rostral extreme of the ACC is activated in emotional states and by affective stimuli. Furthermore the two portions of the ACC, one mediating cognitive and the other affective operations, are mutually inhibitory (Drevets and Raichle 1998; Bush, Luu and Posner 2000). Our model proposes that it is the interaction between these two halves of the ACC that is responsible for the slow emotional Stroop effect as will be discussed below.

1.3. *Slow Emotional Stroop Effect*

The emotional Stroop task involves presenting an emotionally charged word that is written in various ink colours. Subjects are asked to ignore the word and respond to the ink colour. The typical finding is one of longer response latencies to name the ink colour when the word is emotional, usually negative, rather than a control neutral word (Watts, McKenna, Sharrock and Trezise, 1986). Such attentional biases have been argued to reflect the capture of attention by emotional stimuli. In general this is thought to reflect a shift in processing resources towards emotional stimuli. In addition, such processing is often described as automatic, fast acting and even preattentive (Mogg and Bradley 1998). Evidence using paradigms other than the modified Stroop task have also been used to support these claims (Ohman, Flykt and Esteves 2001).

More recently there has emerged evidence in support of the conclusion that there are relatively slow effects of emotionally negative stimuli. One such data set comes from emotional Stroop studies that have mixed negative and neutral trials (McKenna and Sharma 2004). They presented seven Stroop trials: a negative stimulus is shown on the first trial followed by six trials of neutral stimuli. They found that emotional trials are performed at the same speed as neutral trials. However, the following trial had a larger reaction time despite being neutral. This effect returned back to baseline by the second post-emotional word trial. This research highlights the idea that negative words do not capture attention in a relatively fast manner but produce their effects more slowly on attentional components subsequent to detection. Slow effects that operate across trials can explain the larger interference found for emotional stimuli when emotional and neutral stimuli are segregated into separate blocks, compared to when they are mixed together randomly (Holle, Neely and Heimberg 1997).

Emotional Stroop effects that depend on inter-trial effects can be called slow effects, while those that depend on intra-trial effects can be called fast effects (McKenna and Sharma 2004). Our model, derived from previous work in the field, demonstrates both fast and slow components of the emotional Stroop effect, by assuming that fast effects are due to automatic vigilance (Mathews and Harley 1996) and that slow effects are the result of a temporary redistribution of attentional resources in response to the emotional input. We call this latter process salience sensitive control, and view it as an automated process by which the mind of the participant is reconfigured for several hundred milliseconds in order to more readily detect threatening information in the environment.

2. Methods

2.1. *Neural Representation*

Our model is an extension and modification of the framework used by the conflict monitoring and cognitive control model of Botvinick et al (2001). Several components are added to allow this network to model emotional Stroop effects, both fast and slow. The recurrent nature of the inter-layer links are changed to a purely feed-forward pattern to simplify the dynamics of the model.

Our implementation (Figure 1), follows the pattern of the connectionist architecture for Stroop models proposed by Cohen et al (1990). Word and colour information are processed in parallel along different streams, which converge on a common set of nodes. Strong weights within the word stream give word reading an inherent dominance over colour naming. This default tendency is overcome through a task-demand system that provides an excitatory bias to the colour naming stream. In our model, this task demand system also suppresses competing information in the word stream. This implementation is suggested by imaging work from Carter, Mintun and Cohen (1995) which shows that a decrease in left extra-striate cortex accompanied increased activation of dorsal cingulate cortex. The former area is thought to be involved in the processing of word form information (Peterson, Fox, Snyder and Raichle 1990).

Our model uses simple rate code representations. Outputs from each neural element are derived from thresholds combined with a sigmoid-like function. Connections between elements are excitatory or inhibitory and static. There are no synaptic delays. The activation function for the membrane potential of all neural elements is given by (1).

$$MP_{(i,j,t)} = MP_{(i,j,t-1)} + DT\_VM_{(j)} * [(Bias_{(j)} + Excite_{(i,j,t-1)}) * (EE_{(j)} - MP_{(i,j,t-1)}) + Inhib_{(i,j,t-1)} * (EI_{(j)} - MP_{(i,j,t-1)}) + Leak_{(j)} * (EL_{(j)} - MP_{(i,j,t-1)})] \quad (1)$$

$MP_{(i,j,t)}$ is the membrane potential for neuron i of layer j at time t, affected by *Bias*, *Excite*, *Inhib* and *Leak* with a time constant *DT_VM*. *EE*, *EI* and *EL* represent reversal potentials.

This membrane potential is used in the following output mapping which simulates a sigmoid function that bounds a neuron's output to the range [0,1], where θ and γ represent threshold and scaling parameters respectively.

$$Out_{(i,j,t)} = \frac{[MP_{(i,j,t)} - \theta_{(j)}]_+ * \gamma_{(j)}}{[MP_{(i,j,t)} - \theta_{(j)}]_+ * \gamma_{(j)} + 1} \quad \text{where } [x]_+ = \begin{matrix} x & if & x > 0 \\ 0 & otherwise & \end{matrix} \qquad (2)$$

2.2. *Connectivity and Functionality*

The model we use is shown in Figure 1. Colour and word processing are located at the bottom left and right respectively. Nodes in these areas are activated directly by corresponding input units and are intended to represent levels of colour and word-form processing. The word naming stream also contains nodes representing neutral and emotional words.

Nodes in these input layers are connected to the category layer, which is intended to represent activity in a semantic workspace. Connections from the word form area to the category layer are 2.5 times the value of those from the colour area. This asymmetry insures a dominance of word reading over colour naming. The relative strengths of the rest of the connections are not of direct theoretical import and will not be discussed in this limited format. All parameters were identical for all data sets addressed by this model.

Interference occurs within the category and response layers through uniform, lateral, inhibitory connections. Depending on levels of cognitive control, this inhibition may or may not be sufficient to ensure winner take-all dynamics within a layer.

Only the colour nodes of the category layer project to the response layer, in which neurons represent activation of motor plans to commit button pressing responses. Activity in the response nodes builds slowly over the trial until one of the two crosses threshold, at which point a response is determined and the trial concludes. Activation of all nodes are reset to starting values, with the exception of task demand units and those representing the two halves of the ACC.

The colour-naming task demand unit in this model excites colour processing nodes, inhibits word form processing, and excites task relevant modality independent colour nodes in the category layer. The word-naming task demand unit also excites category layer units and has similar but opposite connections to the ink colour and word form nodes as the colour-naming task demand unit. Input from the cognitive portion of the ACC, as well as putatively frontally mediated constant bias selects one of the two task demand units to be active, resulting in suppression of the other.

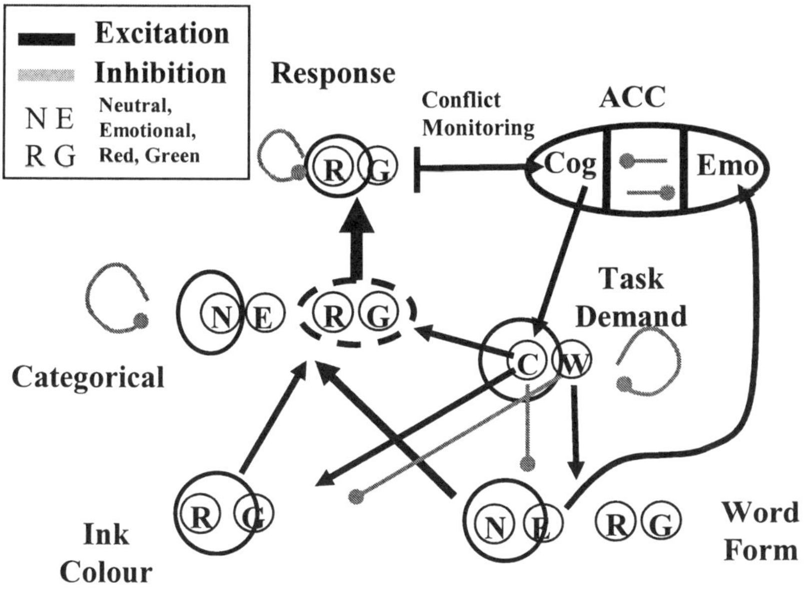

Figure 1. Diagram of the model used in these simulations. Large ovals indicate layers while circles represent units. The dotted line in the categorical layer indicates the semantic grouping of colour.

As in the model of Botvinick et al (2001), and in accord with studies that show ACC conflict monitoring at response production stages (Carter, Macdonald, Botvinick, Ross, Stenger, Noll and Cohen 2000) a Hopfield energy product equation (3) over the two response nodes provides this measure of conflict as the multiplicative product of the two response nodes. Only when both nodes are simultaneously active to some degree is the value of E (i.e. the measure of conflict) greater than zero.

$$E_{(i,j)} = \sum a_{(i)} a_{(j)} \quad \text{where } a_{(i)} \text{ is the output of response unit i.} \qquad (3)$$

The output of this equation is multiplied by a weight and then added as an excitatory input to the node representing the cognitive division of the ACC.

Presentation of an emotional word is capable of activating the affective division of the ACC, which inhibits the cognitive portion of the ACC. This effect is slow relative to the performance of a single trial such that task demand fluctuates gradually across trials.

3. Results

3.1. *Standard Stroop Data*

This model is able to recreate the standard Stroop effects. Interference caused by incongruent trials produces a larger change in reaction time than does the facilitation caused by congruent trials. This is the classical pattern of asymmetry inherent in most Stroop studies. Additionally, word reading tasks are faster than colour naming tasks, and exhibit minimal interference or facilitation.

These effects are illustrated in Figure 2, for which human data from Dunbar and MacLeod (1984) are compared to the results from the model.

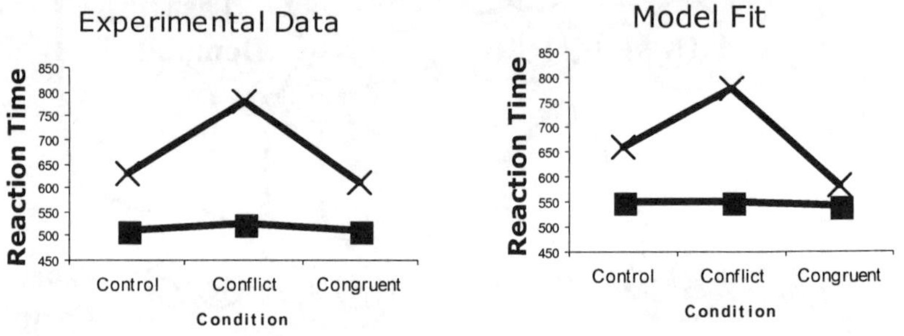

Figure 2. Comparison of simulated to empirical results for the standard Stroop paradigm. The X marked points represent colour naming, while black squares represent word naming.

3.2. *Cognitive Control Data*

The data from Tzelgov, Henik and Berger (1992), among others (Carter et al 2000), make it clear that Stroop effects cannot be considered invariant across blocked and mixed experimental designs. The composition of a trial block affects how the trials within that block are processed.

In Tzelgov et al (1992), subjects were given blocks with different proportions of colour trials and non-colour neutral words in steps of 25%. Of the colour words, half were congruent and half were incongruent. So a 75% colour word block was composed of 37.5% congruent, 37.5% incongruent, and 25% neutral trials. What this work demonstrated is that as the proportion of colour words was increased within a block, the amount of interference caused by an incongruent trial decreased, while facilitation remained essentially unaltered. Likewise in our model, facilitation of congruent trials decreased slightly from 51 to 57 to msec across conditions, while interference decreased from 120 to 87 msec as the frequency of colour trial increased from 25% to 75%. Our model exhibits this effect by virtue of the conflict monitoring system.

An incongruent trial causes simultaneous activation of both colour nodes in the response layer, resulting in a large conflict signal that activates the cognitive division of the ACC. This activation results in an increase in top down cognitive control, further exciting the colour-naming area and suppressing the word-form area as the system becomes strongly biased towards colour naming. Therefore, as the proportion of incongruent trials increases from 12.5% to 25% to 37.5%, the chance of any incongruent trial being preceded by another incongruent trial increases, thereby reducing the average amount of interference.

3.3. *Emotional Stroop Effect*

In our framework of interference, fast and slow emotional Stroop effects are qualitatively distinct phenomena, representative of intra and inter-trial time scales respectively.

3.3.1. Fast Effects

In this model, fast effects are obtained by automatic vigilance, similar to the implementation of Mathews and Harley (1996). Emotional words corresponding to a particular anxiety can be given the ability to strongly activate the category layer representation and thereby interfere in the current trial in a similar manner as incongruent colour words. This can be done by including a threat monitoring unit that adds excitatory bias to a given emotional word form unit, or by increasing the connection strength between these units and the category layer. The degree of interference obtained can vary by adjusting these parameters, so there is not much to be gained in claiming a fit of a particular data set.

3.3.2. Slow Effects

Negative emotional word input in this model has a dual effect by exciting both category layer representations and the affective division of the ACC. Even if the emotional input is insufficient to interfere at the category layer, the resulting activation of the affective portion of the ACC will suppress the cognitive division of the ACC, resulting in a slow reduction in cognitive control. On the following trial, the word-form area will be partially released from suppression allowing even neutral words to interfere at the category layer. It is important to note that the emotional word is not causing any direct interference itself, but rather is enabling the subsequent neutral word to interfere. This is the key point in the qualitative distinction between fast and slow emotional Stroop effects.

This model is able to fit the results of (McKenna and Sharma, 2004) both in time course and in effect size, as shown in figure 3. Note that this effect is specific to words of a negative emotional valence, rather than a positive emotional valence.

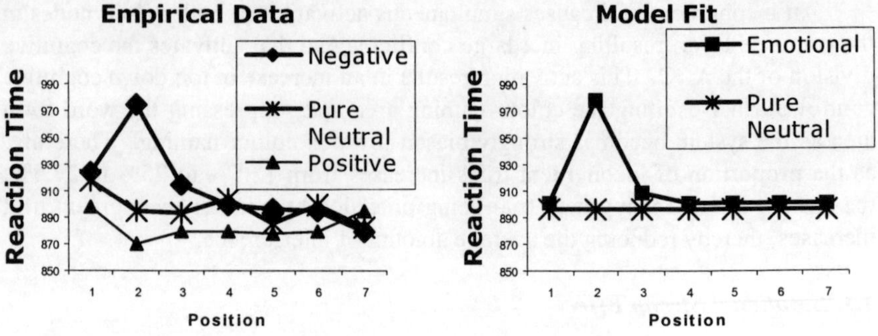

Figure 3. Data from McKenna and Sharma (2004) presented along with the model fit.

4. Predictions

4.1. *Salience Sensitive Control*

The primary prediction of this model and the theory that drives it concerns the fate of information presented to an unattended stream of input following emotional input of any kind. Our conceptualisation of salience sensitive control posits a temporary redistribution of attentional resources across multiple areas of processing following threatening input so as to facilitate the detection of further threatening information. This redistribution comes at the expense of the ongoing task, and the slow emotional Stroop effect is a hallmark of this process.

We predict that if subjects are given a word reading task, and a concurrent audio stream to be ignored, an emotional word presented within the word reading list should create a temporal window of several hundred milliseconds in duration during which input in the unattended audio stream is more deeply processed. This window should give the unattended audio stream an opportunity to prime the word naming task.

4.2. *Cognitive Control of Affective Input*

In this model, processing of word-form information is subject to cognitive control, and this includes processing of emotional salience. Consequently, elevating cognitive control by presenting incongruent trials immediately prior to an emotional word should reduce its ability to interfere with task performance on the subsequent neutral trial. Our model predicts that for the sequence [Incongruent, Emotional, Neutral], there will be no significant slowing of the neutral word compared to the data from McKenna and Sharma (2004).

5. Discussion and Conclusions

We present a theoretical framework that describes an automated response to a threatening stimulus that exists to reorient attentional and cognitive resources towards detecting further threatening information. This position explains the slow emotional Stroop effect (McKenna and Sharma 2004), but also fits with data describing the putative inhibitory interaction of the cognitive and affective portions of the ACC (Bush et al 2000; Drevets and Raichle 1998).

The authors of Algom Chajut and Lev (2004) make a strong empirical case that emotional and colour Stroop effects result from fundamentally distinct processes. Our theoretical position on the role of the ACC and the model which embodies it, support their stance for the case of slow (inter-trial) effects but not necessarily for fast effects. It should be stated that we are not intent on generalizing this phenomenon to all stressful situations, rather specifically to those in which the stressor is not a source of severe anxiety and is task irrelevant.

We propose that this salience sensitive control is the cortical counterpart of the amygdalar and thalamic fear circuitry (LeDoux 2003). Complementing the pre-attentive automatic vigilance towards threatening stimuli and the generalized arousal reaction following such stimuli is a cortical mechanism mediated by the affective portion of the ACC that temporarily suppresses ongoing cognitive activity to further augment the ability to detect and process threatening stimuli.

Acknowledgements

The work presented here was partially supported by EPSRC Grant Number GR/S15075/01.

References

Algom, D., Chajut, E., & Lev, S. (2004). A rational look at the emotional stroop phenomenon: a generic slowdown, not a stroop effect. *J Exp Psychol Gen, 133*(3), 323-338.

Botvinick, M. M., Braver, T. S., Barch, D. M., Carter, C. S., & Cohen, J. D. (2001). Conflict monitoring and cognitive control. *Psychol Rev, 108*(3), 624-652.

Bush, G., Whalen, P. J., Rosen, B. R., Jenike, M. A., McInerney, S. C., & Rauch, S. L. (1998). The counting Stroop: an interference task specialized for functional neuroimaging--validation study with functional MRI. *Hum Brain Mapp, 6*(4), 270-282.

Bush, G., Luu, P., & Posner, M. I. (2000). Cognitive and emotional influences in anterior cingulate cortex. *Trends Cogn Sci, 4*(6), 215-222.

Carter, C. S., Mintun, M., & Cohen, J. D. (1995). Interference and facilitation effects during selective attention: an H215O PET study of Stroop task performance. *Neuroimage, 2*(4), 264-272.

Carter, C. S., Macdonald, A. M., Botvinick, M., Ross, L. L., Stenger, V. A., Noll, D., et al. (2000). Parsing executive processes: strategic vs. evaluative functions of the anterior cingulate cortex. *Proc Natl Acad Sci U S A, 97*(4), 1944-1948.

Cohen, J. D., Dunbar, K., & McClelland, J. L. (1990). On the control of automatic processes: a parallel distributed processing account of the Stroop effect. *Psychol Rev, 97*(3), 332-361.

Drevets, W. C., & Raichle, M. E. (1998). Reciprocoal Suppression of Regional Cerebral Blood Flow during Emotional versus Higher Cognitive Processes: Implications for Interactions between Emotion and Cognition. *Cognition and Emotion, 12*(3), 353-385.

Dunbar, K., & MacLeod, C. M. (1984). A horse race of a different color: Stroop interference patterns with transformed words. *J Exp Psychol Hum Percept Perform, 10*(5), 622-639.

Holle, C., Neely, J. H., & Heimberg, R. G. (1997). The effects of blocked versus random presentation and semantic relatedness of stimulus words on response to a modified Stroop task among social phobics. *Cognitive Therapy and Research, 21*, 681-697.

LeDoux, J. (2003). The emotional brain, fear, and the amygdala. *Cell Mol Neurobiol, 23*(4-5), 727-738.

Mathews, G., & Harley, T. A. (1996). Connectionist Models of Emotional Distress and Attentional Bias. *Cognition and Emotion, 10*(6), 561-600.

McKenna, F. P., & Sharma, D. (2004). Reversing the emotional Stroop effect reveals that it is not what it seems: the role of fast and slow components. *J Exp Psychol Learn Mem Cogn, 30*(2), 382-392.

Mogg, K., & Bradley, B. P. (1998). A cognitive-motivational analysis of anxiety. *Behav Res Ther, 36*(9), 809-848.

Ohman, A., Flykt, A., & Esteves, F. (2001). Emotion drives attention: detecting the snake in the grass. *J Exp Psychol Gen, 130*(3), 466-478.

Petersen, S. E., Fox, P. T., Snyder, A. Z., & Raichle, M. E. (1990). Activation of extrastriate and frontal cortical areas by visual words and word-like stimuli. *Science, 249*(4972), 1041-1044.

Stroop, J. R. (1935). Studies of interference in a letter naming task. *Bulletin of the Psychonomic Society, 10*, 31-34.

Tzelgov, J., Henik, A., & Berger, J. (1992). Controlling Stroop effects by manipulating expectations for color words. *Mem Cognit, 20*(6), 727-735.

Watts, F. N., McKenna, F. P., Sharrock, R., & Trezise, L. (1986). Colour naming of phobia-related words. *Br J Psychol, 77 (Pt 1)*, 97-108.

EVIDENCE OF MODULARITY FROM PRIMATE ERRORS DURING TASK LEARNING

JOANNA J. BRYSON
University of Bath, Department of Computer Science
Bath, BA2 7AY; United Kingdom
E-mail: j.j.bryson@bath.ac.uk

The last two decades have seen a great deal of theorising and speculation about the modular nature of human intelligence, as well as a rise in use of modular architectures in artificial intelligence. Nevertheless, whether such models of natural intelligence are well supported is still an issue of debate. In this paper, I propose that the most important criteria for modularity is specialised representations. I present a modular model of primate learning of the transitive inference task, and propose an extension to this model which would explain task-learning results in other domains. I also briefly relate this work to both neuroscience and established AI learning architectures.

1. Introduction — Modularity and its Criteria

The 1980s saw the publication of two books which drew attention to modular theories for the organisation of intelligence. The first, *The Modularity of Mind* (Fodor, 1983) came from psychology and dwelt primarily on specialised modules for perception. Fodor's full theory proposed that a central processor analyses the output of these modules and chooses some action, then further modules generate expressed behavior from these decisions. The second book, *The Society of Mind* (Minsky, 1985) came from artificial intelligence (AI). It described a system where control is divided and negotiated between agents with different goals and skill bases. Like Fodor, Minsky also proposed some cross-module shared resources, such as long-term memory and a module maintenance system.

Now, many leading psychologists and philosophers of mind take as given that natural intelligence has some sort of modular structure (Carruthers, 2003). Similarly, modular architectures have come to dominate the areas of artificial intelligence that deal with problems most similar to humans': autonomous robotics and virtual reality characters (Bryson, 2000). The relevance of the preferences of AI engineers to theories of natural intelligence is this: These engineers face the same problems as nature, such as dealing with uncertainty in perception and action, and searching enormous, combinatorially-explosive spaces of possible actions. Design features which have proved useful for engineering may therefore be treated as reasonable hypotheses for aspects of natural intelligence. It may be that evolution has had the same result of its search for successful control strategies as we have.

Yet the modularity hypothesis in natural intelligence is still controversial (Karmiloff-Smith, 1992; Carruthers, 2003). This is not least because the *definition* of modularity is still not agreed on. Fodor proposed many possible characteristics for identifying a modular system, such as innateness, localisation and domain specificity. Innateness is extremely difficult to determine, however, because of the nature of development (Elman et al., 1996). Development always begins with a uniform (though modular) embryo of a few cells. The shift into a highly specialised adult moves along a continuum. Selecting a single point when all 'innate' modules might be expressed might be impossible. Certainly no modules will express themselves without an environment containing oxygen. It's quite likely that other modules might not be expressed without the sorts of intellectual stimulation that our species typically receives, such as social interaction. Further, there's no reason to use birth as the defining moment for what attributes have to be present to be considered innate. Given the developmental continuum, other organismic milestones such as puberty might be far more appropriate.

Localisation, on the other hand, is in some sense a given. We know that the brain is divided into organs with different roles and representations. Yet most proposed cognitive modules, e.g. face recognition or language, utilise disjoint sections of brain simultaneously. Similarly, domain specificity has proved difficult. Even V1, the part of the brain identified with 'early visual processing' has lately been indicated for performing other spatial tasks, such as reading braille (Kauffman et al., 2002). This is hardly surprising given the nature of the evolutionary process and neural computing — any cell system found producing orderly and useful computations in a particular context is likely to be exploited and encouraged regardless of whether that is the context in which it first proved useful.

Software engineering has also struggled with the correct way to define modularity, where here 'correct' means 'most useful' — the way most likely to be clear, consistent, maintainable and scalable. Although there is no single universally-accepted strategy for modular decomposition in software engineering, one standard approach is to define a module in terms of its variable state — that is, what it learns or stores, what it brings from past experience in order to generate behavior in the present. Although many AI systems attempt to store all knowledge in a single representation in order to maintain generality, this ignores the intractability of learning in the most general case Bryson (2001). Modularity provides an opportunity for specialised representations which can greatly increase the probability for any agent (animal or artificial) to successfully acquire the skills and knowledge it needs. This theory is derived from a transition the field of psychology made in understanding learning, from the general-purpose learning of the behaviourists to emphasis on specialised learning and modularity (Gallistel et al., 1991; Roper, 1983).

Differences in representation in animals are not only a good indication of modularity, but should in fact be taken as the primary criteria for modularity. While specialised representations are correlated with many of the traditional criteria for modules (such as innateness, domain specificity and localisation) they can also allow for a more flexible definition of 'module' that addresses the difficulties described above. A specialised representation can emerge (be learned) over development as neural systems tune to regularities useful in the agent's context. This representation may apply to more than one domain, but still not to most. It might be encoded across multiple brain sub-systems that are routinely activated in a particular context through reciprocal connections. It may only exist in certain contexts when activation of its component parts is synchronised (or bound (von der Malsburg, 1995)), while in other contexts its component parts perform other functions independently. The concept of module defined in these terms still explains why some capacities of an individual are present in some contexts and and not others. It also still makes the complexity of the overall system more manageable by encapsulating some behavior into a single system.

The remainder of this paper assumes this definition of modularity, which implies that to demonstrate modularity one must only demonstrate multiple distinct representations. We now move to examine whether task learning in primates is modular, beginning with the task of transitive inference.

2. Evidence of Modularity in Transitive Inference Learning

Transitive inference (TI) is the process of reasoning whereby one deduces that if, for some quality, $A > B$ and $B > C$ then $A > C$. This 'reasoning' capacity has been highly researched in psychology. Piaget described transitive inference as an example of concrete operational thought (Piaget,

1954). That is, children become capable of doing transitive inference when they become capable of performing mentally the physical manipulations that would determine the correct answer — in this case, visualising and comparing where two items appeared in a sequence. However, Bryant and Trabasso (1971) showed that pre-operational children could perform TI, given enough training. McGonigle and Chalmers (1977) also showed that monkeys (in this case squirrel monkeys *Saimiri sciureus*), although able to perform well on any two items (e.g. B, D) drawn from a trained series of pairs $\{A > B,\ B > C,\ C > D,\ D > E\}$, if presented with *three* trained items from that set (e.g. B, C, E) performed significantly worse. This last result, which also proved true of children (Chalmers and McGonigle, 1984), indicates that the subjects are not drawing their answers from a simple linear representation.

2.1. *Modelling TI with a Production-Rule Stack*

Harris and McGonigle (1994) presented a model that fit the McGonigle and Chalmers (1977) trigram data extremely well, based on production rule stacks. This model matches the performance of the monkeys very well whether they are modelled as a group and as individuals.

The production-rule stack model requires the following assumptions:

(1) The subject learns a set of rules of the nature "if A is present, select A" or "if D is present, avoid D".
(2) The subject learns a prioritising of these rules.

This process results in a *rule stack*, where the first rule is applied if its trigger finds the context appropriate. If not, the second, and so on.

For an example, consider a subject that has learned this stack:

1. (A present) $\Rightarrow$ select A
2. (E present) $\Rightarrow$ avoid E
3. (D present) $\Rightarrow$ avoid D
4. (B present) $\Rightarrow$ select B

Here the top item (1) is assumed to have the highest priority. If the subject is presented with a pair CD it begins working down its rule stack. Rules 1 and 2 do not apply, since neither A nor E is present. However, rule 3 indicates the subject should avoid D, so consequently it selects C. Priority is critical. For example, for the pair DE, rules 2 and 3 give different results. However, since rule 2 has higher priority, D will be selected.

Adding the assumption that in the trigram test cases, an 'avoid' rule results in random selection between the two remaining items, Harris and McGonigle (1994) model the conglomerate monkey data so well that there is no significant difference between the model and the data. For example, over all possible trigrams, the stack hypothesis predicts a distribution of 0, 25% and 75% for the lowest, middle and highest items. Binary sampling predicts 3%, 35% and 63%, and logic of course 0%, 0% and 100%. The monkeys showed 1%, 22% and 78%. Further, individual performance of most monkeys were matched to a particular stack.

Without trigram data, there would be no way to discriminate which rule set the monkeys use. However, with trigram data, the stacks are distinguishable because of their errors. For example, a stack like

$$\begin{aligned} &1'.\ (A \text{ present}) \Rightarrow \text{select } A\\ &2'.\ (B \text{ present}) \Rightarrow \text{select } B\\ &3'.\ (C \text{ present}) \Rightarrow \text{select } C\\ &4'.\ (D \text{ present}) \Rightarrow \text{select } D \end{aligned}$$

would always select B from the trigram BCD by using rule $2'$, while the previous stack would select B 50% of the time and C 50% because it would base its decision on rule 3.

2.2. *A Two-Tiered Modular System for Learning TI*

Production-rule models are actually common in cognitive modelling (e.g. Newell, 1990; Anderson, 1993), but are not satisfying, perhaps partially because they don't make a clear connection to brain mechanisms (though see Fincham et al., 2002). In an attempt to better understand how the Harris and McGonigle (1994) model could so accurately represent animal behavior, I built a modular neural learning version of their system. I broke the learning into two problems: the association of a stimulus (an item, e.g. B) to an action (select or avoid, as above), and the prioritisation of those perception/action pairs (see Figure 1).

This two-tiered model consists of a list of items seen, each associated with a weight and with a unique 2-item list of possible rules (select or avoid) also associated with weights. When the agent sees a new item, it adds it to its list of stimuli. The biological plausibility of these assumptions is addressed in the discussion.

All of the weights in a single list are normalised (they sum to 1) — new items receive the weight $1/N$ (where N is the current number of items). New items in the stimulus list are also associated with a rule-list where both

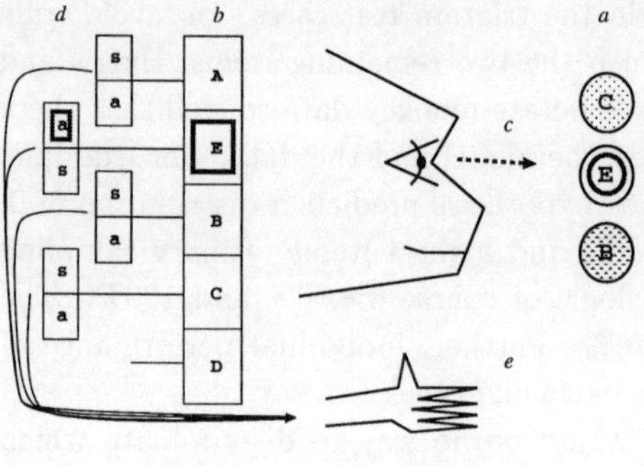

Figure 1. The two-tier model. When the agent observes a set of stimuli (a), a weight vector (b, the first tier) determines which item present is most salient. This attracts visual attention (c) and determines which rule vector (d, the second tier) selects the appropriate action (select or avoid) which controls the monkeys grasp (e). The two vectors that were most recently active (a and one of d) are then updated as determined by the result and the learning rule (Equation 1).

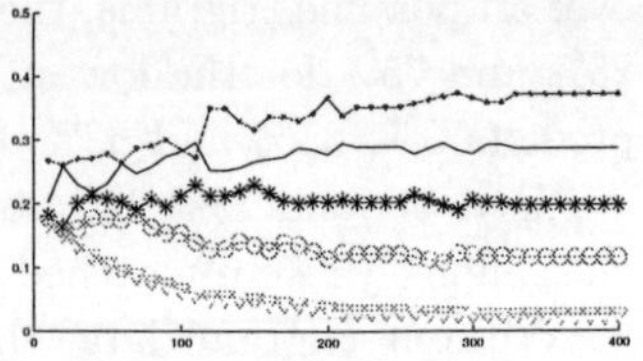

Figure 2. Typical results for single-vector learning (only one rule — select). X-axis: trial number; Y-axis: weights of stimuli vector (sum to one). Free variables are set to Parameters: $\Xi = .08$, $\delta = .02$. Key: $A\bullet$, $B-$, $C*$, $D\bigcirc$, $E\bigtriangledown$.

select and *avoid* have weight 0.5. provided. When the agent is given a trial, it examines its list for the rules associated with the stimuli present. For that set of rules, the highest-priority stimulus determines the focus of the agent's attention. Then the highest priority rule in the rule-list associated with that stimuli determines whether the agent selects or avoids the object it has fixated on. In the case of a tie, outcome is arbitrary.

Weights are updated after every trial, where a trial is a presentation of a pair or trigram, a selection by the agent, and a reward. The training rule is a simple step function. The subject learns a weight vector $\mathbf{v}$. For the pair XY, where X is selected by the subject and $\mathbf{v}_X$ and $\mathbf{v}_Y$ are the weights associated with X and Y respectively:

$$\begin{aligned} &\text{If X is correct and } (\mathbf{v}_X - \mathbf{v}_Y < \Xi)\text{, add } \delta \text{ to } \mathbf{v}_X; \\ &\qquad \text{else, if X is incorrect, subtract } \delta \text{ from } \mathbf{v}_X. \end{aligned} \tag{1}$$

where Ξ and δ are free parameters. Ξ is a threshold over which reward is so expected that it no longer prompts learning. δ is the amount a weight is changed by a single bout of learning. If a weight-change occurs, $\mathbf{v}$ is subsequently renormalised.

The results of this model are an even better match to nature than the original Harris and McGonigle (1994) model. First, when the model learns it exactly replicates the production-rule model, so in that sense it is equivalent. But second, like primates (both children and squirrel monkeys) it *doesn't* always learn successfully. Chalmers and McGonigle (1984) document a training regime that is necessary for getting a majority of children to successfully learn the initial pairs ($A > B$, $B > C \ldots$). This task is difficult because most items are sometimes rewarded and sometimes not.

Figures 2 shows a typical learning of weights in a more conventional, single-representation model of TI. These results are based on the same learning rule given above, but applied only to a single vector. Here learning occurs very reliably — the only way to disturb it is to constrain the learning resources such that the individual cannot find a stable set of weights, in which case the subject occasionally errs.

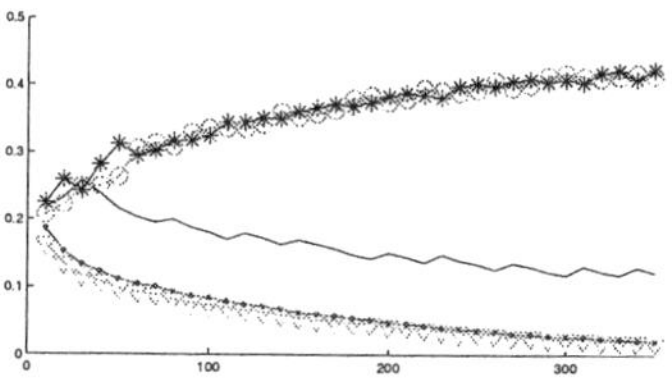

Figure 3. Rule learning with no training regime. Rules learned in descending order of priority: select $D\bigcirc$, avoid $C*$, avoid $B-$. The system cannot stabilise because it is far from a complete solution, but it behaves correctly for every training pair except CD. Parameters: $\Xi = .08$, $\delta = .02$.

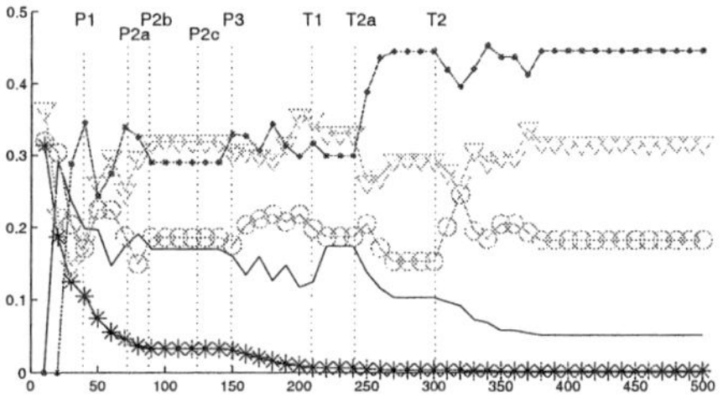

Figure 4. Rule Learning with Phased Training. Labelled lines indicate the *end* of training and testing phases (taken from Chalmers and McGonigle (1984)). This agent arrives at different stable solutions at different points, but they are all correct. The final rules set is: select $A\bullet$, avoid $E\bigtriangledown$, avoid $D\bigcirc$. The agent succeeds even with very 'stupid' parameters: $\Xi = .12$, $\delta = .06$.

Figures 3 and 4 show outcomes with the two-tier model. These diagrams only show the weights of the first tier, the priorities between rules. The rules themselves converge very quickly and then remain stable over the trial; the captions indicate whether 'select' or 'avoid' dominates for each stimuli. Unless the agents are put through a training regime, only roughly a quarter of the agents successfully learn the task. Most of the rest display the pattern shown in Figure 3, where they have learned what to do in the case of the endpoints successfully. Interestingly, the dominant tendency is to neglect the endpoints themselves, and focus attention on their neighbours, which are after all the more difficult learning task. Unfortunately for the agents, no correct rule set exists without an endpoint as the highest priority rule. Learning fixates on finding a correct ordering for two adjacent non-endpoint items, but no such ordering exists. In Figure 3, the agent's behavior is identical whether it selects D or avoids C: either is right in all training cases except when presented with CD, in which case either is wrong.

On the other hand, under the sort of training a child receives when being tested on the TI task, the agents learn successfully almost 9 times in 10, and across a broad range of parameter values Bryson and Leong (2004). Thus the two-tier model can not only successfully learn a model which conforms to Harris and McGonigle (1994), but it also *fails* to learn in an appropriate way: occasionally with training, and very frequently without it.

3. Modularity and the Two-Tier Model

Despite the quality of its results, the two-tier model is unattractive for several reasons. First, why when a single-vector model of TI learning would be not only simpler but more effective would nature choose something so complex? The answer of course is that TI is almost certainly not being learned on a special-purpose mechanism — if it had, perhaps it would also have applied itself to trigrams. This of course begs the question, what is this learning system, and what is it for?

One possible hint to this answer lies in another set of questions about the model. Why are the two rules 'select' and 'avoid'? Why isn't avoid better informed? Why are there the same number of rules as there are stimuli? My guess is that TI is actually revealing the deeper structure of its learning mechanism, a more general purpose one for task learning. I suspect that for any given problem, the agent attempts to determine a minimum set of important discriminative stimuli and also a minimum set of appropriate actions. When subjects are first taught this task, they are taught to 'select' for a reward. Then they are presented with two stimuli, and penalised for selecting incorrectly. This forces them to learn another behavior, 'avoid', a partially-inhibited form of select. 'Select, but not the thing you are attending to.'

Within this ontology, the animal then has two additional things to learn: in which contexts to apply which behavior, and what the prioritisation is between contexts when more than one applies.

The clearest aspect of the two-tier model is that it splits the problem of task-learning into at least two parts:

(1) learning to associate actions with environmental cues
(2) learning to prioritise these cues when more than one association holds.

Neuroscience indicates that long-term rule learning, including categorisation for action selection, seems to be dependent on both the prefrontal cortex (Freedman et al., 2001; Wallis et al., 2001) and the hippocampus. In particular, the entorhinal cortex seems necessary for forming episodic or event memory necessary for such tasks as paired association, while the hippocampal formation proper is necessary for learning more complex relations (Alvarado and Bachevalier, 2000; Baxter and Murray, 2001).

The hippocampus has long been implicated in memory consolidation. Its very sparse, redundant coding system seems ideally suited to recording memories in real time which can then be consolidated — saved to long term storage in an integrated representation — using a slower learning mechanism (McClelland et al., 1995; Louie and Wilson, 2001). Thus the following might be a reasonable neurobiological explanation of the model presented in this paper. The basic association between stimuli and behaviors would be performed by the prefrontal cortex generalising from associations learned by the entorhinal cortex. The hippocampal formation would learn prioritisations between these associations.

4. Conclusion

The two-tier model demonstrates a clearly modular decomposition of human-like intelligence. Associations and priorities are learned separately as different representations and in different parts of the brain. Further refinements on the model, such as explanations of TI-related timing effects, will show further modularity in complex process of learning new tasks.

References

Alvarado, M. C. and Bachevalier, J. (2000). Revisiting the maturation of medial temporal lobe memory functions in primates. *Learning and Memory*, 7(5):244–256.

Anderson, J. R. (1993). *Rules of the Mind.* Lawrence Erlbaum Associates, Hillsdale, NJ.

Baxter, M. G. and Murray, E. A. (2001). Opposite relationship of hippocampal and rhinal cortex damage to delayed nonmatching-to-sample deficits in monkeys. *Hippocampus*, 11(1):61–71.

Bryant, P. E. and Trabasso, T. (1971). Transitive inferences and memory in young children. *Nature*, 232:456–458.

Bryson, J. J. (2000). Cross-paradigm analysis of autonomous agent architecture. *Journal of Experimental and Theoretical Artificial Intelligence*, 12(2):165–190.

Bryson, J. J. (2001). *Intelligence by Design: Principles of Modularity and Coordination for Engineering Complex Adaptive Agents.* PhD thesis, MIT, Department of EECS, Cambridge, MA. AI Technical Report 2001-003.

Bryson, J. J. and Leong, J. C. S. (2004). Representations underlying transitive choice in humans and other primates. submitted.

Carruthers, P. (2003). The cognitive functions of language. *Brain and Behavioral Sciences*, 25(6).

Chalmers, M. and McGonigle, B. O. (1984). Are children any more logical than monkeys on the five term series problem? *Journal of Experimental Child Psychology*, 37:355–377.

Elman, J. L., Bates, E. A., Johnson, M. H., Karmiloff-Smith, A., Parisi, D., and Plunkett, K. (1996). *Rethinking Innateness. A Connectionist Perspective on Development.* MIT Press, Cambridge, MA.

Fincham, J. M., Carter, C. S., Veen, V., Stenger, V. A., and Anderson, J. R. (2002). Neural mechanisms of planning: A computational analysis using event-related fMRI. *Proceedings of the National Academy of Sciences, USA*, 99(5).

Fodor, J. A. (1983). *The Modularity of Mind.* Bradford Books. MIT Press, Cambridge, MA.

Freedman, D. J., Riesenhuber, M., Poggio, T., and Miller, E. K. (2001). Categorical representation of visual stimuli in the primate prefrontal cortex. *Science*, 291:312–316.

Gallistel, C., Brown, A. L., Carey, S., Gelman, R., and Keil, F. C. (1991). Lessons from animal learning for the study of cognitive development. In Carey, S. and Gelman, R., editors, *The Epigenesis of Mind*, pages 3–36. Lawrence Erlbaum, Hillsdale, NJ.

Harris, M. R. and McGonigle, B. O. (1994). A model of transitive choice. *The Quarterly Journal of Experimental Psychology*, 47B(3):319–348.

Karmiloff-Smith, A. (1992). *Beyond Modularity: A Developmental Perspective on Cognitive Change.* MIT Press, Cambridge, MA.

Kauffman, T., Theoret, H., and Pascual-Leone, A. (2002). Braille character discrimination in blindfolded human subjects. *Neuroreport*, 13(5):571–574.

Louie, K. and Wilson, M. A. (2001). Temporally structured replay of awake hippocampal ensemble activity during rapid eye movement sleep. *Neuron*, 29(1):145–156.

McClelland, J. L., McNaughton, B. L., and O'Reilly, R. C. (1995). Why there are complementary learning systems in the hippocampus and neocortex: Insights from the successes and failures of connectionist models of learning and memory. *Psychological Review*, 102(3):419–457.

McGonigle, B. O. and Chalmers, M. (1977). Are monkeys logical? *Nature*, 267:694–696.

Minsky, M. (1985). *The Society of Mind.* Simon and Schuster Inc., New York, NY.

Newell, A. (1990). *Unified Theories of Cognition.* Harvard University Press, Cambridge, Massachusetts.

Piaget, J. (1954). *The Construction of Reality in the Child.* Basic Books, New York.

Roper, T. J. (1983). Learning as a biological phenomena. In Halliday, T. R. and Slater, P. J. B., editors, *Genes, Development and Learning*, volume 3 of *Animal Behaviour*, chapter 6, pages 178–212. Blackwell Scientific Publications, Oxford.

von der Malsburg, C. (1995). Binding in models of perception and brain function. *Current Opinion in Neurobiology*, 5:520–526.

Wallis, J. D., Anderson, K. C., and Miller, E. K. (2001). Single neurons in the prefrontal cortex encode abstract rules. *Nature*, 411:953–956.

ACCOUNTING FOR EPISODIC, SEMANTIC AND PROCEDURAL MEMORY IN THE RECOMMENDATION ARCHITECTURE COGNITIVE MODEL

L. ANDREW COWARD

Department of Computer Science, Australian National University, ACT 0200, Australia
andrew.coward@anu.edu.au

A qualitative explanation for experimental observations of episodic, semantic and procedural memory and the differences between them is provided based on the different ways in which information derived from the senses is recorded and accessed in the recommendation architecture cognitive model.

1. Introduction

The recommendation architecture (RA) is an information architecture which attempts to bridge the gap between psychological descriptions of cognitive phenomena and physiology (Coward, 1990). It represents a radically different approach from artificial neural networks, dynamical systems or computational modeling of symbolic systems. The different approach is needed because of the requirement to recognize and address the severe architectural constraints experienced by any system which learns large numbers of features with limited information handling resources. Information handling resources include resources needed for information recording, information manipulation (or "processing") and internal connectivity. Theoretical arguments demonstrate that as the ratio of features to resources increases, the system will be constrained more and more tightly into the RA form (Coward, 2004a). If a cognitive model fails to take these constraints into account, it may be able to model one or a few features, or many features if there are no limits to available resources, but will break down when requirements similar to those on real biological brains are imposed.

There are strong resemblances between the physical forms of the RA and mammal brain physiology (Coward, 1990; 2001). Electronic implementations of the RA demonstrate bootstrapping of behaviour analogous with cognition with minimal a priori guidance (Coward, 2001).

This paper demonstrates that the phenomena and observed dissociations between semantic, episodic and procedural memory find a straightforward explanation in terms of the available, electronically implemented information recording and access mechanisms in the RA cognitive model.

2. The Recommendation Architecture (RA)

In the RA there is a primary architectural separation between a modular hierarchy called clustering and a component hierarchy called competition. Clustering defines conditions within the information available to the brain and detects the occurrence of any defined condition. Modules in clustering detect sets of similar conditions, where the similarity between conditions within a higher module is somewhat less than within the more detailed modules that make up the higher level module. A subset of the conditions detected at any point in time is communicated to competition. Different components within competition correspond with general behaviour types, more specific types, and individual behaviours and sequences of behaviours. Any one condition detection by clustering is communicated to a wide range of components, where it is interpreted as a recommendation in favour of the corresponding behaviours, each recommendation having a different weight. Competition adds the weights of each recommended behaviour across all currently detected conditions, and implements the most strongly recommended behaviours. Consequence feedback following a behaviour can change the recommendation weights of recently active conditions into recent behaviours, but cannot change the definitions of the conditions.

Devices in clustering learn and respond in radically different ways from conventional connectionist device algorithms. A clustering device permanently records a set of conditions. In other words, once a condition has been recorded, an exact repetition of that condition will always be detected. Systems based on this type of device algorithm have been implemented electronically and demonstrated cognitive like capabilities (Coward, 2001). To be recorded, a condition must actually occur within the information available to the device, be sufficiently similar to conditions already recorded on the device, and the device must also be receiving signals encouraging it to record conditions. These signals come from special purpose devices indicating the general level of condition detection in other modules. A device is activated by the recording of and by any subsequent repetition of a condition. The output of an activated device is a series of spikes. The average rate of spike production indicates the number of conditions which are currently present, and a frequency modulation of the spike rate (i.e. bunching of spikes towards the peaks in a modulation frequency) indicates the input population within which the condition was detected. This modulation is determined by the modulation of the inputs to the device. For example, if modulation is imposed only on the subset of the outputs from a visual input domain which correspond with an area within a closed boundary (i.e. a visual object), only conditions within the object will be detected by devices processing outputs from the whole domain, and recommended behaviours will be in response to this object. Conditions detected by the same physical set of devices within two different objects can be kept separate if different phases of frequency modulation are imposed on inputs from the two objects (Coward, 2004b).

Careful management of when and where conditions will be recorded is essential. Additions are driven by the requirement that an adequate range of behavioural recommendations be available in response to every input state, which is equivalent to a requirement that the number of conditions detected in response to every input state must be above a minimum level. Additional conditions are recorded until this level is reached. Recording of conditions takes place in the modules containing the most similar conditions. New modules are created if conditions dissimilar to those in any existing module conditions are required. The location of condition recording is determined by the relative activity in a range of alternative locations (Coward 2001).

Devices in clustering are arranged in layers in which the first layer receives sensory inputs, and condition defining inputs to each subsequent layer come from just the preceding layer. All conditions detected within one layer are within the same range of complexity, where the complexity of a condition is the number of sensory inputs (including duplicates) that contribute to the condition either directly or via intermediate conditions. Parallel arrays of columns extend through several layers. Columns manage when and where additional conditions will be recorded, and in addition they detect and communicate the detection of conditions within a specific range of complexity to competition. These conditions are called the portfolio of the column, and arrays add conditions until some portfolios are detected within every input state.

These device algorithms and change management processes have all been implemented in electronic form (Coward, 2001; 2004b).

3. Definition of Conditions and Portfolios

A vast range of inputs containing information about the external environment and the internal state of both the brain and the body are available to the brain from the senses etc. A somewhat oversimplified way of understanding the definition of conditions is that one condition corresponds with each input in a specific set of inputs being present to an individually specified degree. Because conditions cannot be specified a priori, there is a random element to the definition of conditions, and because conditions are not changed after being recorded, any one condition or portfolio is perceptually, cognitively and behaviourally ambiguous. Unambiguous meanings are only achieved in competition, using populations of portfolios. However, conditions with complexities of the same order of magnitude as visual object perceptions will tend to be less ambiguous with respect to categorization of visual objects than conditions on other levels of complexity. Although no conditions on any level correlate unambiguously with such categories, conditions on such levels of complexity will provide better discrimination between such categories.

This simple view of conditions is made much more complex because a column can be activated not only by the presence of its conditions within sensory inputs, but also indirectly by two types of mechanism. One mechanism is activation if a number of other columns are already active which were often active in the past at the same time as the column. The other mechanism is activation if a number of other columns are already active which recorded conditions at the same time in the past as the column. An available variation is if the past activity or condition recording of one portfolio was shortly before or after that in the other. Such indirect activations are behaviours which must be recommended by already active columns to competition and accepted.

When there is simultaneous activity or condition recording in two columns, there is a strong recommendation weight created initially in favour of the indirect activation, but such weights decline fairly rapidly with time. However, if an indirect activation occurs shortly before a behaviour which is followed by positive consequences, the decline is reduced. Frequent such occurrences stabilize or increase the weight.

These indirect activation mechanisms can be viewed as supplementing the conditions present in current sensory inputs with other conditions which have a significant probability of being relevant to determining the most appropriate current behaviour.

A newly recorded condition is made up of a set of currently active component conditions. Some of these component conditions may be combinations of currently present sensory inputs, and others could have been activated by one of the two indirect mechanisms. Both the definition of conditions in terms of sensory inputs and the relationship between sensory inputs and the resultant pattern of condition activation can therefore be very complex.

4. The RA cognitive model

In the RA, information can be therefore accessed by four qualitatively different mechanisms. Firstly, the actual presence of a condition within current sensory inputs activates portfolios containing the condition. Secondly, an activated portfolio can recommend activation of other portfolios which were often been active at the same time in the past. Thirdly, an activated portfolio can recommend activation of other portfolios which recorded conditions at the same time in the past. A variant of these mechanisms is activation on the basis of activity by a portfolio somewhat before or somewhat after activity by the other portfolio. Indirectly activated portfolios can in turn recommend activation of yet other portfolios. The fourth mechanism is comparison of recommendation weights. The weights of all active behaviours into each recommended behaviour are totaled, and the behaviours with the strongest weights are implemented.

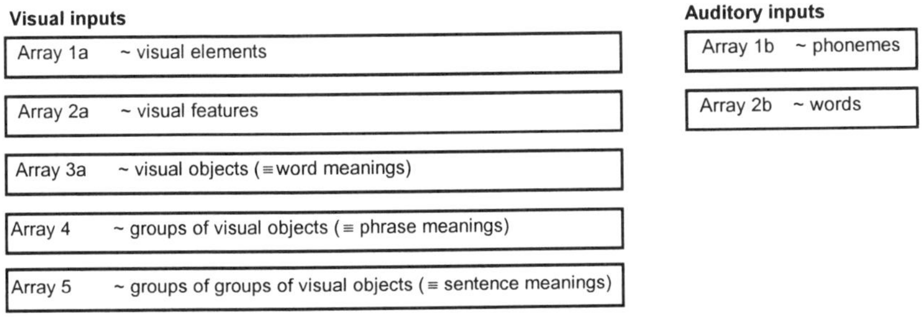

Figure 1. Architecture to support cognitive processes. Arrays detect conditions on different levels of complexity, with condition defining information passing sequentially from top to bottom. The outputs of an array indicate the detection of conditions within roughly the same range of complexity as the indicated cognitive category (features, objects, groups of objects etc.) but conditions do not correlate unambiguously with such categories. Conditions are detected independently in the visual and auditory domains.

The simplest arrangement of clustering arrays to support cognitive behaviour is illustrated in figure 1, and examples of competition subsystems associated with one array are illustrated in figure 2. In figure 1, different array outputs indicate the detection of portfolios with a complexity able to discriminate between cognitive features, objects, groups of objects etc. but individual portfolios do not correlate unambiguously with such objects etc. Portfolios able to discriminate between different groups of objects record conditions containing information derived from each member of groups.

Behaviour types which could be recommended by a portfolio through connectivity to the corresponding component include: prolong the activity of some currently active portfolios; activate portfolios active at the same time in the past as the currently active portfolio; activate portfolios containing conditions recorded at the same time in the past as some conditions in the currently active portfolio; activate portfolios containing conditions recorded just before or just after some conditions in the currently active portfolio; synchronize the activity (i.e. phase of modulation frequency) of several different groups of currently active portfolios; release outputs from one array to the next array; perform a general sequence of attention behaviours; perform a specific sequences of attention behaviours; perform an individual attention behaviour; speak a word; and say a phrase.

Competition receives outputs from portfolios currently being detected by clustering. If a portfolio has been present in the past at the same time as the performance of a number of different behaviours, it will have acquired recommendation weights in favour of or against those behaviours in the component corresponding with the behavior, depending on the consequence feedback from those behaviours. Competition adds the weights of all currently recommended behaviours and selects the behaviour with the largest weight. New portfolios are given an initial weight similar to the weights of the most similar previously existing portfolios, or genetically defined initial weights. If portfolios are new and no significant recommendation strength has been assigned in these ways, target behaviours are selected randomly. Such random selection can be limited to behaviours within a behaviour type which has already been selected. Alternatively, a behaviour can be selected by imitation of an externally observed behaviour.

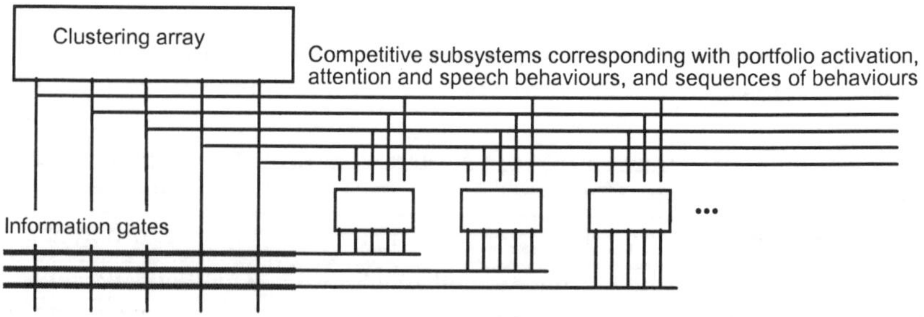

Figure 2. Competitive components receiving outputs from one array. Different behavioural interpretations are placed upon the same clustering outputs by different components. Competitive inhibition between and within competitive components limits selected behaviours to a small, consistent set. The behaviour accepted by a competitive subsystem is implemented by release of the outputs from clustering which correspond with the behaviour to a clustering level or to motor behaviour. This release behaviour is indicated by the information gates.

5. Different Types of Memory

The permanent recording of conditions means that the system has the capability to learn from single experiences. The model implies that there will be a level of condition recording in response to every experience, with higher levels for experiences with higher degrees of novelty. The level of condition recording therefore correlates with the degree of novelty of the experience, accounting for the high human capability to detect the novelty or otherwise of an experience (Standing, Conexio & Haber, 1970).

5.1. *Episodic Memory*

Episodic memory is memory of the past with a context of what else happened at the time, in contrast with semantic memory in which memories of facts are detached from memories of where those facts were learned (Tulving, 2002). Various types of recall experiments measure episodic memory.

In targeted recall, subjects are asked to recall particular past events (Neisser & Libby, 2000). In cued recall, subjects are given a cue which may be a word (Robinson, 1976) or a type such as "vivid memories" (Rubin & Kozin, 1984). In involuntary recall, some specific environmental stimulus such as a smell or a taste brings a memory to mind unsought (Berntsen, 1996).

The starting point for targeted recall in the RA model is hearing words which describe an event. Portfolios are activated which contain conditions within the sounds of the words. Secondary populations of portfolios containing conditions which occurred within visual and other sensory inputs are activated on the basis that the secondary portfolios have often been active in the past at the same time as the primary "auditory" portfolios. A significant proportion of the portfolios in these secondary populations were also active during the target event, and a somewhat smaller proportion recorded conditions during that event. Because of the words used, the proportions are larger for the target event than for any other event.

All active portfolios have recommendation strengths in favour of activating other portfolios which recorded conditions at similar times in the past. Active populations at higher levels derived from the presence of words like "recall" have recommendation strength in favour of accepting these types of recommendations. Because the target event has the highest proportion of activated portfolios in the secondary population, acceptance of such recommendations will tend to result in an active tertiary population with an even higher proportion of portfolios which recorded conditions during the target event. This process is self reinforcing, especially if a large number of conditions were recorded during the target event. The resultant population will be experienced as a re-perception of the original event, although the portfolios closest to sensory input are not reactivated, and there is therefore no visual hallucination of the event. The activated portfolios in this population have recommendation strength in favour of, for example, generating verbal descriptions of the event. Use of recommendation strength in favour of activating portfolios which recorded conditions somewhat before or somewhat after condition recording in currently active portfolios allows the re-perception to be set at the beginning of the event and to be moved through the event.

Recommendation strengths will always also be present in favour of activating portfolios on the basis of simultaneous past activity and simultaneous past recording during other events. The activated population is therefore unlikely to be an exact match for the original, although in general the higher the level of condition recording during the event, the greater the probability of a close match.

Cued recall operates in a very similar fashion. However, the initial secondary population may contain portfolios which recorded conditions during a number of past events. In the absence of specific indication of one event in the verbal cue, the tertiary population will evolve towards the event which happened to be represented by the highest degree of condition recording in the initial population. Events which resulted in a high degree of condition recording across many portfolios will tend to be the end points of this process.

Involuntary recall is the result of strong condition recording in portfolios activated in response to a sensory stimulus (for example, a novel smell or taste) at the same time as strong condition recording in other portfolios in response to some event. A later repetition of the sensory stimulus activates the portfolios which originally responded to that stimulus. These portfolios in turn activate portfolios which recorded conditions at the same time in the past, resulting in a re-perception of the event.

5.2. *Semantic Memory*

A typical way of measuring semantic memory in the laboratory is sentence and category verification. Sentence verification experiments measure the time for subjects to respond with the correctness of sentences like "Is a robin a bird" or "Is a penguin a bird". Category verification experiments are essentially equivalent and measure such times for simple category-exemplar pairs like bird-robin or bird-tree. It is found that for different members of the same category paired with the correct category, responses are faster for more typical category exemplars. For example, the response to bird-robin is faster than to bird-chicken. However, responses to clearly incorrect category-exemplar pairs like "Is canary an animal" is also fast (Rips, Shoben & Smith, 1973).

The RA model for category verification experiments can be understood by considering the portfolio populations activated in response to the words indicating category and exemplar. The portfolios activated in response to hearing the name of the category are portfolios which have often been active in the past when the category name was also present. Hence they will be portfolios also present when exemplars of the category were present, since this is the way the category is learned. Hence the active population is the set of portfolios which have most often been present when different category exemplars have been present. There will therefore be an overlap between the population activated in response to the name of the category and that activated in response to the name of the exemplar. This overlap will be greater for more typical exemplars, and very small if the exemplar is not a member of the category. The degree of overlap is itself a condition which can be detected and recommends for or against identifying the exemplar as a category member. If the exemplar is typical, overlap is substantial and its detection is rapid. Atypical exemplars have more moderate overlap, and more time may be required to expand the portfolio populations to include portfolios active at the same time in the past but slightly less often to achieve an overlap adequate to generate the appropriate verbal response. However, objects which are not in any way members of the category will have negligible overlap which again is detected rapidly. The model is therefore in agreement with the observations of Rips, Shoben & Smith (1973).

In contrast with the spreading activation model of Collins & Loftus (1975), there are no units which correspond with concepts like categories or the features of categories. Portfolios are groups of conditions in which there has been a degree of randomness in the definition of individual conditions, but conditions within one portfolio have some similarity with each other and have tended to occur at similar times in the past. A portfolio may therefore have a probabilistic correlation with many different features and categories, with the probabilities expressed, for example, as recommendation weights into naming the features or categories.

5.3. *Procedural Memory*

Procedural memory is defined as the ability to acquire skills. Observations of amnesics indicate that such memory is at least partially dissociated from semantic memory, since amnesics can acquire such skills at apparently normal rates. Thus amnesics can acquire motor skills such as mirror tracing tasks (Corkin, 1968).

In the RA model for learning a skill, portfolios activated within clustering in environments where the skill is relevant must acquire weights in competition associated with skilled behaviours. These portfolios will generally include both new information elements resulting from novelty in the environments and information elements recorded in prior experiences. Although the new elements may be particularly useful for recommending the new behaviours, some skill learning would be possible using only previously recorded elements which happen to occur in the new environments. Thus skill acquisition could proceed in the absence of condition recording.

6. Conclusions

The RA provides a qualitative account for the phenomena of episodic, semantic and procedural memory and the differences between them. This capability results from the different mechanisms by which information derived from sensory inputs is recorded and accessed, with different combinations of mechanisms corresponding with the different memory systems.

References

Berntsen, D. (1996). Involuntary autobiographical memories. *Applied Cognitive Psychology* 10, 435 - 454.

Collins, A. & Loftus, E. (1975). A spreading-activation theory of semantic processing. *Psychological Review* 82(6), 407 – 428.Corkin, S. (1968). Acquisition of motor skill after bilateral medial temporal-lobe excision. *Neuropsychologia* 6, 225 - 264.

Coward, L. A. (1990). *Pattern Thinking*. New York: Praeger.

Coward, L. A. (2001). The RA: lessons from the design of large scale electronic systems for cognitive science. *Journal of Cognitive Systems Research*, 2, 2, 111-156.

Coward, L. A. (2004a). The RA Model for Human Cognition. Proceedings of the Conference on Brain Inspired Cognitive Systems, University of Stirling, Scotland.

Coward, L. A. (2004b). Simulation of a Proposed Binding Model. Proceedings of Workshop on Information Coding in Early Sensory Stages, BICS 2004, University of Stirling, Scotland.

Neisser, U., & Libby, L. K. (2000). Remembering life experiences. In E. Tulving and F. I. M. Craik (Eds.).*The Oxford Handbook of Memory*. Oxford: Oxford University Press.

Rips, L., Shoben, J. & Smith, E. (1973). Semantic Distance and Verification of Semantic Relations. *Journal of Verbal Learning and Verbal Behaviour* 12, 1 – 20.

Robinson, J. A. (1976). Sampling autobiographical memory. *Cognitive Psychology* 8, 578 - 595.

Rubin, D. C. & Kozin, M. (1984). Vivid memories. *Cognition* 16, 81 - 95.

Standing, L., Conexio, J., & Haber, R.N. (1970). Perception and memory for pictures: single-trial learning of 2500 visual stimuli. *Psychonomic Science* 19(2), 73 - 74.

CAN SELF-CONTROL BE EXPLAINED THROUGH GAMES?

GAYE BANFIELD[A] AND CHRIS CHRISTODOULOU

School of Computer Science and Information Systems, Birkbeck College, University of London, Malet Street, London WC1E 7HX, UK [A]gaye@dcs.bbk.ac.uk

It has been suggested that self-control can be represented as an internal conflict between the higher and lower centres of the brain. Based on this viewpoint we implemented a two-player model, simulating a top-level conscious processor for effortful reasoning, and an intuitive processor for heuristic problem solving, competing in a game theoretical situation. The results show that reducing the number of patterns of play promotes cooperation, which supports the theory that self-control can be explained in terms of learning to co-operate with oneself where behaviour variability is a factor.

1. Introduction

In the viewpoint of modern cognitive neuroscience, self-control as an internal process can be represented as network architecture of two interacting networks of neurons, to reflect the higher and lower centres of the brain (Rachlin, 2000). Based on this viewpoint we used game theory for its realization. More specifically our model framework is implemented as two players, simulating a top-level conscious processor for effortful reasoning and an intuitive processor for heuristic problem solving, competing in a game-theoretical situation. Two artificial neural networks, representing the two players play the general-sum game Iterated Prisoner's Dilemma (IPD). A general-sum game is one where the players' payoffs are neither totally positively nor totally negatively correlated. Reinforcement Learning is considerably more difficult to implement in such games (Banfield and Christodoulou, 2003). IPD has been used to model the evolution of human cooperation (Axelrod and Hamilton, 1981). The IPD is appropriate as it has been suggested that there is a relationship between self-control and social cooperation (Brown and Rachlin, 1999). Brown and Rachlin suggest that self-control can be explained in terms of the probability of reciprocation of defection or co-operation. In self-control the probability of reciprocation is with oneself. In the IPD game, defection yields a higher immediate payoff. If both players defect however, they would both do worse rather than if both players co-operate. In terms of self-control this can viewed as learning to co-operate with oneself in order to maximise the total accumulated payoff. In this study we make the theoretical premise, that the higher and the lower systems of the brain are largely independent and are competing in a form of prisoner's dilemma for optimal control. The higher and the lower systems are represented as artificial neural networks. In the first phase of the study, both networks implement reinforcement learning in multi-layer feed forward architecture but with different update rules. The Selective Bootstrap rule (Widrow et al., 1973) is used to simulate the lower brain system and the Temporal Difference update rule (Sutton, 1988) to simulate the higher brain system. In the second phase of the study we implemented the network representing the higher centre of the brain as a simple recurrent network (Elman, 1990), which is in line with current research on goal directed behaviour (Carver and Scheier, 1998). The results provide an explanatory framework for this complex behaviour.

The remainder of the paper is orgainsed as follows: Section 2 describes a model for self-control behaviour; Section 3 describes the development and testing of the model and Section 4 discusses the results and future research directions.

2. A model of Self-Control

S*elf-control* can be defined as choosing a large delayed reward over a small immediate reward (Rachlin, 1995). Increasing the delay of the later reward increases our discounting of the reward (Mischel et al., 1989). The greater the discounting of the larger-later reward, the sooner its current value sinks below the value of the smaller-sooner reward. This is defined as a reversal of preferences as in hyperbolic discounting (Green et al., 1994). To give an example of self-control, let the larger-later reward represent obtaining good grades. At the start of an academic year, most students value achieving good grades above that of going to the pub (smaller-sooner reward). When invited to the pub, some time later, the value of going to the pub exceeds their long term reward of getting good grades. If the student exercises self-control s/he will choose studying over the pub. Research to date suggests that we recognise that we have self-control problems and we try to solve them by precommitment behaviour (Rachlin, 2000; Ariely, 2002). *Precommitment* is defined as making a choice with the specific aim of denying oneself future choices (Rachlin, 1995). An example of precommitment is putting an alarm clock away from your bed, to force you to get up to turn it off. Precommitment removes the smaller-sooner reward (sleep) by restricting the choice at an earlier point in time to the larger-later reward. Precommitment can be viewed as an example of the internal conflicts that arise from the "higher" centre of the brain (rational thought) competing with the "low-level" centre (instinctive behaviour) (Nesse, 2001), as illustrated in Figure 1.

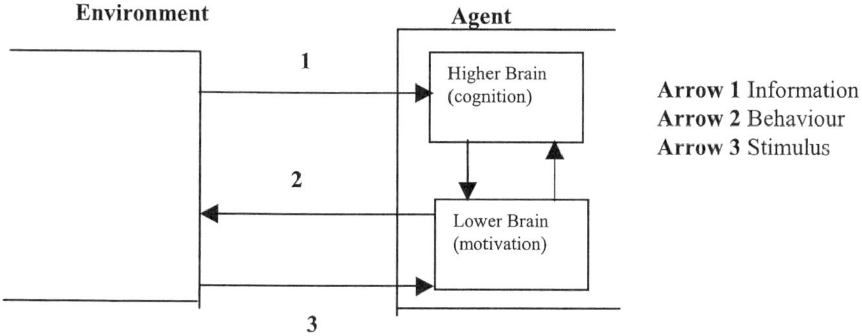

Figure 1. Self-control as an internal process in the viewpoint of modern cognitive neuroscience. Arrow 1 summarises information both past and current about the environment, Arrow 2 is the emergent behaviour of the Agent, and Arrow 3 is the reinforcement signal, which can either be a global reward or penalty as a response to Arrow 2 (based upon Rachlin, 2000).

Arrow 1 in Fig. 1 denotes information coming into the cognitive system located in the higher centre of the brain, which represents the frontal lobes associated with rational behaviour such as planning and control. This information combines with messages from the lower brain, representing the limbic system (including memory from the hippocampus) that is associated with emotion and action selection (O'Reilly and Munakata, 2000; Rachlin, 2000). This travels back down to the lower brain and finally results in behaviour (arrow 2 in Fig. 1), which is rewarded or punished by stimuli entering the lower brain (arrow 3 in Fig. 1). In this simple model, the modules can be represented as a network architecture of two interacting networks of neurons. This follows on from the ideas proposed above where precommitment behaviour is an example of some internal conflict between the lower and the higher centres of the brain. From this viewpoint a computational model of the neural cognitive system of self-control through precommitment behaviour has been developed. This simple model of Fig. 1 is implemented as two Artificial Neural Networks (ANNs) simulating two players competing against each other in general-sum games using reinforcement-learning. The *State* (corresponding to arrow 1 in Fig. 1) summarises information both past and current about the environment; the *Action* (corresponding to arrow 2 in Fig. 1) is the emergent behaviour of the combined networks and the *reinforcer* (corresponding to arrow 3 in Fig. 1) is a global reward or penalty signal as appropriate to the action.

Research on self-control suggests that there is a relationship between cooperation and self-control (Brown and Rachlin, 1999). As human cooperation has been modelled as a game of Iterated Prisoner's Dilemma (Axelrod and Hamilton, 1981), it seemed an appropriate game to use in our current study. In the final phase of this study we implement precommitment by adding signalling information to the two-player model.

3. Development and Testing of The Model

3.1. *Selective Bootstrap vs. TD Feed Forward Networks Playing Iterated Prisoner's Dilemma (IPD)*

The IPD consists of two players who compete with each other repeatedly. Each player can either co-operate or defect. Defection is the higher payoff for the individual player, however if both players defect then the resulting payoff for both is worse. The network is configured with an input to represent the opponent's previous action (to defect or co-operate) and an output representing a response. A game consists of one or more rounds. The goal is to maximise the total payoff. The payoff matrix is adapted from the definition of the Prisoner's Dilemma given by Maynard Smith (1982) summarised in Table 1. In this study Temptation has a numerical value of 2, Reward a value of 1, Punishment a value of zero and Sucker's payoff a value of (–1).

Table 1. Maynard Smith's Payoff Matrix for the Iterated Prisoner's Dilemma (Maynard Smith, 1982) where T: Temptation, R: reward, P: punishment, S: sucker's payoff. Rule 1 (T>R>P>S) defines the game. Rule 2 (2R>T+S) ensures that the payoff is greater to two players who co-operate, than a pair who alternately co-operate and defect.

	Co-operate	**Defect**
Co-operate	R	S
Defect	T	P

Rules:
1. T>R>P>S
2. 2R>T+S

The ANN simulating the lower brain functions was implemented with the Selective Bootstrap weight update rule (Widrow et al., 1973) and the ANN simulating the higher brain functions with the Temporal Difference weight update rule (Sutton, 1988). The Selective Bootstrap weight update rule has no mechanism for discounting future rewards; it simply learns the value of each action. For this reason it can be viewed as being concerned with immediate gratification and therefore is used in our model to represent the lower brain system. The Temporal Difference update rule maintains a history of previous rewards and includes a discount factor in determining the value of future rewards; this is consistent with planning and control of higher brain functions (Carver and Scheier, 1998) and therefore we use it to represent the higher brain system. In the Selective Bootstrap rule no explicit reinforcement signal is used; instead the actual output is used as the "desired output" and learning follows in a supervised format (Widrow et al., 1973). In the case of a success, this "desired output" is used to reward the network by Eq. (1):

$$\Delta w(t) = \eta\ [a(t) - s(t)]x(t) \tag{1}$$

where time (t) is represented in terms of number of completed rounds, a(t) is the actual output at time t used as "desired output", s(t) is the sum of the weighted inputs, w(t) is the weight at time t, η is the learning rate and x(t) is the input at time t. For penalisation of the network if the action was deemed a failure, the weights are updated as if the actual output "desired output" was not produced as shown by Eq. (2):

$$\Delta w(t) = \eta\ [1 - a(t) - s(t)]x(t) \tag{2}$$

where a(t) is the actual output (used as desired output at time t, s(t) the sum of the weighted inputs, w(t) the weight at time t, η is the learning rate and x(t) the input at time t.

Temporal Difference (TD) learning is used as a model of classical conditioning in psychology (Barto and Sutton, 2002). TD is implemented in our study with a lookup table, which is in effect the Value function V(St). The Value function is the probability of receiving the highest payoff, given the current state of the environment and the agent's action. Table 2 shows the initial values for the Value function for each state/action combination.

Table 2. A table of initial values for the learned value function, where D stands for Defect represented by a value >= 0.5 and C co-operate represented by a value < 0.5. The initial values are based on the probability of gaining the highest payoff derived from the payoff matrix by Maynard Smith (1982).

State D/C	Action D/C	Value Function $V(S_t)$
D	D	0
C	C	0.5
D	C	0
C	D	1

We use Eq. (3) to update the table based on the probability of winning from the previous state as taken from Barto and Sutton (2002):

$$V(S_p) = V(S_p) + \alpha[V(S_c) - V(S_p)] \tag{3}$$

where $V(S_p)$ is the value function of the state for the previous round (p) defined by the opponent's previous action to either defect or co-operate and the network matching action (either to defect or co-operate). Similarly $V(S_c)$ is the value function of the state for the current round (c). α is the stepsize parameter (or learning rate). The table is updated at the end of each turn of the game. The Value function and the reinforcement signal are used to calculate the temporal difference error given by Eq. (4):

$$\delta_t = r_{t+1} + \gamma V(S_{t+1}) - V(S_t) \quad (4)$$

where time t is represented by the number of rounds completed, r_{t+1} is the reinforcement signal at time t+1, γ is the discounting factor of future rewards, $V(S_{t+1})$ is the Value function at time t+1 and $V(S_t)$ the value function at time t. The reinforcement signal is adapted from the payoff matrix from Maynard Smith (1982) in Table 1. The temporal difference error is used to update the weights as in Eq. (5):

$$\Delta w(t) = \alpha \delta_t x(t) \quad (5)$$

where α is the stepsize parameter and x(t) is the input at time t to the neuron.

For evaluating the game, at each round we tracked the payoff for each network and at the end of the game the average payoff was calculated for both networks over the number of rounds played. The game was repeated 3 times and the average payoff for the three trials for each network was recorded. To avoid any first player advantage/disadvantage the starting network was selected at random. The networks are rewarded or penalised at the end of each round and at the end of the game. The TD Feed Forward network achieved a slightly higher accumulated payoff (3 points per game), however the average payoff did not increase to 1 (reward for co-operation in Table 1); in fact the average payoff became worse with the number of rounds played as shown in Fig. 2.

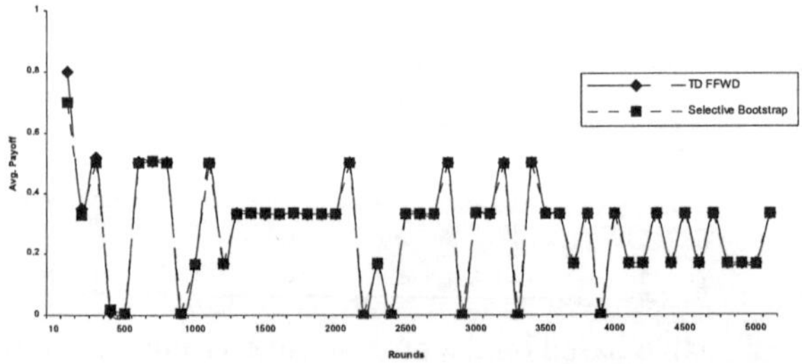

Figure 2. Average payoff for a Temporal Difference Feed Forward Network competing against a Selective Bootstrap Network in the Iterated Prisoner Dilemma's game played to 5000 rounds.

In addition, the networks did not follow any specific pattern of behaviour, i.e. both networks could choose to defect every round or one network could play a round of cooperation followed by a round of defection. This is indicated by the average payoff fluctuations observed in Fig. 2. The Selective Bootstrap had a tendency to defect.

3.2. *TD SRN vs. Selective Bootstrap Feed Forward Network Playing IPD*

For reducing the variability of behaviour and for improving the average payoff we included feedback by changing the network architecture of the higher brain to be the Simple Recurrent Network (SRN) of Elman (1990). In the implementation of the SRN, the neurons in the context layer do not interact with the outside world (environment); they interact exclusively with neurons internal to the network. The hidden units feed back to activate the context units. At time t+1 the context layer contains values that are exactly the same as the values of the hidden layer at time t. The neurons at the context layer have an initial output value of 0.5 with the activation function between 0.0 and 1.0; hence there is a 1:1 correspondence between the context and hidden layers. We played a version of IPD with TD lookup implemented as a SRN to simulate the higher brain, competing against Selective Bootstrap in a Feed Forward network to simulate the lower brain. In this case, the number of patterns of play were reduced although no specific patterns of behaviour emerged. The average payoff did not increase to 1 (representing co-operation in Table 1), however it did not decrease significantly as shown in Fig. 3 (as opposed to the case as shown in Fig. 2). TD SRN gained a slightly higher accumulated payoff (3 points per game).

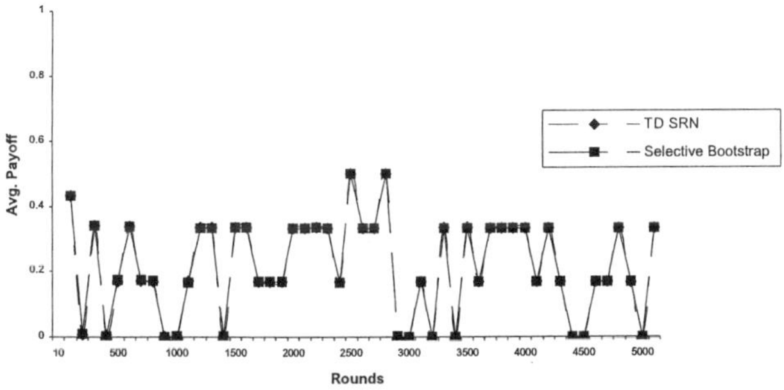

Figure 3. Average payoff for a Temporal Difference Simple Recurrent Network competing against a Selective Bootstrap Network in the Iterated Prisoner Dilemma's game played to 5000 rounds.

3.3. *TD SRN v TD Feed Forward Playing IPD*

For improving the performance of the model we played a version of IPD with TD lookup implemented as a SRN with different discount factors to simulate the higher brain, competing against TD lookup in a Feed Forward network with a lower discount factor, representing immediate gratification simulating the lower brain. In this case the average payoff was higher for both networks (as opposed to the cases shown in Fig. 2 and 3), moving closer to a value of 1 for co-operation (refer to Table 1) as shown in Fig. 4. With this arrangement (TD Lookup SRN versus TD Feed Forward), in games of symmetric play, i.e. when both networks co-operate, the accumulated payoff was the same for both networks. In games of asymmetric play, i.e. one network defects and the other co-operates, whichever defects, achieves a slightly higher payoff.

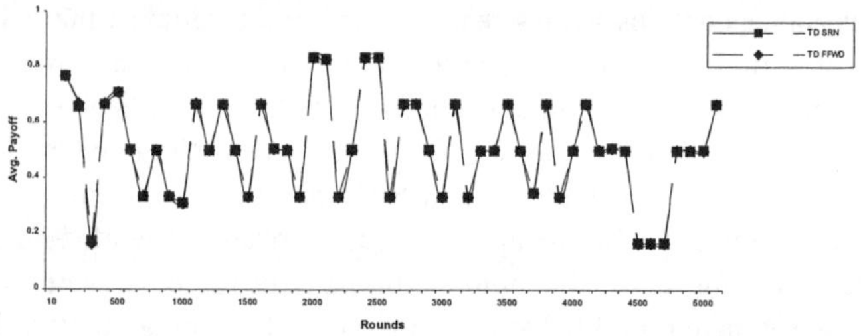

Figure 4. Average payoff for a Temporal Difference Simple Recurrent Network with a discount factor of 0.9 competing against a Temporal Difference Feed Forward Network with a discount factor of 0.1 in the Iterated Prisoner Dilemma's game played to 5000 rounds.

3.4. *Modelling Precommitment*

We implemented precommitment as a parameter with the following assumptions. Precommitment once announced must be irrevocable. This distinguishes personal precommitment and precommitment as viewed in game theory. In personal precommitment one has moral and other dilemmas to deal with. There maybe different levels of commitment. For example moving the alarm clock away from the bed biased our later choice of getting up when the alarm clock rings. A more extreme form of commitment however, would be to rig up some contraption which would physically prevent ourselves from going back to sleep. We have modelled these different levels of commitment as input values from zero (no precommitment, e.g. the alarm clock stays by the bed), to one (fully committed, e.g. there is no way we are going back to sleep). A version of IPD was played with the TD Feed Forward network competing against the TD SRN network (as in section 3.3). Precommitment was modelled as an additional input node whose value ranged from zero, indicating that the network has no commitment to any behavioural pattern, to one where the network is fully committed to a behavioural pattern. A behavioural pattern could be for example to defect then co-operate. The value was fixed for the duration of the game. The aim was to reduce behaviour variability from the network by increasing the level of commitment. A series of experiments were run to test the effect of precommitment on the networks' emergent behaviour when (i) both networks signal commitment and (ii) when just one network signals commitment. The results showed that when both networks signal commitment, play tends to be asymmetric, increasing commitment on both networks promotes co-operation. When only the TD SRN network commits, play is symmetric with a tendency to defect, however when only the TD Feed Forward network commits, play is symmetric with a tendency to co-operate.

4. Conclusion

The results from the two Feed Forward ANNs (TD representing the higher brain and Selective Bootstrap representing the lower brain) can be explained as follows: the Selective Bootstrap network has no history of rewards; it simply learns the value of each action. The Selective Bootstrap network can be viewed as myopic as it holds no value to future rewards. The TD network even in this simple implementation maintains a history of previous rewards and includes a discount factor in determining the value of future rewards. Learning with the Selective Bootstrap rule in the IPD had a tendency to defect, which is consistent with myopic behaviour. In an attempt to reduce the behaviour variability from the networks, we played a version of IPD with TD Lookup implemented in a SRN architecture while keeping the other network as a Feed Forward with Selective Bootstrap. The number of patterns of play did reduce, however the average payoff did not increase as expected. With TD SRN representing the higher brain and TD Feed Forward representing the lower brain the results showed that both networks tended to co-operate as indicated by the higher average payoff from both networks. The TD SRN however had a tendency to defect for longer. As the TD SRN was implemented with a lookup table learning to co-operate sooner was expected. A possible explanation is that TD SRN has also to learn which history of play to keep track of, therefore could require a longer learning period than that of the TD Feed Forward network. When we implemented a high level of precommitment on both networks we saw the behaviour variability reduce further with both networks tending to co-operate. Self-control has been explained in terms of learning to co-operate with oneself where behaviour variability is a factor (Rachlin, 2000). The results of the IPD with precommitment provide an explanatory framework for future work which will explore this theory further by evolving the networks playing IPD.

References

Ariely D. (2002). *Procrastination, Deadlines, and Performance: Self-Control by Precommitment*. Cambridge, MA: MIT Press.

Axelrod R. and Hamilton W.D. (1981). The Evolution of Cooperation, *Sci.* 211, 1390-1396.

Banfield G. and Christodoulou C. (2003) On Reinforcement Learning in two player "real-world" games. *Proceedings International Conference on Cognitive Science* (ICCS ASCS 2003), Sydney, Australia, 22.

Barto A. G. and Sutton R. S. (2002). *Reinforcement Learning: An Introduction*. Cambridge, MA: MIT Press.

Brown J. and Rachlin H. (1999). Self-control and Social Cooperation, *Behavioral Processes* 47, 65-72.

Carver C. S. and Scheier M. F. (1998). *On the Self Regulation of Behavior*. Cambridge: Cambridge University Press.

Elman, J. L. (1990). Finding Structure in Time. *Cognit. Sci.* 14, 179-211.

Green, L., Fry A. F. and Myerson J. (1994). Discounting of Delayed Rewards: A Life Span Comparison. *Psychological Science* 5, 33-36.

Maynard Smith, J. (1982). *Evolution and the Theory of Games*. Cambridge: Cambridge University Press.

Mischel, W., Shoda, Y. and Rodriguez, M. (1989). Delay of Gratification in Children. *Sci.* 244, 933-938.

Nesse R. M. (2001). Natural Selection and the Capacity for Subjective Commitment. In R. M. Nesse (Ed.), *Evolution and the Capacity for Commitment*, New York: Russell Sage (1-44).

O'Reilly R. and Munakata Y. (2000). *Computational Explorations in Cognitive Neuroscience*. Cambridge, MA: MIT Press.

Rachlin H. (1995). Self-Control: Beyond commitment. *Behavioural and Brain Sciences* 18, 109-159.

Rachlin H. (2000). *The Science of Self-Control*. MA: Harvard University Press.

Sutton, R. S. (1988). Learning to predict by the method of temporal differences, *Machine Learning* 3, 9-44.

Widrow B., Gupta N. K. and Maitra S. (1973). Punish/reward: Learning with a critic in Adaptive Threshold Systems. *IEEE Trans. on Sys., Man and Cyber.* 5, 455-465.

AN INVESTIGATION OF THE MYOPIA FOR FUTURE CONSEQUENCES THEORY OF VMF PATIENT BEHAVIOUR ON THE IOWA GAMBLING TASK: AN ABSTRACT NEURAL NETWORK SIMULATION

KIRAN KALIDINDI, HOWARD BOWMAN AND BRADLEY WYBLE

Centre for Cognitive Neuroscience and Cognitive Systems, Computing Laboratory, University of Kent, Canterbury, Kent CT2 7NF, England.
kk49@kent.ac.uk, H.Bowman@kent.ac.uk, B.Wyble@kent.ac.uk

'Somatic marker' theory proposes that body states act as a valence associated with potential choices based on prior outcomes; and thus aid decision-making. The main supporting evidence for this theory arose from clinical interviews of subjects with ventromedial prefrontal cortex (VMF) lesions and their performance on the Iowa 'Gambling Task' (IGT). VMF patient behaviour has been described as 'myopia' about future consequences. The aim of this paper is to investigate the implications of this description using an abstract simulation of the neural mechanisms that could underlie decision-making in this type of reinforcement learning task.

1. Introduction

Over the past decade there has been an increasing interest in the role of emotions in everyday decision-making. A theory particularly well represented in the literature is A. Damasio's 'somatic marker' theory (Damasio, 1994), which proposes that body states act as a valence which can be associated with potential choices based on prior outcomes; and thus aid decision-making. The main supporting evidence for this theory arose from clinical interviews of subjects with ventromedial prefrontal cortex (VMF) lesions and their performance on the Iowa 'Gambling Task' (IGT) (Bechara, 1999), compared to normal controls and those with lesions in other brain areas. The IGT consists of four decks that subjects can pick from; two decks, A and B, which yield high wins but higher losses (Disadvantageous) and the other two, C and D, that yield low wins with lower losses (Advantageous). Normal subjects start by picking from the disadvantageous decks but learn to pick from the advantageous decks, unlike the VMF patients who, as in their real social and personal lives, continue to pick disadvantageously.

A description that accounts for this behaviour in VMF patients' is 'myopia' for future consequences, in that they are driven by immediate reward and are less interested in uncertain future loss or gain (Bechara et al, 2000a). The main aim of this paper is to investigate the implications of this 'myopia' with an abstract neural network simulation that could characterize decision-making in this type of task. The current literature lacks a model that accurately reproduces these experimental results, and is abstracted from specific anatomical details, which underlie other models e.g. Wagar and Thagard (2004).

This model focuses on a time averaging parameter τ, which dictates the relative influence of current and past information in decision-making. A 'myopia' for future consequences could be caused by a similar 'myopia' for past events, where the experience from those events is not well integrated into current decision-making. Exploration of the state space for τ, from constant values to linear or logarithmic growth over the 100 trials in each example of the task, found that certain values of τ provided a good fit to the experimental data for normal controls and VMF subjects. This investigation informs us about how these different groups may utilise past information in present decision-making, especially in relation to situational novelty.

2. Design

In the design of the neural network, attempts were made to make the least number of assumptions and still produce an abstract network that can replicate the human data. The model makes the following basic assumptions:

1. That pleasure and pain are equally valued by Normal controls and VMF patients of the IGT.
2. An activation-based memory is sufficient to hold the required information about the task.

Based on the above assumptions, the network shown in Figure 1 was designed and implemented as a possible solution.

For each choice, there are two memory units which represent a measure of expected value, with one unit recording positive information (the positive unit) resulting from a choice and the other unit negative information (the negative unit). The results of a particular choice are passed on to the relevant Memory Layer units by having the unit associated with the current choice active in the Result Layer, while all other units in that layer are inactive. By passing the activation from the Result layer multiplicatively with activation in the Value Input Layer through the gating neurons, only the Memory Layer units connected with the current choice are reinforced on any particular trial, all other units only decay. Activation in the Value Input Layer (See Figure 1) is calculated using the current win and/or current loss divided by the starting balance of $2,000 or the maximum card seen so far. This gives a value between 0 and 1, which, once gated (as discussed above), becomes the input into the Memory Layer, y_α (See Equation 1) which is passed through a time-averaging equation. The time-averaging equation (Equation 1) represents how information decays over time in the memory layer and how present information is valued against past knowledge.

$$act(t) = \tau.act(t-1) + (1-\tau).y_{\alpha}(t) \tag{1}$$

where $y_{\alpha}(t)$ is the input activation from the Value Input Layer gated to the relevant Memory Layer unit at time t; and $act(t)$ is the units activation at time t.

Finally, on each trial, the activation from the negative memory units is subtracted from the activation in the corresponding positive memory units and then passed through a sigmoidal function onto the connected Response Layer units. Once all Response Layer units have received their inputs from the memory layer, and additional random noise has been added, a winner-takes-all system ensures that the unit with the highest activation becomes the choice for the next trial. The addition of random noise allows for increased exploration, preventing action choice being completely greedy. Note also that, after each trial units in the Value Input Layer and the Response layer are set to zero. Whereas the Memory Layer units (which hold a measure of past experience related to a particular choice) and those units in the Result Layer, holding the Response Layer result from the current trial, to be used as the choice in the next trial, retain activation.

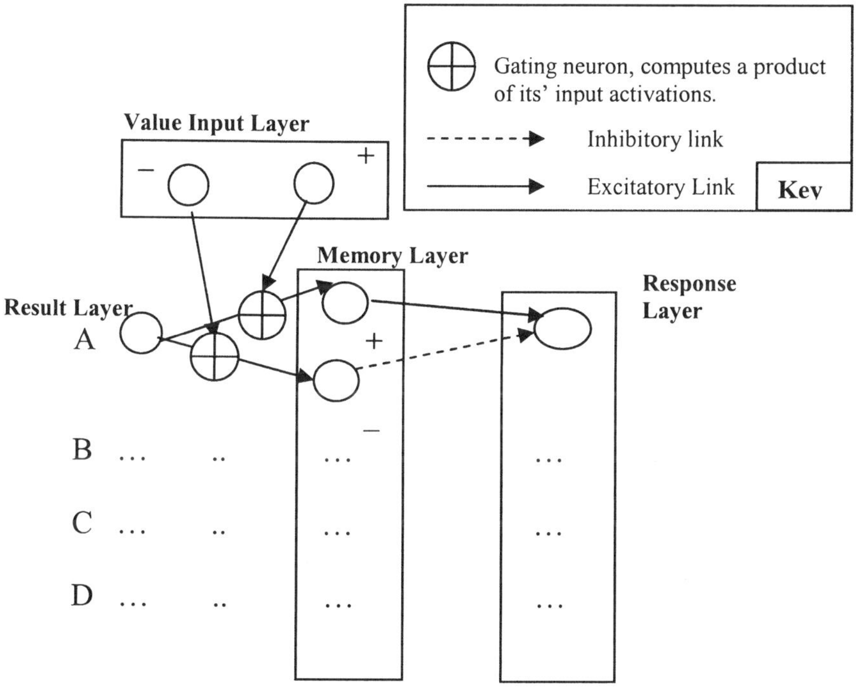

Figure 1. Diagram of Neural Network used for the Iowa Gambling Task. It shows the repetition of the basic architecture for each choice in the problem space. In the IGT there are four choices/decks.

3. Results

In the exploration of the parameter space for τ (See equation 1), it was found that the normal control deck choice profiles could only be reproduced with a rising value of τ over the course of the IGT. However, the closest to the deck choice profiles for the VMF patients was best replicated using a constant $\tau = 0.52$. (This gave an average variance in data points between the simulation and the real data of 2%). The important factor in reproducing the normal controls data was not a particular growth function of τ across trials, but that it began at $\tau = 0.52$ or below and rose to 0.9 by the 80th trial and then stayed constant at 0.9 for the remaining 20 trials. The 80th trial was chosen as the limiting point as this is when many subjects, both patients and controls, gained conceptual knowledge of the advantageous and disadvantageous decks and therefore a conscious strategy (Bechara et al, 2000b). The essential quality was an increasing τ over time/trials: with logarithmic growth from 0-0.9 in 80 trials; sigmoidal growth from 0.3-0.9 in 80 trials and linear growth from 0.52-0.9 in 80 trials, each giving an average variance of approximately 6% between the real data and that from the simulation. Interestingly, the starting point for linear growth of τ for normal subjects is the same as the constant value most accurate in reproducing the VMF patient data.

4. Conclusions

The required growth in τ suggests that normal controls build up a knowledge of the task in hand by valuing new information more early in the task than later. This has the consequence that previous learning, as activation, is built up and not lost as rapidly as with a constant τ where new information can never have more than the initial $1\text{-}\tau$ influence. Therefore, if τ is initially low and then increases over trials, new information has a greater impact early in the task. These simulation results are largely consistent with the myopia for future consequences theory of VMF patient behaviour. Although the model fleshes out the theory somewhat, by suggesting that what is missing in VMF patients is the capacity to adjust the trade-off between responding to new reward/punishment information and allowing past experience to affect the current choice. That is, in post-initial phases of the Iowa Gambling Task VMF patients are unable to switch to a strategy in which past experience is emphasized and preserved. By its nature such a strategy would focus more on future consequences, since learning that pays heed to past experience is more likely to enable decision-making that is more tuned to longer-term future consequences. This leads to the suggestion that the ventromedial prefrontal cortex is important in sustaining activation for goal directed behaviour by integrating relevant past experience. Perhaps VMF patients are unable to hold onto or access this information. It has been proposed that this retention of goal relevant information is a key role of the prefrontal cortex. "How does the PFC (Prefrontal Cortex) 'latch' onto goal-relevant information and maintain it without disruption? Several ideas have been suggested. These typically employ a form of gating signal that instructs the PFC network when to maintain a given activity state. This gating signal may come from dopaminergic (DA) neurons in the midbrain and basal ganglia." (p1370, Squire, 2003). This could be related to the growth in τ over trials.

Damasio (1994) has suggested that VMF patients' decision-making processes are no longer properly supported by 'somatic markers'. The authors of this paper only argue that VMF patients have a limited capacity in integrating relevant past information into their decision processes, without going as far as stating what form this information takes, e.g. affect or 'cold' cognitive. It is known that information is passed onto and analyzed in some way within consciousness as some VMF patients (50% by 80 cards) report 'conceptual' knowledge of the best decks, 70% of controls reach this state by 80 cards (Bechara et al, 2000b). But unlike controls VMF patients do not reflect this knowledge in their subsequent behaviour and still pick more or less randomly. It seems that conceptual conclusions for these patients are insufficient for behavioural control. This could be similar to when some addicts 'know of what not to do', but lack the behavioural control to prevent repeated use. Do theories of dual systems, provide a potential solution, where the non-conscious System 1 (Evans, 2003) in VMF patients is making decisions containing affect, with little or no influence from conscious conclusions (System 2)? The current simulation is considered to be at a level of abstraction above that describing a dual systems theory, but it could be modified to become the basis of a representation of 'somatic-makers' in the System 1 part of a dual-process model, with the relative influence of System 2 over System 1 controlling the growth of τ across trials.

References

Bechara A., Damasio A.R., Damasio H. and Lee G.P. (1999). Different contributions of the human amygdala and ventromedial prefrontal cortex to decision-making. *The Journal of Neuroscience*, 19(13):5473-5481.

Bechara A., Tranel D. and Damasio H. (2000a). Characterization of the decision-making deficit of patients with ventromedial prefrontal cortex lesions, *Brain* 123: 2189-2202.

Bechara A., Tranel D. and Damasio H. (2000b). Emotion, Decision Making and the Orbitofrontal Cortex, *Cerebral Cortex* 10: 295-307.

Damasio A.R. (1994). Descartes Error: Emotion, Reason and the Human Brain. New York: Grosset/Putnam.

Evans J.St.B.T. (2003). In two minds: dual-process accounts of reasoning, *Trends in Cognitive Sciences,* Vol 7, No 10: 454-459.

Squire L.R. (2003). *Fundamental Neuroscience.* Amsterdam; London : Academic.

Wagar B. and Thagard P. (2004). Spiking Phineas Gage: A Neurocomputational Theory of Cognitive-Affective Integration in Decision Making, *Psychological Review*, Vol 111, No 1: 67-79.

Face Processing and Classification

RECOGNITION OF PAIN EXPRESSIONS

P.J.B. HANCOCK[A] C.D. FROWD[A] E. BRODIE[B] C.A. NIVEN[C]

[A] *Department of Psychology, University of Stirling, Stirling, FK9 4LA, UK*
pjbh1@stir.ac.uk

[B] *Department of Psychology, Glasgow Caledonian University, George Moore Building, Cowcaddens Road, Glasgow G4 0BA, UK*

[C] *Department of Nursing and Midwifery, University of Stirling, Stirling, FK9 4LA, UK*

The automatic recognition of facial expressions of pain has potential medical significance: some patients are unable or unwilling to ask for analgesia, but it has been rather little studied. We report some preliminary work. The expressions were produced by actors. Human participants were able to distinguish the expressions intended to be pain from others with an average accuracy of around 80%, with disgust the most difficult to distinguish. A simple computer model consisting of a Gabor filtering front end and a linear discriminator network performed similarly well, suggesting that quite a simple model is able to emulate human performance.

1. Introduction

Much work has been done on classifying facial expressions by computer, but this has mostly been confined to Ekman's six basic emotions: anger, happiness, fear, disgust, sadness and surprise. Performance on these six expressions has been quite good (e.g. Dailey et al. (2000) report average performance of 86%). Pain, however, appears to have been largely neglected. It has potential medical significance: there are patients who either cannot or do not wish to report their pain (e.g stroke victims, post operative patients, premature babies or just not wanting to make a fuss, e.g. LeResche and Dworkin (1988)). There are some patients that medical staff are supposed to monitor, to check they are not in pain, but they may either forget to do it or fail to recognise it. In addition, clinicians regard patients' verbal reports of pain as unreliable and may therefore rely on non-verbal cues (Hill and Craig 2002). Unfortunately, medical staff tend to be worse than lay public at detecting pain expressions (Prkachin, 1997). Indeed, it has even been suggested that cancer patients be taught how to express pain in ways that medical practitioners will recognise (Keefe and Dunsmore, 1992). An alternative approach to helping with this problem would be to build a computer system capable of recognising when someone is in pain by looking at them. Other possibilities for automation exist, for example attempting to classify the cry made by babies, to detect when they are in pain (Petroni et al., 1995).

Ekman developed the Facial Action Coding Scheme, that identifies the individual facial muscle movements, termed Action Units (AUs), which combine to make up expressions (Ekman and Friesen, 1978). Each expression can therefore be categorized in terms of the AUs that are displayed. For pain, there are 11, including AU4, brow lowering, AU12, lip corner puller and AU 43, eyes closed (Williams, 2002). However, it is to be expected that the combination of AUs displayed will vary with the type of pain: a sudden, sharp pain differing from more chronic suffering. They will also differ in terms of dynamics, with sharp pains producing a rapid change in expression while chronic pain results in a more fixed, drawn appearance.

There appears to be little previous work on computer classification of pain expressions (e.g. Mori et al., 2000), though Pantic and Rothkrantz (2002) speculate about the feasibility of doing so. However, there have been a number of attempts to build general expression recognition systems, reviewed by Pantic and Rothkrantz (2000). One method of doing this is to attempt to identify the individual AUs: if this can be done, then the appropriate combinations can be used to identify, in principle, any expressions using Ekman's classification. Pantic and Rothkrantz (2004) use a very rule-based approach, first locating key fiducial points on a face and then classifying the individual AUs in terms of relationships between them. For example, if the position of the eyebrow "arcade" viewed from profile moves up or down by more than some threshold amount, AU1 is indicated. The individual AUs can then be combined to identify expressions. Bartlett et al (2004) use a more holistic approach. Following an automatic face finding and alignment stage, the faces are filtered by multi resolution and orientation Gabor filters. Individual features are selected by an Adaboost procedure, before individual support vector machines classify each AU. However, Dailey et al (2000) reported quite good performance on the six basic expressions by using the same Gabor filtering stage, followed by a simple network-based classifier. This simple approach is used in the pilot experiments reported here.

2. Human Experiments

2.1. *Stimuli*

Our stimulus set derived from 23 actors from a local drama school, who were asked to simulate expressions for the six basic emotions (anger, disgust, fear, happiness, sadness and surprise), plus pain. A neutral face was added to make eight classes. The actors were videoed, and ten different examples of pain and two of each other expression were captured to give still frames. These were 720x576 pixels, of which the face occupied approximately the central 300x400 pixels.

2.2. *Study 1*

The task here was two alternative forced choice (2AFC): participants had to decide which of two displayed faces depicted pain. One example of each of the non-pain expressions was selected from each of eight actors, four male and four female. Each expression was paired with a different example of pain from the same actor, so no images were repeated. This gave a total of 56 pairs, with pain appearing on the left and right equally often for all the other expressions. The participants were first given four practice trials, using images from an actor unused in the main experiment. They were asked to press the 'z' key if pain was on the left and 'm' if it was on the right. For the practice trials, participants received feedback on their answer; for the main experiment they received none.

Participants were 24 students from the university, mean age 24, who took part voluntarily.

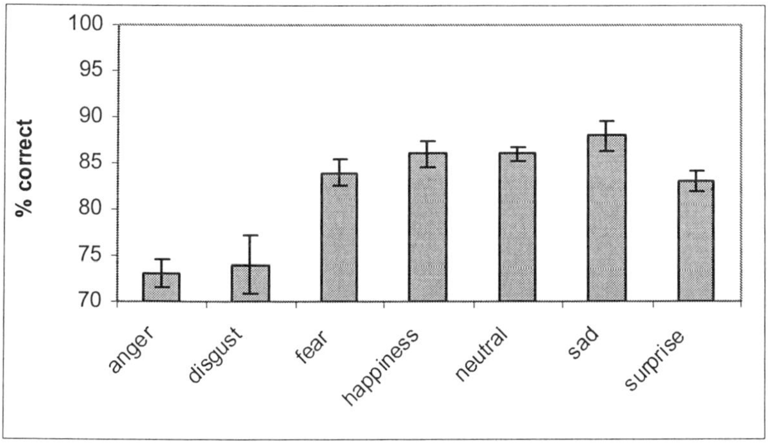

Figure 1 Accuracy with which pain is selected for each alternative expression on the 2AFC task in study 1. Chance is at 50%

2.3. *Results*

Overall average accuracy was 82%. A breakdown of accuracy for each expression is shown in Figure 1. Each bar shows the accuracy with which pain was differentiated from the particular expression, since it is a 2AFC task, chance is at 50%. Anger appears to be the hardest to distinguish, with the pain expression being chosen 73% of the time, implying that the Anger expression was incorrectly chosen as being pain on 27% of occasions. A repeated measures Anova confirmed an effect of expression ($F_{6,132}=13.4$, $p< 0.001$), i.e. there is a significant difference between the accuracy rates for the different expressions. Testing these differences with pairwise comparisons showed that anger and disgust produced the lowest accuracy rates, significantly lower than the others, but still significantly above chance. Anger and disgust are therefore most confused with pain, but the pain expression is still correctly chosen about three times out of four.

2.4. *Study 2*

The task for this study was 4AFC, with a pain expression placed among fear, disgust and sadness from the same actor. These three were chosen as being three similarly negative emotions. The faces were positioned in a 2x2 array on the screen, with a number, 1-4, next to each. Participants were asked to respond with the number of the face they considered to be displaying pain. An initial four practice trials were displayed, using images not used in the main experiment. Feedback was given for practice, but not for the main set of trials. In the main set, each actor featured twice, using different examples of each of the expressions. Actors were displayed in a randomised order.

Participants were 20 university students, mean age 26, who took part voluntarily.

2.5. *Results*

Overall accuracy, i.e. correctly choosing pain, was 79%. Figure 2 shows the distribution of choices. Since this is a 4AFC task, with each of the four expressions being presented on each trial, the bars represent the percentage of trials on which that expression was picked as being the one depicting pain. Most of the errors were caused by selecting the disgust expression, 13.5% of the time. Paired t-tests indicated that disgust was chosen significantly more often that the other two, which did not differ significantly from each other (e.g. disgust-fear $t_{19}=5.14$, $p< 0.001$).

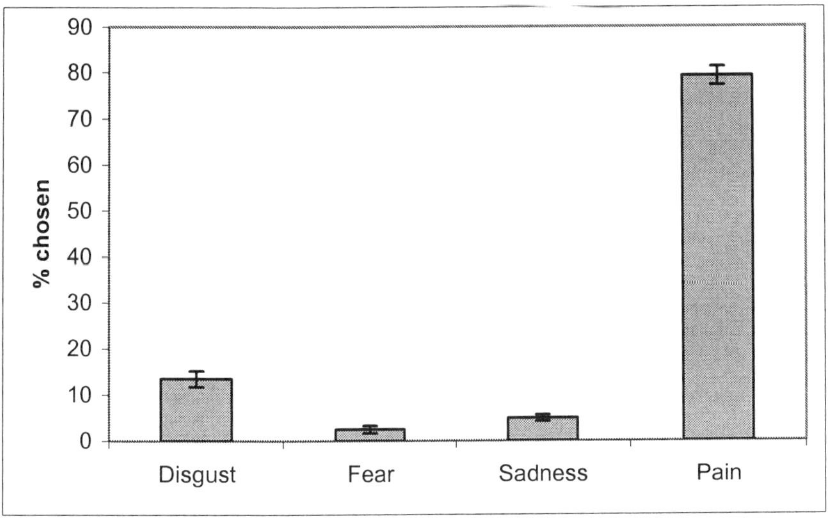

Figure 2 Results for the 4AFC task: percentage of trials on which each expression was chosen as being the one best depicting pain. Chance performance would be 25%

2.6. *Discussion*

The first aim of these studies was to provide some verification of the stimuli. Since they were performed by actors, it was possible that they would not be very clear. In the event, our participants recognized the pain expression about 80% of the time for both tasks. Since the second experiment always included a comparison with disgust, this suggests that they are quite recognisable as pain, even if perhaps not quite the same as real pain expressions. Pain was most commonly mistaken for disgust in a study using health professionals as participants (Kapesser & Williams, 2002). They reported an error rate of 18% for pain misclassified as disgust, though only 3% the other way round, disgust being more often mistaken for anger in their experiment.

3. Computer simulation

The simulation followed the design of Dailey et al. (2000), with images processed by multi-resolution Gabor filtering, followed by a linear network classification stage. We used 10 male and 10 female faces each with 2 examples of the 8 expressions for 320 images in total – 2 (gender) x 10 (individuals) x 8 (emotions) x 2 (examples). These were formed into a training and a test set, each with one of the examples from each face. The faces were rotated and scaled to put the centre of the eyes in a common location, then cropped to 180x240 pixels, in monochrome. Figure 3 shows some examples of the edited images.

Each image was convolved with Gabor filters, at up to five different spatial scales, each at seven orientations (0, 30, 60, 90, 120, 150 degrees). Initial tests used five scales, of sd from 1-5 pixels, this was varied to obtain the best results. The filtered images were sampled at eight pixel intervals to give a 21 x 29 array of intensity values for each scale and orientation. All the samples for each original image were concatenated to give a vector of size 21315 (7 x 5 x 21 x 29). These long vectors were then compressed by a principal components analysis, taking the first 160 components.

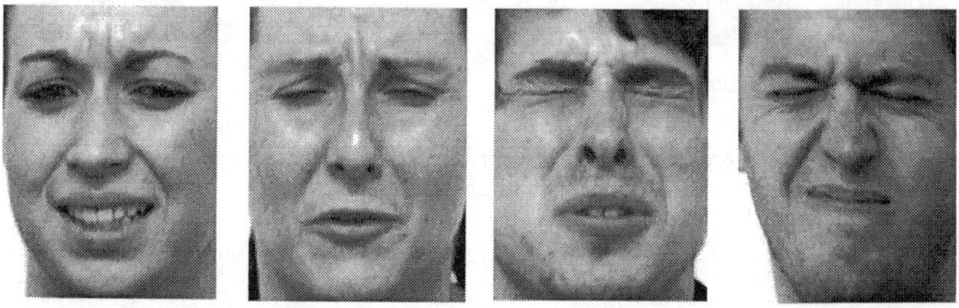

Figure 3 Four examples of pain expressions, as cropped for image processing.

The PCA reduced vectors were then classified using a linear neural net, with 160 inputs and 8 outputs. One set of images was used as a training set, the other for testing and the two sets then reversed for cross validation. Training used a learning rate of 0.0005 for 10000 epochs (although correct classification on the training set was generally obtained after 500). We experimented with the number of inputs used and found that training was unsuccessful with fewer, for example 90 inputs gave only 70% correct classification.

Testing indicated that the initial rather fine scale filtering used (a sd range from 1-5) gave rather poor test performance, but that this improved with larger spatial scales in the range of 10-15. Marginally the best overall performance was given by the use of three spatial scales 10, 11 and 12, at 71.6%, averaged across both test sets. However, performance varied rather little with the precise choice of spatial scales and the average of four different runs, using a variety of combinations of scale from 10-15, averaged 69.5% overall. However, pain is relatively poorly classified, averaging only 46.3% overall.

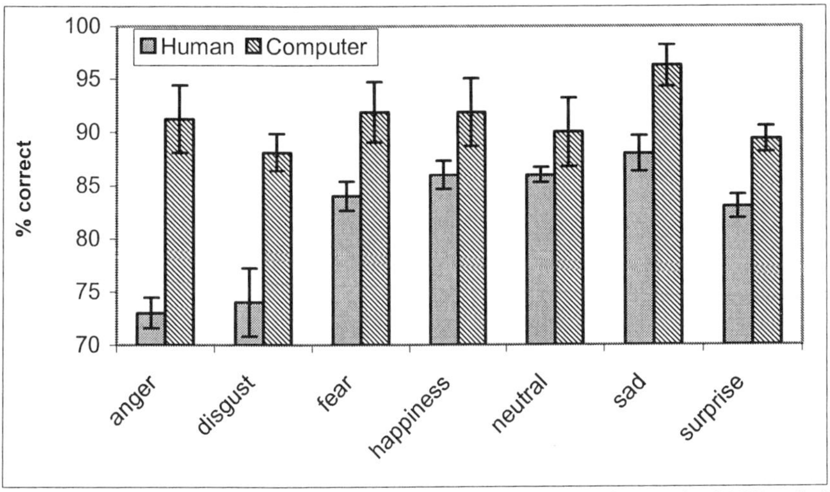

Figure 4 Accuracy of pain identification for each competing expression on the 2AFC task for humans in study 1 and for the computer simulation.

3.1. *Comparison between computer and human performance*

While the computer classification performance is not especially good, it is being asked to do a different task from the human participants. They were asked only to decide which of 2 or 4 images was pain, not to decide what expression a given image portrayed. We therefore asked the same of the computer simulation: for each identity in the set, compare the pain image with the others to see which gave the higher signal on the pain output unit. Note that while the images used in the computer and human studies are from the same set, for a variety of reasons they are not identical, e.g. in the 2AFC task, 8 actors were used for the human study to limit the length of the task, while 20 were used in simulation.

Average performance on the 2AFC task, across the four different spatial scale sets, is shown in Figure 4. The scores are much higher than for thc overall classification task: the pain expression only has to produces a higher signal on the pain output than the one alternative image, rather than a higher output on pain than for any other output. That is, the system may classify the pain image as anger, for example, but still more like pain than the alternative image, which would therefore count as a correct answer. Performance is also higher than humans on the same task, though it is interesting that both find sadness the easiest to discriminate from pain and disgust one of the hardest.

Average performance on the 4AFC task is shown in Figure 5. On this task the performance of the computer system is almost identical to that of the human participants. Both correctly pick the pain expression about 80% of the time, while making most errors for disgust.

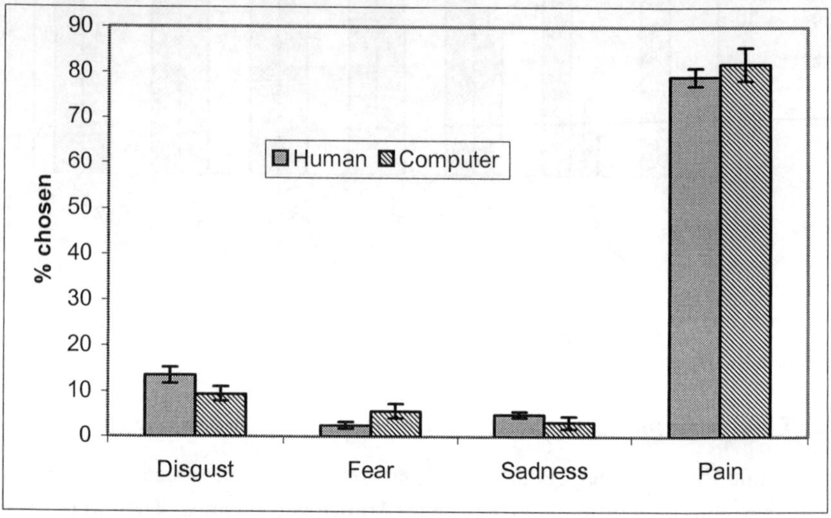

Figure 5 Results for the 4AFC task: the second experiment for human participants, and for the computer simulation.

4. Discussion

The work described here is very much pilot work, intended as a proof of concept, both to get an indication of the validity of the posed expression set and to see how easy it might be to classify pain. The computer model is very simple, though based on one that successfully classified the six basic expressions (Dailey et al., 2000). The classification method, though implemented as a neural net, is linear, and therefore equivalent to discriminant analysis. It is assumed that a more sophisticated classifier might perform better, though a larger training set might be required to constrain it adequately. The net was only asked to generalize over different example images from individual actors that were also in the training set: it was not asked to generalize across identities. This would obviously be needed in a useful system and is one focus for future development. This system also required entirely manual intervention to identify and normalize the position and size of each face. Accurately locating a face is a surprisingly difficult task to automate, even with quite consistent images such as these, though good progress is being made (e.g. Bartlett et al 2004).

The match between computer and human participants on the 4AFC task is striking but should not be over-interpreted. Although the early spatial filtering stage of the computer system is loosely modelled on human visual processing, the rest of it is very simple and the match may well be coincidental: certainly the 2AFC results are not so obviously similar. In any case, emulating human performance is not the target of this work: ideally we need to improve upon it. However, comparison with human performance provides a good benchmark for any computerized system, and more extensive similarities might provide some circumstantial evidence as to how humans perform the task, which would be of more general interest.

Acknowledgments

This work was funded by a grant from the University of Stirling Human Sciences Faculty research development fund.

References

Bartlett M., Littlewort G., Lainscsek C., Fasel I., Movellan J., (2004). Machine learning methods for fully automatic recognition of facial expressions and facial actions. In *IEEE Conference on Systems, Man & Cybernetics*.

Dailey M.N., Cottrell G.W., Adolphs R. (2000). A six-unit network is all you need to discover happiness. *Proceedings of the Twenty-Second Annual Conference of the Cognitive Science Society*.

Ekman P., Friesen W.V. (1978). *Facial Action Coding System* Consulting Psychologists Press.

Hill M.L., Craig D.M. (2002). Detecting deception in pain expressions: the structure of genuine and deceptive facial displays. *Pain*, 99, 135-144.

Kapesser J., Williams A.C.de C. (2002). Pain and negative emotions in the face: judgements by healthcare professionals. *Pain*, 99, 197-206.

Keefe F.J., Dunsmore J. (1992). Pain behavior: concepts and controversies. *American Pain Society Journal*, 1, 92-100.

LeResche L., Dworkin S.F. (1988). Facial expressions of pain and emotions in chronic TMD patients. *Pain*, 35, 71-78.

Mori T., Okazaki Y., Kawai T., Sato T. (2000). Objective estimation of pain based on facial expression, vital signs and body movements *Advanced Robotics* 14, 413-414.

Pantic M., Rothkrantz L.J.M. (2000). Automatic analysis of facial expressions: the state of the art. *IEEE Transactions on Pattern Analysis and Machine Intelligence,* 22, 1424-1445.

Pantic M., Rothkrantz L.J.M. (2002). Machine understanding of facial expressions of pain. *Behavioral and Brain Sciences*, 25, 469-470.

Petroni M., Malowany A.S., Johnston C.C., Stevens B.J. (1995). Classification of infant cry vocalizations using artificial neural networks (ANNs). *Proceedings of IEEE International Conference on Acoustics and Speech Signal Processing*, Vol 5, 3475-3478.

Prkachin K.M. (1997). Afterword: the consistency of facial expressions of pain. In Ekman P. and Rosenberg E. (Eds.), *What the face reveals*. Oxford: Oxford University Press.

Williams A.C.de C. (2002). Facial expression of pain: An evolutionary account. *Behavioral and Brain Sciences*, 25, 439-488.

GLOBAL AND FEATURE BASED GENDER CLASSIFICATION OF FACES: A COMPARISON OF HUMAN PERFORMANCE AND COMPUTATIONAL MODELS

SAMARASENA BUCHALA[A] TIM M.GALE[A,B] NEIL DAVEY[A] RAY J.FRANK[A]
KERRY FOLEY[B]

[A] *Department of Computer Science, University of Hertfordshire, College Lane, Hatfield, AL10 9AB, UK*
{S.Buchala, N.Davey, T.Gale, R.J.Frank}@herts.ac.uk

[B] *Department of Psychiatry, QEII Hospital, Welwyn Garden City, AL7 4HQ, UK*
kbmf@btopenworld.com

Most computational models for gender classification use global information (the full face image) giving equal weight to the whole face area irrespective of the importance of the internal features. Here, we use a global and feature based representation of face images that includes both global and featural information. We use dimensionality reduction techniques and a support vector machine classifier and show that this method performs better than either global or feature based representations alone.

1. Introduction

Most computational models of gender classification use whole face images, giving equal weight to all areas of the face, irrespective of the importance of internal facial features. In this paper we evaluate the importance of global and local information in a series of gender recognition experiments. Global processing of faces is assumed to encode coarse information like shape and configuration of internal features, while featural processing utilises more detailed representations of facial features (e.g. eyes, mouth etc). In psychological terms, the latter implies an attentional component whereby salient features are processed in more detail than the coarse image. In this study we present these two kinds of representation and use a Support Vector Machine (SVM) to classify gender. Since face image data has very high dimensionality, we also implement dimensionality reduction techniques before classification.

The remainder of the paper is organised as follows. Related work is discussed in the next Section. Section 3 discusses the methodology used for this study. Sections 4 and 5 present the computational and human experimental results. We conclude with some discussion of the results in Section 6.

2. Related Work

Issues in gender classification have stimulated a great deal of research by psychologists and computer scientists. While research in Psychology (Bruce et al., 1993, Burton et al., 1993, Abdi et al., 1995) has largely been within the context of human visual processing, and identifying key featural differences in males and females, Computer Science research (Golomb et al., 1991, Brunelli & Poggio, 1992, Moghaddam & Yang, 2000, Sun et al., 2002) has been geared more towards specific face identification. The computational models range from using pixel-based information to representations derived from geometric measurements. Studies also vary considerably in the size of training sets used and in the type of features present or absent (for example, some studies use hair information while others do not). Nevertheless, most models, and specifically those that are pixel-based, have used whole face images, where the salience of specific facial features is not captured. These can be termed as global models.

3. Methodology

3.1. *Face Representation*

Hair, especially for females, forms a major part of a facial image and has a dominating effect on classification. Abdi et al (1995) reported gender classification accuracy of 80% for hairless faces against 91.8% for the same faces with hair information included. However, classification rate on hairless faces was better than that on faces with hair information in our previous work (Buchala et al., 2004). The performance degradation for face images with hair in our experiments was due to the variability of hairstyles in the dataset. Despite these disparate results, hair can certainly be an important visual cue for gender identification. The first image in Figure 1 shows a pictorial view of the *difference* in means of female and male face images. The lighter the pixel luminance, the larger is the difference and the darker the luminance, the smaller is the difference between means. This pictorial view suggests that regions around the face outline, chin, mouth, and above the eyes carry discriminatory information. However, the region around the face outline, with much brighter luminance, carries greater discriminatory information. This region signifies the presence or absence of hair. The second and third images of Figure 1 are the pictorial views of the standard deviations within the female and male face images respectively. Again, the lighter the pixel luminance, the larger is the standard deviation. These images, however, indicate that the discriminatory information of the regions around neck and face outline is variable to a large extent in females and to a certain extent in males. From this simple analysis, it can be said that hair information is important. However, a psychologically plausible face-representation should overcome the problem of variable hairstyles.

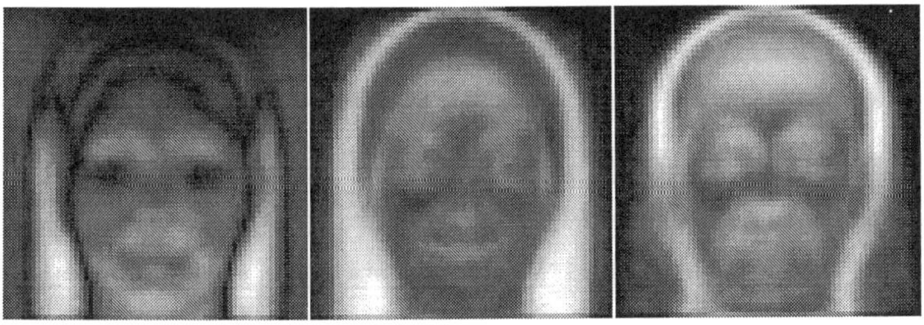

Figure 1. The first image is the pictorial representation of the difference of the means, of female and male face images. The second image is the standard deviation within the female face images. The third image is the standard deviation within the male face images.

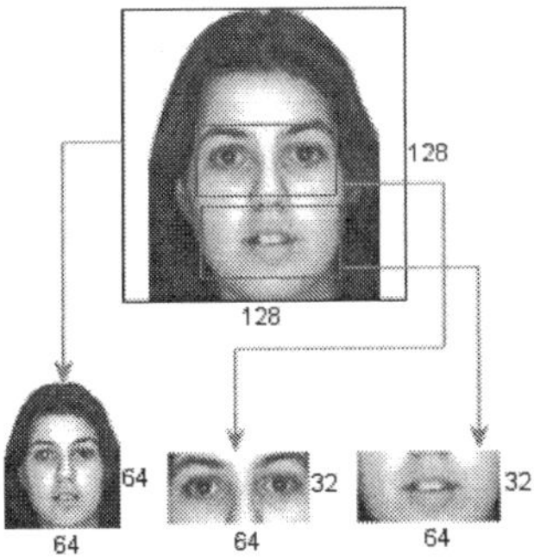

Figure 2. Three sub-images are obtained from the original 128 × 128 image. A 32 × 64 image pertaining to the eye region and a 32 × 64 image pertaining to the mouth region are extracted from the original image. The third sub-image is a 64 × 64 reduced resolution version of the original image.

In this study we use a global and feature based representation of face images which embodies both global and featural information. From a 128 × 128 face image, three sub-images are obtained as illustrated in Figure 2. A 32 × 64 pixel strip pertaining to the eyes region, taking the midpoint between the two eyes as a reference point, and a 32 × 64 pixel strip pertaining to the mouth region, taking midpoint of the mouth as a reference point, are extracted from each face image. These sub-images represent salient featural information. The third sub-image is a 64 × 64 reduced resolution version of the original image and this represents global information. In this study, the quantity of pixel information is identical for featural and global representations. A similar type of face representation was also used by Luckman et al (1995) for their computational model of familiar face recognition.

3.2. *Dimensionality Reduction*

Face image data has very high dimensionality and owing to the "curse of dimensionality" (Bellman, 1961), we apply dimensionality reduction techniques before using an SVM for classification.

Principal Component Analysis (PCA) (Jolliffe, 1986) is a popular dimensionality reduction technique that linearly transforms a D dimensional dataset X to a d dimensional dataset Y, without significant loss of information, where $d \leq D$.

Self Organising Map (SOM) (Kohonen, 2001) is a nonlinear method that learns a mapping from a D dimensional input space X to a d dimensional output space Y by using principles of Vector Quantization and Topological Mapping.

Curvilinear Component Analysis (CCA) (Demartines & Herault, 1997), a recent technique, has the ability to reduce the dimensionality of strongly-nonlinear data. The output is a free space that assumes the shape of the submanifold of the data. CCA minimizes the following error function:

$$E = \frac{1}{2}\sum_{i=1}^{N}\sum_{j=1}^{N}\left(d_{i,j}^{X} - d_{i,j}^{Y}\right)^2 F_\lambda\left(d_{i,j}^{Y}\right) \quad \forall\, j \neq i \tag{1}$$

Where $d_{i,j}^{X}$ and $d_{i,j}^{Y}$ are the Euclidean distances between points i and j in the input space X and output space Y respectively. $F_\lambda\left(d_{i,j}^{Y}\right)$ is the neighbourhood function. The idea of CCA is to match distances in the input and output spaces. However, preservation of larger distances may not be possible in the case of nonlinear data. In this case, it is important that at least local (smaller) distances should be preserved. For this reason CCA uses the neighbourhood function that ensures the condition of distance matching is satisfied for smaller distances while it is relaxed for larger distances.

3.3. *Support Vector Machine*

The classification is performed using an SVM. The SVM (Cortes & Vapnik, 1995) is a recently developed learning method, for pattern classification and regression. The basic idea of the SVM is to find the optimal hyperplane that has the maximal margin of separation between the classes, while having minimum classification errors.

Given a set of examples and their labels $\{(\mathbf{x}_1, \mathbf{y}_1), (\mathbf{x}_2, \mathbf{y}_2), \ldots (\mathbf{x}_N, \mathbf{y}_N)\}$ where $\mathbf{y}_i \in \{-1,1\}$, the optimal hyperplane is given as:

$$f(X) = \sum_{i=1}^{N} \alpha_i y_i k(X, X_i) + b \tag{2}$$

Constructing the optimal hyperplane is equivalent to finding α_i with nonzero values. The examples corresponding to the nonzero α_i are called support vectors. $k(X, X_i)$ is a kernel function, which implicitly maps the example data points into a high dimensional feature space, and takes inner product in that feature space. The potential benefit of a kernel function is that the data is more likely to be linearly separable in the high dimensional feature space, and also the actual mapping to the higher-dimensional space is never needed.

4. Computational Experiments

Experiments are carried out using 400 frontal face (200 female and 200 male) greyscale images. The faces are from the following databases: FERET (Phillips et al., 1998), AR (Martiniz & Benavente, 1998), and BioId (Jesorsky et al., 2001). All face images are aligned based on their eye-locations. Three sub-images, as explained in the previous Section, are extracted for each of the 400 faces. Histogram equalization is then applied on all three sub-images to reduce lighting effects. We use five-fold cross validation, with 320 faces (160 females and 160 males) for each training set and 80 faces (40 females and 40 males) for each test set, and report average classification rates using an SVM classifier, with RBF kernel. Before applying classification, dimensionality reduction techniques discussed in Section 3 are applied on the sub-image data. For PCA reduction we use the first few principal components, which account for 95% of the total variance of the data. Since CCA has the ability to reduce the dimensionality of strongly-nonlinear data, we use an Intrinsic Dimension[a] estimation technique, the Correlation Dimension (Grassberger & Proccacia, 1983), and reduce the data dimension to this Intrinsic Dimension. For SOM reduction, the subspace dimensionality is chosen as 64 (8×8 output grid) for the whole face and 36 (6×6 output grid) for eyes and mouth sub-images.

---

[a] Due to correlations, linear and nonlinear, a D dimensional data may actually lie in a d dimensional space. This true dimension d is called Intrinsic Dimension, where $d \leq D$. As PCA accounts only linear correlations, it is unable to reduce the data dimension to its intrinsic dimension when the correlations are nonlinear.

First we present classification results on the sub-images data. As shown in Table 1, all three sub-images produced high classification rates, indicating a surprisingly high amount of gender information in each of them. The figures in parentheses indicate the subspace dimensionality. Classification is performed on the composite data, obtained by applying dimensionality reduction on the sub-images individually and combining the resultant data. It can be seen from Table 2 that PCA performed marginally better than CCA and SOM. However, CCA uses far fewer variables (70) than PCA (759). For a comparison, we also report the classification rates of the data of the original 128 × 128 faces. It can be seen from Table 2 that the composite data, which includes both global and featural information, performed significantly better than the global model. Figure 3 shows that the composite data outperformed all other data representations.

Table 1. Average classification rates of the sub-images by an SVM. Figures in parentheses are the number of variables obtained after dimensionality reduction.

Feature	PCA	CCA	SOM
Eyes	85.5% (250)	82.75% (22)	80.25% (36)
Mouth	81.25% (253)	81.55% (22)	80.25% (36)
Full Face	87.5% (256)	87% (26)	83.25% (64)

Table 2. Classification rates of the composite data and original image data by an SVM. Figures in parentheses are the number of variables obtained after dimensionality reduction.

Feature	PCA	CCA	SOM
Composite	92.25% (759)	91.5% (70)	89.75% (136)
Original Full Face	86.5% (283)	85.5% (36)	83.25% (81)

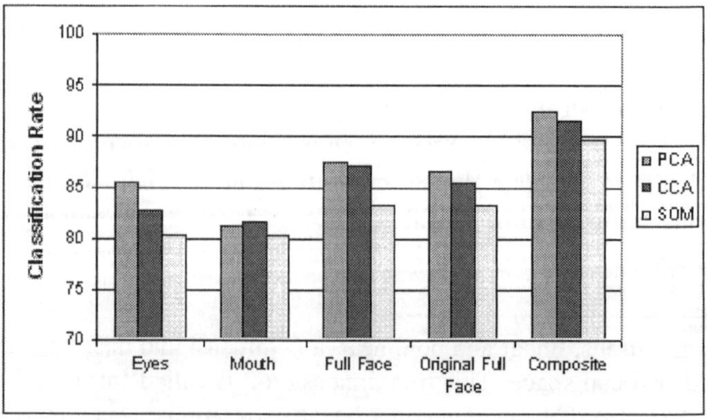

Figure 3. Average classification rates on different features.

5. Human Experiments

We recruited 80 participants (40 Male, 40 Female, mean age 40.1 years) to undertake a gender classification task using exactly the same sub-images that were used to test the dimensionality reduction techniques. Each participant viewed 80 eye and 80 mouth sub-images and were asked to record their best guess for the gender for each image. Each set (eyes and mouths) comprised 40 male and 40 female images, presented in random order. Data were collated by subjects and by items (the latter for the purpose of error analysis). We compared human and model performance on gender classification (N.B. in this context, the model performance was that for the sub images only, not the composite data).

5.1. *Eye Images*

Mean performance accuracy for eye classification was 77.25% (standard deviation = 5.42%). Chance performance on this task would be 50% so participants performed well above chance. There was no difference between male and female participants in terms of their accuracy. In the items analysis, gender recognition accuracy varied considerably across the 80 eye images (range 13–100% correct). Interestingly, there were very few sets of eyes that elicited chance levels of recognition performance. Rather, they tended to be correctly classified by the majority or incorrectly classified by the majority.

A major focus of interest with this work is whether the classification errors of human participants are associated with those of the computational models (PCA, CCA, and SOM) under generalization. We subdivided the 80 eye images into 2 groups based on whether each model had classified the gender correctly. We then investigated whether those items that were erroneously classified by the model were less accurately classified by the 80 human participants. This analysis is summarized in the table below.

Table 3. A comparison of classification accuracy rates by human participants for eye images classified incorrectly and correctly by the 3 computational models

Model	No. Incorrect Items	No. Correct Items	Mean human accuracy for items incorrectly classified by the model	Mean human accuracy for items correctly classified by the model
PCA	13	67	72.7 %	78.4%
CCA	13	67	71.9 %	78.6%
SOM	19	61	57.1 %	83.8%

Although the accuracy of humans was always higher for items that had been correctly classified by the models, this difference was statistically significant only for the SOM ($p < 0.005$). Since the data were not normally distributed, differences were analysed non-parametrically (with Mann-Whitney's U Test).

It is notable that the SOM made more classification errors than the two other models and this may be why it predicts the human data more correctly. The other two models made few errors overall and hence the sample size is small.

5.2. *Mouth Images*

Mean performance accuracy for gender classification of mouth images was 75.4% (standard deviation = 5.7%). The fact that, once again, participants scored well above chance level suggests that information useful for gender recognition can be derived from specific facial features, even when represented at a fairly low level of resolution.

The overall accuracy rate of the models and human participants is very similar. As with the eye data, we compared human performance on those mouth images that the model had classified incorrectly and correctly. These data are presented in table 4.

Table 4. A comparison of classification accuracy rates by human participants for mouth images classified incorrectly and correctly by the 3 computational models

Model	No. Incorrect Items	No. Correct Items	Mean human accuracy for items incorrectly classified by the model	Mean human accuracy for items correctly classified by the model
PCA	15	65	57 %	79.7 %
CCA	20	60	54.6 %	82.4 %
SOM	21	59	59.2 %	81.2 %

The differences were significant at $p < 0.001$ or less for all 3 methods, showing that those items which the models fail to categorise correctly are more likely to elicit gender recognition errors in humans.

6. Discussion and Conclusion

Hair, especially for females, forms a major part of the image and has a dominating effect on the classification. Many males with long hair and females with short hair were misclassified when the original full face images are used. The global and feature based model largely solved this problem, by reducing the effect of misleading hairstyles, while not removing important hair information. Figure 4 shows examples of individual faces that are misclassified when the original full face images are used and classified correctly by the global and feature based model.

Figure 4. Examples of the faces that are misclassified due to hair style of the individuals.

The global and feature based model for gender classification presented here performs significantly better than the global and featural models individually. This model allows inspection of facial data at various component levels and the results presented suggest that all components carry high levels of gender information. We believe that this type of representation also acts as a weighting factor of information, where highly variable discriminatory information (like hair) alone does not affect classification. Importantly, the global and feature based model captures an attentional component of human face recognition, whereby a human observer may use specific face feature cues to aid gender identification. Our experiments with human subjects showed that impressive levels of gender recognition accuracy were obtained from low resolution representations of single facial features (i.e. eyes and mouths). This underscores the importance of these specific features and supports the psychological plausibility of the global and feature based model discussed in this paper. Moreover, there was some association between the errors made by the models and those made by human observers. This, again, supports the psychological plausibility of these models although we will need to replicate this in some new sets of feature images that reflect a greater number of classification errors by the 3 models. We hope that this approach will also facilitate a useful comparison between the different dimensionality reduction techniques.

Finally, we note that the performance of CCA, a nonlinear technique, is comparable to PCA, with the added advantage that it uses far fewer variables than PCA.

References

Abdi, H., Valentin, D., Edelman, B., O'Toole, J. A. (1995). More about the difference between men and women: evidence from linear neural networks and the principal component approach. *Perception, 24*, 539-562.

Bellman, R. E. (1961). *Adaptive control processes: A guided tour*: Princeton University Press.

Bruce, V., Burton, A. M., Hanna, E., Healy, P., Mason, O., Coombes, A., Fright, R., Linney, A. (1993). Sex discrimination: how do we tell the difference between male and female faces? *Perception, 22*, 131-152.

Brunelli, R., Poggio, T. (1992). *HyperBF networks for gender classification.* Paper presented at the DARPA Image Understanding Workshop.

Buchala, S., Davey, N., Frank, R. J., Gale, T. M. (2004). *Dimensionality reduction of face images for gender classification* (Technical Report 408): Department of Computer Science, The University of Hertfordshire, UK.

Burton, A. M., Bruce, V., Dench, N. (1993). What's the difference between men and women? Evidence from facial measurement. *Perception, 22*, 153-176.

Cortes, C., Vapnik, V. (1995). Support-vector networks. *Machine Learning, 20*, 273-297.

Demartines, P., Herault, J. (1997). Curvilinear component analysis: A self-organizing neural network for nonlinear mapping of data sets. *IEEE Transactions on Neural Networks, 8*(1), 148-154.

Golomb, B. A., Lawrence, D. T., Sejnowski, T. J. (1991). Sexnet: A neural network identifies sex from human faces. *Advances in Neural Information Processing Systems, 3*, 572-577.

Grassberger, P., Proccacia, I. (1983). Measuring the strangeness of strange attractors. *Physica D, 9*, 189-208.

Jesorsky, O., Kirchberg, K., Frischholz, R. (2001). *Robust face detection using the hausdorff distance.* Paper presented at the International Conference on Audio- and Video-based Biometric Person Authentification, Halmstad, Sweden.

Jolliffe, I. T. (1986). *Principal component analysis*. New York: Springer-Verlag.

Kohonen, T. (2001). *Self organizing maps* (3rd ed.): Springer-Verlag.

Luckman, A., Allinson, N. M., Ellis, A. M., Flude, B. M. (1995). Familiar face recognition: A comparative study of a connectionist model and human performance. *Neurocomputing, 7*, 3-27.

Martiniz, A. M., Benavente, R. (1998). *The AR face database* (Technical Report 24): CVC.

Moghaddam, B., Yang, M.-H. (2000). *Gender classification with support vector machines* (Technical Report TR-2000-01): Mitsubishi Electric Research Laboratory.

Phillips, P. J., Wechsler, H., Huang, J., Rauss, P. (1998). The FERET database and evaluation procedure for face recognition algorithms. *Image and Vision Computing, 16*(5), 295-306.

Sun, Z., Yuan, X., Bebis, G., Louis, S. J. (2002). *Neural-Network-based gender classification using genetic search for eigen-feature selection.* Paper presented at the IEEE international joint conference on neural networks.

Developmental Processes

A NEURAL NETWORK INVESTIGATION OF THE HEAD PREFERENCE: PROBLEMS EXPLAINING EMPIRICAL RESULTS BY BOTTOM-UP PROCESSES ALONE

MARTIAL MERMILLOD[A], ROBERT M. FRENCH[B],
HUGUES SMEETS[B] & JANINE SPENCER[C]

[A]*Laboratory of Psychology and NeuroCognition, University of Grenoble, France*
[B]*Cognitive Science department, University of Liège, Belgium*
[C]*Centre for Research in Infant Behaviour, Brunel University, United Kingdom*
martial.mermillod@upmf-grenoble.fr

Abstract

Quinn & Eimas (1996a) and Spencer, Quinn, Johnson & Karmiloff-Smith (1997) have shown that 3- to 4-month-old infants mainly use head information to categorize basic-level exemplars like cats and dogs. The question raised in this paper is the following: could this preference for head information be explained by the statistics of the input related to the head or was an attentional component necessary? In other words, it might be possible that the statistical information related to the head of the animals is more useful for a distributed model of visual categorization. We used the distributed model of visual categorization proposed by Mareschal & French (1997) & Mareschal, French & Quinn (2000) to confirm i) that the infants may well use mainly head information to create their category representations and ii) that this model, based on purely bottom-up processing, does not seem to be able to explain the preference for head-related information reported in 3- to 4-month-old infants. Finally, we suggest that an attentional focus on the head information might be necessary to simulate available behavioral data.

1. Introduction

In the past decade, a distributed model of cognition was proposed to simulate infant visual categorization skills (Mareschal & French, 2000; Mareschal, Quinn, & French, 2002; French, Mareschal, Mermillod & Quinn, 2004). This connectionist model, a back-propagation autoencoder, was able to simulate bottom-up processes that were thought to occur in 3- to 4-month old infants. Specifically, the neural network reproduced the behaviour of the infants familiarized on a visual category and then exposed to new exemplars of the same versus other categories. Because infants focus more attention on an exemplar coming from a new category, the autoencoder model produced greater error on output when exposed to novel category exemplars. This simple model succeeded in reproducing a visual categorization asymmetry shown by Quinn, Eimas & Rosenkrantz (1993) and assumed to be based on the perceptual variance of the stimuli (French et al., 2004).

In other words, the autoencoder has proven to be a good model of visual categorization used to investigate bottom-up processes occurring in 3- to 4-

month olds (French et al., 2004). This model was used in the present paper to test another developmental hypothesis put forward by Johnson, Dziurawiec, Ellis & Morton (1991) and Morton & Johnson (1991). These authors have shown that, eseentially from birth onwards, infants direct more attention toward face attributes than to stimuli that have other configural properties. More generally, Quinn & Eimas (1996a) and Spencer, Quinn, Johnson & Karmiloff-Smith (1997) have demonstrated that 3- to 4-month-old infants mainly use head information to categorize basic-level categories, such as Cat and Dog. To explain these results, it could be that there is an innate system directing more attention to face information (CONSPEC) and associated with a learning system (CONLERN). The face-preference found in newborns was assumed to be based on structural information extracted by CONSPEC, whereas CONLERN was for learning about the visual characteristics of conspecifics (Morton et al., 1991). An alternate explanation could be that the face-preference effect might be generated in a purely bottom-up manner based on the statistics of the perceptual input. In other words, the perceptual information about the head might contain better correlations between the different features and thus preferentially catch the attention of the infants. We decided to test this latter hypothesis by means of the connectionist model proposed by Mareschal & French (1997, 2000) and French et al. (2004).

2. Empirical experiments

In the connectionist simulation, we are simulating the experimental design used by Spencer et al. (1997). This experiment was based on an earlier experiment proposed by Quinn & Eimas (1996a). In this study, the authors were investigating the type of perceptual cues used by 3- to 4-month-olds to categorize basic level categories, such as Cat and Dog. There were three experimental groups: Whole Animal, Face Only or Body Only. The Whole Animal group was familiarized with 12 cats and then given a preference test that paired a novel cat with a novel dog. The Face Only and Body Only groups were familiarized and tested with the same cats as the Whole Animal group, but with their bodies (Face Only group) or faces (Body Only group) occluded. Reliable preferences for the novel dog stimuli were observed in the Whole Animal and Face Only groups, but not the Body Only group. These findings suggest that young infants rely more on head/face information than on body information to categorically differentiate among animal species.

The experiments proposed by Spencer et al. (1997) aims at generalizing these findings by means of a better control of the input. The authors do not simply use a part of the animal, but, rather, a mixture of the head and body information. This is achieved by means of hybrid stimuli (Figure 1). As in the Quinn et al. (1993), Quinn & Eimas (1996a) experiments, infants were familiarized with 12 cats or dogs, but here, they were then tested on two sets of test trials that paired a novel cat with a novel dog (Whole Animal test) and novel hybrids of cats and dogs (Hybrid test).

Figure 1. Hybrid stimuli by Spencer et al. (1997).

The results obtained by Spencer et al. (1997) confirm, as suggested by Quinn et al. (1994), that 3- to 4-month old infants rely preferentially on head information. The theoretical explanations proposed by the authors were: "First, the head and face area may simply be the more diagnostic of category membership (i.e., within-category similarity and between-category dissimilarity may be greater for the cat and dog heads than it is for the cat and dog bodies), and the attention of both infants and adults is pulled to that region of the stimuli." This is the bottom-up hypothesis. However, the authors also formulated a top-down hypothesis: "Second, infants (as well as adults) may possess an innate mechanism that pushes their attention to the facial information in the stimuli (i.e., the CONSPEC mechanism hypothesized by Johnson and Morton, 1991…)." The autoencoder model proposed by Mareschal & French (1997, 2000) relies on bottom-up processes based on the perceptual distribution of these stimuli (French et al., 2004) and is, therefore, appropriate to test the bottom-up hypothesis.

3. Connectionist simulation

3.1. *Neural network architecture*

The simulation reported here was done on a standard feedforward backpropagation 10-8-10 autoencoder network. The learning rate was: 0.1, momentum: 0.9, and Fahlman offset: 0.1 and all weights were initialised between 2.0 and –2.0. In the original simulation, there were 3 features coding for the body and 7 features coding for the head. These characteristics of the input vector might create a bias for the purpose of the present study. In the present study, we have selected 7 features for the body (vertical extent, horizontal extent, neck width, lengths of front leg, left rear leg, right rear leg, and tail) and 7 features for the head (head width, head length, distance between eyes, ear width, ear length, nose width, and nose length). We used a 14-10-14 autoencoder network in all other respects identical to the 10-8-10 autoencoder used in Mareschal & French (1997, 2000) with a comparable input-hidden compression ratio.

Stimuli and training algorithm

Training was identical to that in Mareschal et al. (2000). Networks were trained in batches of 2 patterns for a maximum of 250 epochs. Results were averaged over 50 runs. This simulated familiarization with pairs of pictures for a fixed period before being presented with a new familiarization pair. The comparison between the empirical results and the connectionist simulations was facilitated by the fact that there were based on the same experimental design (i.e, the familiarization/preference for novelty paradigm). For this reason, we maintained the same training procedure used in Mareschal & French (1997, 2000), Mareschal et al. (2002) and French et al. (2004) for the work presented here.

3.2. *Results*

Figure 2 compares the results obtained by Spencer et al. (1997) with the results of our current connectionist simulation. Spencer et al. show that infants attend preferentially to a novel head in the hybrid stimuli compared to a novel body. For example, when infants are familiarized with exemplars from the Cat category, they were more interested in the hybrid stimulus having a novel dog head rather than the hybrid having a novel dog body. The effect is significant ($t(11)=3.17$, $p<0.01$). On the other hand, our autoencoder simulation did not produce the corresponding increase in output error. When trained on a particular category, there was no significant difference between the output error for novel heads compared to novel bodies on the hybrid stimuli.

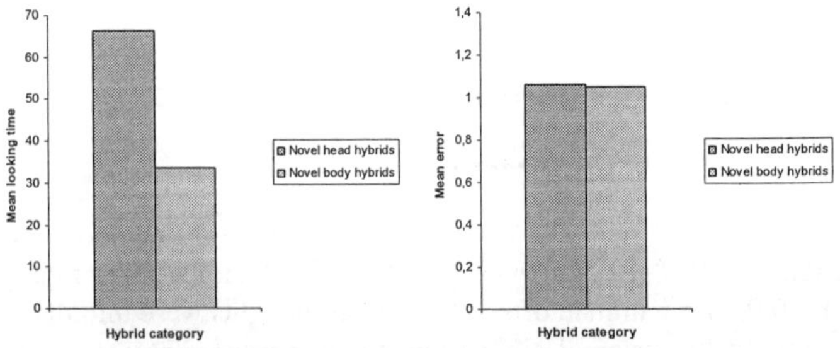

Figure 2. On the left, the empirical results obtained by Spencer et al. (1997). On the right, the results of the autoencoder simulation.

4. Conclusion

Despite the reliability of the neural network to correctly simulate previous empirical results, we were unable to reproduce the head preference effect by means of pure bottom-up processes using the autoencoder model. This might be due to feature-based input encoding not being able to account for the structural complexity of head information. However, the head preference effect was also obtained on very simple stimuli made of line drawings (Johnson et al., 1991), meaning that the effect requires only a simplified description of structural information.

Different aspects could be investigated in producing the face preference effect in a simple back-propagation network. *First*, we could test a quantitative effect at the level of the input features. For example, the head might convey more physical attributes (nose, ears, eyes, cheeks, chin, forebrain, etc.) than body part. Therefore, it would be interesting to study the difference, at a quantitative level, between head and body preference with respect to the amount of information provided by head or body components. We could speculate that a distributed classifier might focus more interest on facial information when the amount of facial information is more available at the level of the input. *Second,* at the level of the processes applied on the inputs, we can imagine that facial information might be more informative in a recognition task than a categorization task. Then, a face-preference effect might emerge in another cognitive task than categorization and subsequently be transferred to categorization in a second stage.

Finally, a top-down attentional component might also be required to simulate Spencer et al.'s empirical results. Further connectionist models will be proposed to test the possibility of such an attentional component in order to reproduce the behavioural data. A model like ALCOVE (Krushke, 1992) might be able to focus attention on structural information. However, as suggested in the present study, head information might not contain sufficient information by itself to focus attention appropriately on the head. Simulating these results might ultimately require positing the existence of an innate preference for head information (Johnson et al., 1991; Morton & Johnson, 1991).

5. Acknowledgments

This work has been supported in part by EC grant HPRN-CT-1999-00065.

References

French, R. M., Mareschal, D., Mermillod, M., & Quinn, P. C. (2004). The Role of Bottom-up Processing in Perceptual Categorization by 3- to 4-month-old Infants: Simulations and Data. *JEP:General*, *133 (3)*, 382–397.

Johnson, M.H., Dziurawiec, S., Ellis, H., & Morton, J. (1991). Newborns' preferential tracking of face-like stimuli and its subsequent decline. *Cognition*, 40, 1-19.

Kruschke, J.K. (1992). ALCOVE: an exemplar-based connectionist model of category learning. *Psychological review, 99*, 22-44.

Mareschal, D., & French, R. M. (1997). A connectionist account of interference effects in early infant memory and categorization. In M. G. Shafto & P. Langley (Eds.), *Proceedings of the 19th annual conference of the Cognitive Science Society* (pp. 484-489). Mahwah, NJ: Erlbaum.

Mareschal, D., French, R. M., & Quinn, P. C (2000). A connectionist account of asymmetric category learning in early infancy. *Developmental Psychology, 36,* 635-645.

Mareschal, D., Quinn, P. C., & French, R. M. (2002) Asymmetric interference in 3- to 4-month-olds' sequential category learning. *Cognitive Science, 26,* 377-389.

Morton, J., & Johnson, M. H. (1991). CONSPEC and CONLERN: A two-process theory of infant face recognition. *Psychological Review, 98*, 164-181.

Quinn, P., Eimas, P., & Rosenkrantz, S. (1993). Evidence for representations of perceptually similar natural categories by 3- and 4-month-old infants. P*erception, 22,* 463-475.

Quinn, P.C., & Eimas, P.D. (1996). Perceptual cues that permit categorical differentiation of animal species by infants. *Journal of Experimental Child Psychology, 63*, 189-211.

Spencer, J., Quinn, P.C., Johnson, M.H., and Karmiloff-Smith, A. (1997). Heads you win, tails you lose: Evidence for young infants categorising of mammals by facial attributes. *Early Development and Parenting, 6,* 113-126.

CHILDREN'S CAUSAL INFERENCES AS REVEALED BY BACKWARDS BLOCKING TASKS: A MEMORY SELF-REFRESHING NEURAL NETWORKS ACCOUNT

SERBAN C. MUSCA

Laboratoire de Psychologie et NeuroCognition — CNRS UMR 5105

Université Pierre Mendès France, BP 47, 38040 Grenoble Cedex 9, France

E-mail: Serban.Musca@upmf-grenoble.fr

Children aged 3½ exhibit less backwards blocking effect than those aged 4½; the latter only are sensitive to probabilities (Sobel, Tenenbaum & Gopnik, 2004). The original account proposed by Sobel *et al.* (2004) is that children develop a mechanism for Bayesian structure learning. This account is problematic because it evades the explanation of the origins of the initial core of knowledge that is used by the posited Bayesian mechanism. I propose here an alternative explanation: Children's differential performance stems from a memory limitation, with retroactive interference stronger in younger children, but adult-like in older children. This claim is supported by simulations with Ans and Rousset's (1997) memory self-refreshing neural networks architecture.

1. Introduction

Recently, in order to explain how the causal structure of the world is learned from interactions with the environment during early childhood, Sobel, Tenenbaum & Gopnik (2004) proposed that children develop a mechanism for Bayesian structure learning. The key experiments in support of this view used backwards blocking (hereafter BB) tasks and showed that children aged 4½ exhibit more BB than children aged 4½ (experiment 1), and that only the latter are sensitive to a manipulation of the probability of the outcome of interest in a pre-experimental phase (experiment 3).

I propose here an alternative explanation. The developmental difference in performance found by Sobel *et al.* (2004) with BB tasks is a by-product of a developing *memory system* in children: that of younger children, still immature, leads to a higher level of retroactive interference.

Backwards blocking is part of the broader class of retrospective revaluation tasks, where changes in the learned response to a cue (X) arise without further presentations (*i.e.* training) of that cue but due to the training of an associate of X (e.g., cue Y) during a second phase of training.

In the present chapter, I first show that the results of experiment 1 of Escobar *et al.* (2002) with *adults* can be simulated with Ans and Rousset's (1997) memory self-refreshing neural networks architecture (hereafter DRSR).

The same architecture (**including parameters**) is then used to simulate the results of the 4½-year-olds in Sobel et al.'s (2004) experiments 1 and 3. Finally, it is shown that DRSR can simulate the results of the 3½-year-olds only when the memory self-refreshing mechanism is made inefficient — which leads to a high level of retroactive interference in the architecture.

2. Simulations

All simulations used Ans and Rousset's (1997) DRSR. DRSR comprises two 20-8-[20+1] feedforward backpropagation networks — 20 input, 8 hidden, 20 auto-associative and 1 hetero-associative nodes — trained with a learning rate of 0.01 and a momentum of 0.7. First, NET1, the "first half" of the architecture, was trained; then NET1 generated reverberated pseudopatterns (hereafter PPs) that were used to train NET2, the "complementary half" of the architecture. After being trained on PPs issued from NET1, NET2 generates PPs that are used to train NET1, which at this point is concurrently trained on new associations.

DRSR has three parameters: the number of PPs (*Pn*) sent from NET1 to NET2 before the latter generates PPs that are used to maintain NET1's memory, the ratio (*Pr*) between the weight chances on PPs and on the new associations in NET1, and *R*, the number of reverberations used to produce a PP. In all the simulations presented here *Pn* was set to 10^4 and *R* to 5.

2.1. *Retrospective revaluation in adults*

The results found by Escobar *et al.* (2002, experiment 1) in adults were simulated with *Pr* set to 10 (an efficient memory self-refreshing). The stimuli and the lists for the 12 replications that were run correspond to those used by Escobar *et al.* (2002): 12 stimuli were generated (10-bit binary vectors made of five *zeros* and five *ones*), 6 of them corresponding to *X* stimuli, the other 6 to *A* stimuli; a list contained 8 items (*i.e.* 2 items, randomly chosen, were used twice per list). Firstly NET1 was trained on the first list (*e.g.*, for both *Compound* conditions: concurrently on $X \rightarrow O_1$ and on $AX \rightarrow O_2$)* to the RMS criterion 0.1, then 10^4 PPs generated in NET1 were used to train NET2. Secondly, NET1 was trained on the second list (*e.g.* $A \rightarrow O_1$ for the *Compound-O_1* condition†) while

---

* The hetero-associative target (*i.e.* O_1, and O_2) was 0, and 1. In $X \rightarrow O_1$, the input was a 20-bit vector: a 10-bit nil vector followed by the 10-bit *X* vector; similarly, *AX* is the concatenation of the two 10-bit vectors *A* and *X*.

† The *Compound-O_2* condition is identical to the *Compound-O_1* condition except for the second phase, where *A-O_2* is presented. The two *Elemental* conditions

its memory was refreshed by PPs generated in NET2. The training ended when the criterion was reached on the items of the second list, or after a maximum of 10^4 epochs. The results are presented in Figure 1.

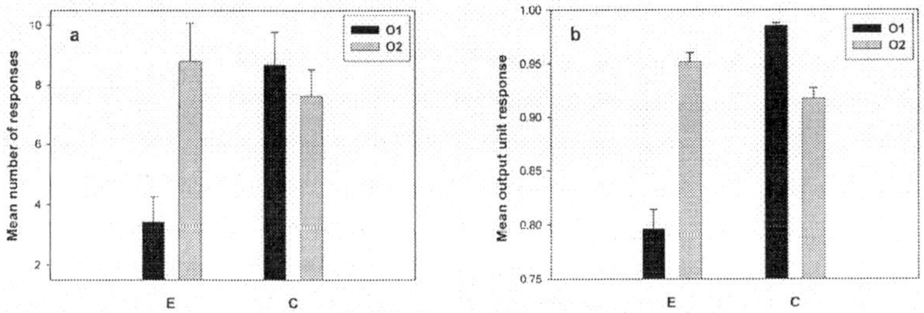

Figure 1. Retrospective revaluation in adults in the Elemental (E) but the Compound (C) condition: a) behavioural results (Escobar *et al.*, 2002, experiment 1); b) simulations with DRSR.

2.2. *Retrospective revaluation in children*

A particular form of retrospective revaluation — backwards blocking — was investigated in children by Sobel *et al.* (2004, experiments 1 and 3). In the next two sections I shall simulate the result patterns of these experiments with the same architecture and parameters as in the previous simulation, except for the *Pr* parameter: to simulate the performance of older children *Pr* will be set to the same value as the one I used to simulate adults' performance (*i.e.* 10), and to a much lower value (*i.e.* 0.25) to simulate younger children's performance. As the latter *Pr* value corresponds to an inefficient memory self-refreshing, this is consistent with the hypothesis that I propose in the introduction.

2.1.1. *Backwards blocking*

The stimuli were created in order to match the most directly possible the experimental situation of experiment 1 of Sobel *et al.* (2004). Twenty 10-bit binary vectors were generated, 16 of which were made up of five *zeros* and five *ones*, 2 comprising four *ones* and 2 comprising three *ones*. These last 4 vectors were used as knobs and tee-joints (cf. *Procedure* of experiment 1 in Sobel *et al.*, 2004). Four of the remaining 16 vectors corresponding to the demonstration items, four corresponding to the training items, and two corresponding to the experimental items were randomly chosen for each of the 16 replications. The training proceeded in 4 steps, each one followed by a self-refreshing involving

---

are identical to the *Compound* conditions, except for the stimuli presented in the first list: $X \rightarrow O_1$ and $A \rightarrow O_2$ (instead of $X \rightarrow O_1$ and $AX \rightarrow O_2$).

NET2: training on the knobs and tee-joints (auto-associatively only), then on the demonstration and training items, on the association of *AB* with the outcome *O*, and finally on the association of *A* with *O*. The results are presented in Figure 2.

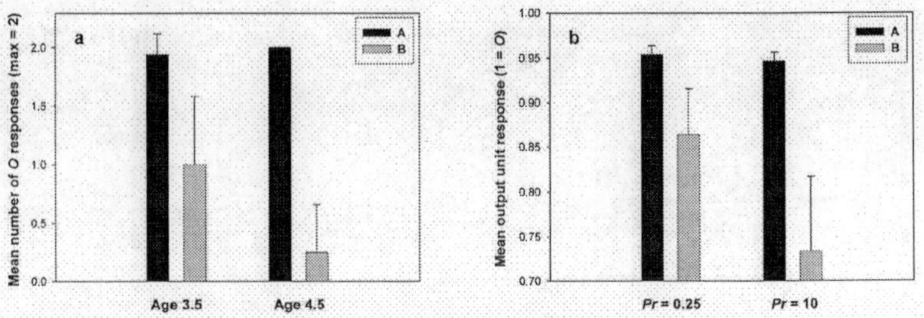

Figure 2. Differential backwards blocking in children simulated: a) behavioural results (Sobel *et al.*, 2004, experiment 1); b) simulations with DRSR, with *Pr* manipulated.

2.1.1. *Sensitivity to prior probabilities*

The simulation of the results of experiment 3 of Sobel *et al.* (2004) is presented here. The stimuli used were the same as in the previous simulation, except for two additional vectors made up of five *zeros* and five *ones*. The training proceeded in five steps, the same first two as in the previous simulation, a probability step — involving the training on 10 associations (either 1 or 9 of which had *O* as the associated outcome) — then the last two steps as in the previous simulation. The results of 25 replications are presented in Figure 3.

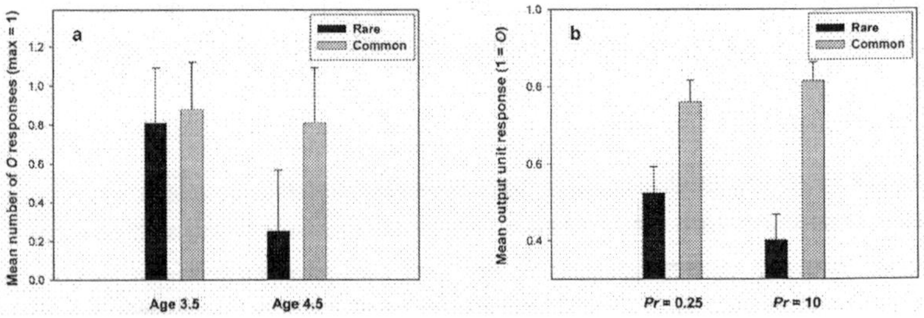

Figure 3. Differential sensitivity to the priors in children simulated: a) behavioural results (Sobel *et al.*, 2004, experiment 3); b) simulations with DRSR, with *Pr* manipulated.

3. Conclusion

Considering the facts that only one parameter was manipulated and the networks were initialized with random weights (unlike humans, whose memory prior to any experimental manipulation is not blank), the behavioural results were simulated surprisingly well. Critically, DRSR with the same *Pr* parameter value simulated both the results of *adults* and of *older children*. The simulation of the behavioural results of the *younger children* was made possible by using a much smaller *Pr* value, which resulted in less consolidation in DRSR.

Taken together, these results tend to show that an explanation of a differential backwards blocking effects in children as a function of their age can be given in terms of memory consolidation effectiveness: older children (and adults) would dispose of a memory consolidation mechanism that is not fully efficient in younger children. If the realism of the DRSR architecture (Ans & Rousset, 1997, 2000) may be called into question on the ground of involving two coupled networks that need to work together, Ans (2004) recently proposed a single neural network that is able to successfully self-refresh its own memory.

The account I propose here is motivated by the fact that Sobel *et al.*'s (2004) proposal suffers from some limitations. Conceptually, it is plagued by a chicken and egg problem, as the theory somehow posits the very initial core of (causal) knowledge that it aims to explain: children are supposed to apply their statistical knowledge in order to enhance some *pre-existing* knowledge "database". In addition, it does not predict revaluation for experiment 1 of Escobar et al. (2002) since the learning (during Phase 2) of the associations of the second list does not modify the probabilities of the X-O_1 associations of the first list (learned during Phase 1); yet such a revaluation is found in the *Elemental-O_1* condition, and is naturally simulated with DRSR.

References

Ans, B. (2004). Sequential learning in distributed neural networks without catastrophic forgetting: A single and realistic self-refreshing memory can do it. *Neural Information Processing - Letters and Reviews*, 4, 27-32.

Ans, B., & Rousset, S. (1997). Avoiding catastrophic forgetting by coupling two reverberating neural networks. *C. R. Acad. Sci. Paris, Sciences de la vie*, 320, 989-997.

Ans, B., & Rousset, S. (2000). Neural networks with a self-refreshing memory: Knowledge transfer in sequential learning tasks without catastrophic forgetting. *Connection Science*, 12, 1-19.

Escobar, M., Pineño, O., & Matute, H. (2002). A comparison between elemental and compound training of cues in retrospective revaluation. *Animal Learning & Behavior*, 30, 228-238.

Sobel, D. M., Tenenbaum, J. B., & Gopnik, A. (2004). Children's causal inferences from indirect evidence: Backwards blocking and Bayesian reasoning in preschoolers. *Cognitive Science*, 28, 303-333.

Neural Network Methodology

EMBRYOLOGICAL MODELLING OF THE EVOLUTION OF NEURAL ARCHITECTURE

CHRIS P. BOWERS & JOHN A. BULLINARIA
School of Computer Science, The University of Birmingham
Birmingham, B15 2TT, UK
{c.p.bowers, j.a.bullinaria}@cs.bham.ac.uk

Attempts in the past to use evolutionary simulations to model the emergence of modular neural architectures has led to conflicting results. Here, we present some preliminary work on the use of computational embryogeny to model the evolution of neural architecture at a less coarse level of description. We believe that such an approach will lead to much more reliable and biologically realistic simulations of brain evolution.

1. Introduction

The architecture of the human brain, in particular the existence of functionally specialised neural modules, is becoming increasingly well mapped out. However, some of the early results from neural computational modelling in this area (Rueckl, Cave & Kosslyn, 1989) have recently been thrown into doubt by more detailed evolutionary simulations (Bullinaria, 2001). There remain a range of feasible computational approaches to understanding neural architecture (Jacobs, 1994), and we believe the best way forward is to look at the evolution of these structures at a much lower level of description than is often used.

Embryogeny is the process by which an embryo structure develops from an initial stem state. This genotype to phenotype mapping process is adaptive in that the rules of growth are dependent upon the state of the system and vice-versa, thus forming a dynamical system. The modelling of such processes in a simulated evolutionary framework is known as computational embryogeny.

In this paper we shall present work in which such an approach is utilised to grow neural architectures from an initial stem state, based upon a simple model encapsulating individual cell states and the diffusion of chemicals in a three-dimensional space. This formulation is sufficiently expressive to account for features such as topology and connectivity, as well as the initial weight and learning parameters. A systematic series of computational experiments is carried out to explore the possibility of using this approach to study the emergence of neural structures appropriate for the simplified 'what-where' task used previously (Rueckl, et al., 1989; Bullinaria, 2001). The resulting structures are then compared against a series of more abstract directly coded models of the type studied previously by Bullinaria (2001), but with a matched evolutionary regime. In this way we can gain a better understanding of the reliability of, and the relationships between, models at different levels of description.

2. Modelling the Emergence of Modularity

We ground our work in the simplified 'what' and 'where' vision tasks, for which extensive investigations into the internal representations learned by simple neural networks have already been carried out (Rueckl, et al., 1989), as have a series of evolutionary simulations (Bullinaria, 2001). These tasks consist of mapping from a 5x5 binary retinal image, to a set of 9 'what' and 9 'where' binary classification outputs. The problem is to identify 9 different 3x3 sub-retinal images that are positioned in the 9 different possible locations upon the 5x5 retina, which yield a total set of 81 possible input patterns.

As in the previous work, the basic neural network structure takes the form of a 3-layer network consisting of 25 input units, 18 output units and various hidden units (Figure 1). The hidden units are all fully connected to each of the inputs, but their connectivity to outputs can be described as belonging to one of three different groups: connected to only the 'what' outputs (*Hid1*), connected to only the 'where' outputs (*Hid2*), or connected to both (*Hid12*). In this respect, a network where all the hidden units belong to the group *Hid12* would be considered totally non-modular, whereas a network where there are no hidden units in *Hid12* would be considered to have totally modular connectivity.

The analysis of Reuckl et al. (1989) appeared to show a clear advantage for modular architectures. However, taking a population of networks, and using simulated evolution by natural selection to determine all the learning and architecture parameters, the emergent architecture was found to depend on the learning algorithm cost function used (Bullinaria, 2001). If the most efficient (cross entropy based) learning algorithm is used, or if the learning algorithm is allowed to evolve, we always end up with a non-modular architecture. These conflicting results have led us to look for more biologically realistic approaches for understanding the emergence of modularity in neural systems.

3. The Embryological Model

Embryogeny is the process by which an embryo develops from a single stem cell. Computational embryogeny refers to the simulation, within computational systems, of the kind of embryonic processes observed in nature. The embryogeny mapping requires that the phenotype consist of discrete processing blocks (cells). In our model the cell environment will be represented as a three-dimensional space that can accommodate various levels of proteins and cell states at any given real valued position.

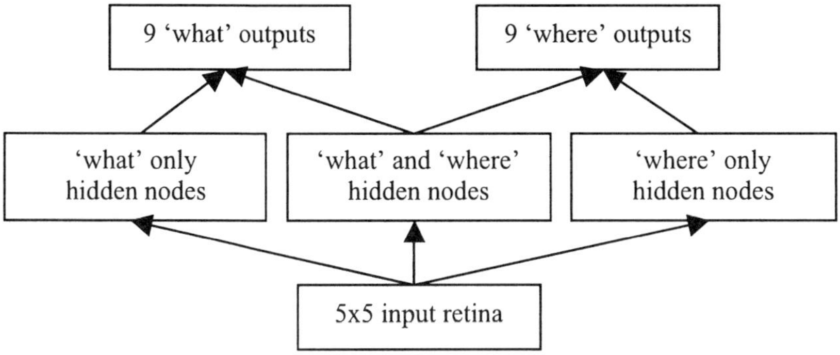

Figure 1: The architecture of the basic neural network model for the 'what' and 'where' task.

In nature, the functionality of genes is determined by the state of a cell and the surrounding environment, i.e. by proteins and other chemicals. The function of a gene may, in turn, be to alter the cell state upon which these genes are dependent. In order to capture the essence of this process in a computer simulation, a number of simplifications over real biological systems must be made to render the computational costs feasible. All cell sizes and shapes are taken to be identical, and are represented as spheres with diameters in proportion to the dimensions of the overall space. Direct cell-cell interaction beyond simple chemical diffusion is prohibited, so there is no cell-adhesion or any other form of force between cells. If a cell is to divide, or move into the location of an existing cell, it simply overwrites the existing cell. This is clearly not biologically plausible, but it is necessary to reduce the computational costs of a more realistic physical model.

In nature, any limits on the size of the physical system are set by physical interaction. Since, in our model, no such forces exist, the environment must be set at some finite size to impose some constraint on the size of the emergent system. This poses the problem of what to do when a cell attempts to move or divide outside of this finite environment. A common approach is to cycle around the edges so that, for example, in two-dimensional space a toroidal shape is formed, removing the need to worry about finite limitations. However, this can often introduce unwanted cyclic behaviour, so a different approach is used in which any cells dividing outside of the modelled space are simply ignored.

The genetic representation of the embryogeny system is based upon a simplified model of the genetic regulatory network (GRN), using the Operon model as a basis for individual gene function (Alberts, et al., 1994). Biologists use this model to describe how groups of individual genes can form complex GRNs. In our work, a much-simplified version of the Operon model is used in which a gene has two dependents, an inducer/repressor and a product. If the inducer/repressor exists, the gene is switched on/off respectively. If the gene is switched on, it releases a product pre-defined by the gene. This product may result in the increase or decrease of the concentration of some protein, or it may lead to some function upon the cell. For example, a gene that promotes axonal growth may cause an axon head to grow in the direction of the protein gradient as defined by the product, as long as no repressor protein is present. The GRN comes into play since the function of a gene may be to produce an inducer or repressor itself, and in this way a gene can control the expression of other genes and so form a network of interaction between genes. In our model, there are 34 gene functions consisting of concepts such as cell division and movement, protein emission, protein absorption, axon growth, etc. These allow the cells to adapt and change their environment and cell states.

The mapping process starts with an initial stem state, consisting of pre-determined input and output neurons and a single stem cell. The growth process is performed in fixed steps, in parallel, across the space, such that the proteins and genome interact to produce some cell functionality. This interaction is the critical part of the model which defines what is expressible in the phenotype.

4. Computational Embryogeny

Computational embryogeny creates a complex mapping between genotype and phenotype spaces that exhibits some key characteristics. First, the mapping is many-to-many in that a single genotype can form a variety of different phenotypes dependent upon the environmental conditions. Similarly, a number of different genotypes can result in identical phenotypes. Second, the neighbourhood structure is not necessarily preserved by the mapping, in that two similar genotypes can result in very different phenotypes, or vice-versa.

From these characteristics, it is easy to assume that such a mapping would be disadvantageous for a simulated evolutionary process, since it would introduce a hugely rugged landscape. In some respects, this is true. However, this is based on the viewpoint that the fitness landscape is a fixed entity through which a population of individuals drift and jump around based on some heuristics, or search operators, biased toward fitter individuals. These search operators are, however, more powerful than this viewpoint suggests, since it is not only the genetic representation that determines the structure of the fitness landscape, but also the search operators themselves. If the effects of the search operators can be adapted in some way during the evolutionary process, then this would effectively cause a restructuring of the fitness landscape. Thus the analogy of traversing around landscapes begins to break down.

A further complication is the huge number of possible individuals with equivalent fitness. The result is that the search space has vast networks of neutrality, where there are threads, planes, or even higher dimensional structures of equal fitness. Traditionally, this has been viewed as a bad thing to incorporate, since it results in convergence of the population or dead time in terms of evolutionary change. However, such neutral networks have been shown to be extremely useful with the right kind of algorithm, and evidence suggests they are also critical to natural evolution (Harvey, 1997). Since there can be no selection pressure in these neutral areas, neutral evolution has a different behaviour to that of adaptive evolution, which relies on selection pressure to drive evolution to fitter individuals. These neutral networks can provide a way to freely traverse the search space, without selection pressure driving toward some fixed convergence point. The result is that individuals converge onto the most optimal neutral network, and then diverge across that network. When an individual finds a new neutral network within its reach, it jumps across, and the rest of the population converges on this new network. This results in staggered jumps in fitness through evolutionary time, and gradual increases in diversity followed by a sudden drop as the population jumps to the new network (Smith, Husbands & O'Shea, 2001).

An important point to note here is that, in the absence of selection pressure, evolution does not simply stop. Since individuals are selected randomly, search operators are still used to produce new offspring during neutral evolution. The evolutionary drive is no longer towards fitter individuals, but individuals that can maintain or improve their fitness through mutations and crossover. In other words, genetic material that is more susceptible to damage will be less likely to propagate to the next generation, regardless of its contribution to fitness. It could be said that adaptive evolution applies evolutionary drive to fitter individuals, whilst neutral evolution drives towards more evolvable individuals.

To take advantage of these elements, our evolutionary algorithm consists of the following processes:

- The mutation operator consists of point mutations applied to each allele of the genome with a given mutation probability, resulting in random replacement with a value from within the range of validity. Crossover involves the selection of two random crossover points which may only fall in between genes. This helps to prevent destruction of genes and ensures the conservation of genome length.
- Recombination is based on a steady state approach, consisting of the replacement of the last individual in a linearly ranked population only if its fitness is equal or less than the individual that replaces it.
- Fitness is defined as the grown network's performance on the training data after gradient descent learning using our chosen cost function.

It has been shown that for a complex mapping, optimal neutral evolution will occur when the population is phenotypically converged. In other words, when the entire population is drifting on the same neutral network there is more chance of finding a bridge to an individual/network of higher fitness (Harvey, 1997). However, in order for adaptive evolution to be successful, it must maintain diversity amongst the population to perform effective search across the search landscape. Obviously, these are two conflicting ideals, and so the solution has been to use an algorithm based on an island population model (Gordon, Whitley & Bohn, 1992). This consists of a number of separate sub-populations being allowed to evolve and swap individuals. Populations swap individuals under the following conditions:

- An individual is chosen as a migrant to the receiving population if it is the best individual from another population having the greatest genetic distance from the best individual in the receiving population.
- The chosen migrant replaces the individual in the receiving population with the closest genetic distance.

Genetic distance is measured as the number of differences between two individuals. This allows a group of individuals to converge on a neutral network, whilst ensuring different groups are effectively forced to diverge from each other. The result is a system that can swiftly adapt to suit either neutral or adaptive evolution, or even a combination of both. An added advantage of the island model, and one of the more common reasons for its use, is that it is easy to distribute across numerous machines and so speed up the evaluation time of the overall population. This is a significant advantage in this case where fitness evaluation can be particularly computationally expensive.

5. Experimental Details

Our aim for this paper is to extend the previous work on evolving modularity (Bullinaria, 2001) by allowing the evolutionary process greater flexibility in the architectures and learning parameters it can express. In particular, we wish to gain a better understanding of the relationship between modularity and learning for the simplified 'what-where' vision task discussed above.

Although the embryogeny model is capable of expressing any of the parameters required for describing and evaluating the performance of a neural network, in our study it does not specify values for all the network weights and biases, only the ranges for their random initial values. As with most biological networks, the actual values are learned from appropriate training data. In our experiments, we use an online gradient descent algorithm, based on either the

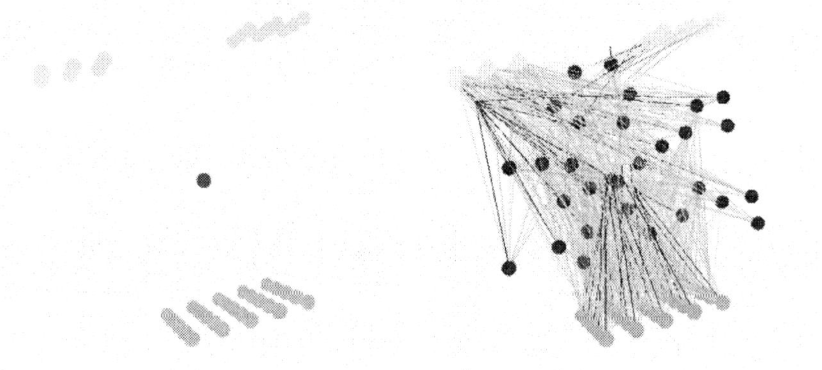

Figure 2: (a) The initial stem state of the model. (b) A typical fully grown evolved neural network.

Sum Squared Error (SSE) or the Cross Entropy (CE) cost function.

The types of architectures that can be found by our evolutionary system are extremely flexible, encapsulating all forms of connectivity, including lateral and recurrent connections. The learning algorithm must account for this possibility, and so a windowed propagation approach is used whereby neural activations can only propagate through a single neuron per iteration. This means that several iterations of the training algorithm are required before activation at the input neurons finally reaches the output neurons, but recurrent looping and output settling problems are eliminated.

The embryogeny process requires a stem state from which to start. This stem state consists of a set of predefined input and output neurons on opposing faces of the three-dimensional environment with a single stem cell positioned at the centre (Figure 2a). The input neuron positioning is consistent with the problem description forming a 5x5 array, whilst the two groups of output neurons are arranged in 3x3 arrays to minimise the connectivity radius required to connect to all 9 outputs. For simplicity, we took the dendrite lengths and overall environment size to be measured in units of cell widths. This removes any need to assign definitive sizes, rather than just proportionality ratios between different aspects of the model. The network growing process is allowed a fixed number of growth steps sufficient to allow any cell to re-position itself to any location in the environment.

Our modeling approach is also flexible enough to accommodate the idea, and consequences for modularity, of constraints on neural connectivity, including those discussed by Jacobs and Jordan (1992). By constraining the distance from which an axon head can create a synaptic connection to a neuron, it is possible to effectively constrain the number of neurons a particular neuron can connect to. This is represented in Figure 3.

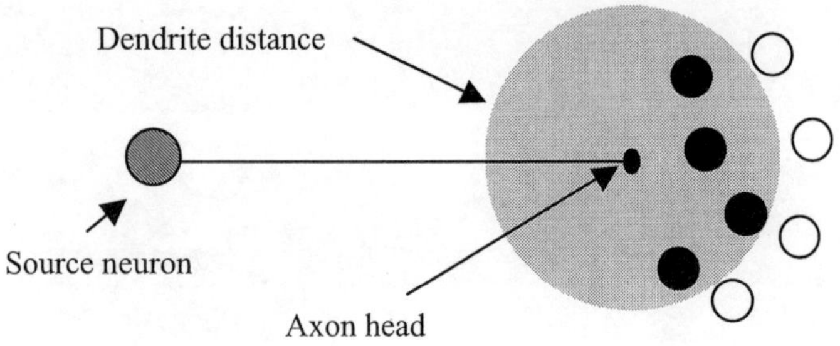

Figure 3: The effect of dendrite distance on connectivity. Only cells within the dendritic radius create synaptic connections with the source neuron.

6. Simulation Results

We now present our results from a series of simulated evolutionary processes obtained for various dendrite distances, and for both the SSE and CE learning algorithm cost functions. Figure 4a shows the hidden to output connectivity plot for CE, where *pHid1*, *pHid2*, and *pHid12* specify the proportion of hidden units connected to each output block. Each point in the plot represents an optimal network found by an evolutionary process. For larger dendrite distances, the results agree with those of the coarser evolutionary simulations discussed above (Bullinaria, 2001), with networks tending to be located in the *pHid12=1* region, indicating that non-modular fully distributed networks are preferred.

However, if we reduce the dendrite distance, and so constrain the connectivity, the results show a distinct difference, with far fewer neurons connected to both output blocks. Two points can be made from these results. First, for larger dendrite distances, the results show non-modular networks emerging, and so, even with greater expressivity in terms of network structure and learning parameters, non-modular networks are still found to be optimal. Second, it is possible to encourage the emergence of less distributed and more modular architectures, simply by decreasing the allowable dendrite distances.

Figure 4b shows the average performance rates throughout the evolutionary process. We see that for shorter dendrite distances it takes more evolutionary time to find an optimal solution. Thus, although we can force less distributed and more modular architectures on networks trained with CE, the evolutionary progress is worse than for networks that use a fully distributed approach.

If we look at the connectivity plot results for the corresponding SSE simulations (Figure 5a), we see that for longer dendrite distances the network connectivity behaves much like those for CE. However, finding optimal networks for smaller dendrite distances becomes extremely hard, and this is reflected by the lack of plotted points representing optimal networks with

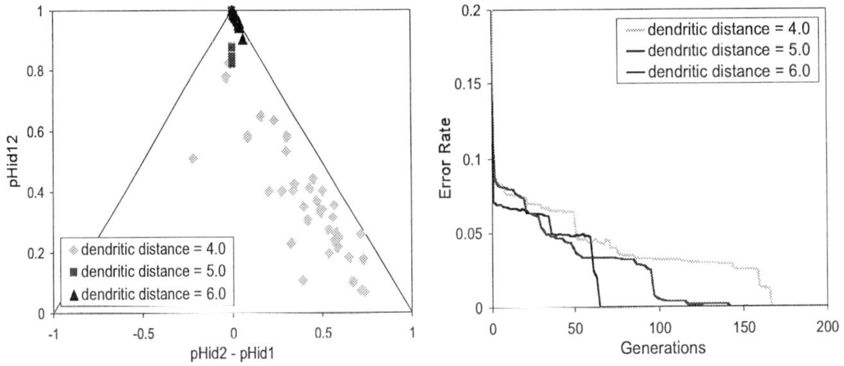

Figure 4: (a) Connectivity plot of optimal networks and (b) average evolutionary learning curves for the CE based learning algorithm.

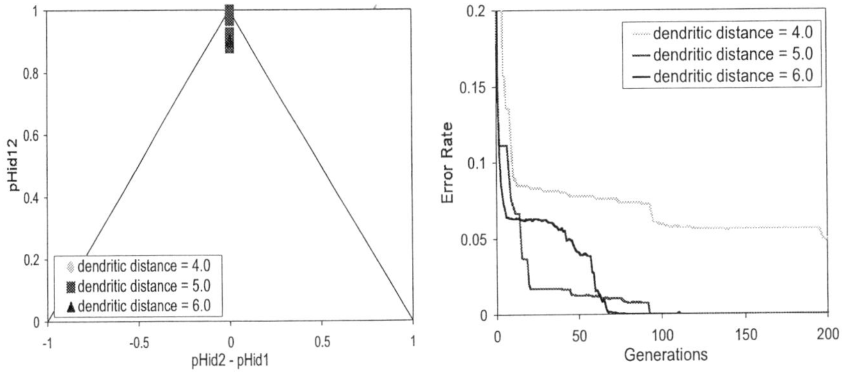

Figure 5: (a) Connectivity plot of optimal networks and (b) average evolutionary learning curves for the SSE based learning algorithm.

dendrite distance of 4.0. This is confirmed in Figure 5b where we can see that the learning behaviour during evolutionary time is similar to CE except that, for smaller dendrite distances, optimal networks are never actually found and evolution gets stuck on poorly performing networks.

Using more directly encoded representations, the CE cost function always produced non-modular optimal networks, whilst SSE tended to produce modular optimal networks, as Bullinaria (2001) found. When comparing this with our embryological modelling results for CE, it is clear that the behaviour is similar until the networks connectivity is constrained by short dendrite distances. Comparison of the results for SSE show the greatest differences, since our work suggests that, under a more expressive representation, non-modular networks are optimal in this case too. This appears to be at odds with all the previous work (Rueckl et al., 1989; Bullinaria, 2001). Moreover, constraining the connectivity by restricting the dendrite distances has a serious effect for SSE networks, in that it results in a failure of the evolutionary algorithm to find networks of any description that can learn to perform the given tasks.

7. Conclusion

We began by reviewing the Rueckl et al. (1989) and Bullinaria (2001) studies into the emergence of modularity in neural systems, and argued that performing more detailed simulations was a sensible approach to gaining a better understanding of the problem. After an outline of a computational embryogeny approach to evolving networks that grow and learn to perform two distinct tasks (the 'what' and 'where' tasks studied previously), we found that the emerging architectures always take on a distinctly non-modular form, unless restrictions are imposed on the dendritic distances. This ties in with the work of Jacobs and Jordan (1992) on the consequences of a natural bias to short neural connections. We clearly have much more work to do to explore fully all the relevant issues, but we believe we have a promising approach for studying the evolution of neural architectures.

References

Alberts, A., Bray, D., Lewis, J., Raff, M., Roberts, K. & Watson, J.D. *Molecular Biology of the Cell.* 3rd edition, Garland. 1994.

Bullinaria, J.A. Simulating the Evolution of Modular Neural Systems. In *Proceedings of the Twenty-third Annual Conference of the Cognitive Science Society*. 146-151. Mahwah, NJ: Lawrence Erlbaum. 2001.

Gordon, V., Whitley, D. & Bohn, A. Dataflow Parallelism in Genetic Algorithms. In Manner, R. & Manderick, B. (Eds), *Parallel Problem Solving from Nature 2*, 553-542. Amsterdam: Elsevier. 1992.

Harvey, I. Artificial Evolution for Real Problems. In *Evolutionary Robotics: From Intelligent Robots to Artificial Life*. 187-220. AAI. 1997.

Jacobs, RA. Computational Studies of the Development of Functionally Specialized Neural Modules. *Trends in Cognitive Science*, **3**, 31-38 1999.

Jacobs, R.A. & Jordan, M.I. Computational Consequences of a Bias Toward Short Connections. *Journal of Cognitive Neuroscience*, **4**:323-336. 1992.

Rueckl, J.G. Cave, K.R. & Kosslyn, S.M. Why are "What" and "Where" Processed by Separate Cortical Visual Systems? A Computational Investigation. *Journal of Cognitive Neuroscience*, 1, 171-186. 1989.

Smith, T. Husbands, P. & O'Shea, M. Neutral Networks and Evolvability with Complex Genotype-Phenotype Mapping. *European Conference on Artificial Life*. 272-281. Springer. 2001.

EVOLVING NEURAL NETWORKS THAT SUFFER MINIMAL CATASTROPHIC FORGETTING

TEBOGO SEIPONE   JOHN A. BULLINARIA
School of Computer Science, The University of Birmingham
Birmingham, B15 2TT, UK
{t.seipone, j.a.bullinaria}@cs.bham.ac.uk

Catastrophic forgetting is a well-known failing of many neural network systems whereby training on new patterns causes them to forget previously learned patterns. Humans have evolved mechanisms to minimize this problem, and in this paper we present our preliminary attempts to use simulated evolution to generate neural networks that suffer significantly less from catastrophic forgetting than traditionally formulated networks.

1. Introduction

Virtually all natural systems gradually forget what they have learned previously, particularly when they learn new information. However, with traditional artificial neural networks, the forgetting is much more catastrophic, and this proves to be a serious limitation for them (McCloskey & Cohen, 1989; Ratcliff, 1990). Some relatively complex systems, involving the interleaving of pseudo-patterns, and/or dual network architectures, have already been shown to deal with this problem quite successfully (French, 1999). However, we wish to explore the possibility of avoiding the problem within much simpler systems. Natural neural networks, such as human brains, have presumably evolved by natural selection to minimize the forgetting, and the aim of this paper is to present a preliminary investigation into using simulated evolution to see how far we can minimize the problem in artificial neural networks. The strategy employed starts by measuring how well traditional networks remember sets of input-output mappings after being trained on new items, and then explores systematically whether these can be evolved into better performing networks.

2. Evolving Neural Networks

We simulate evolution by maintaining a population of individual neural networks, each specified by a number of 'innate' parameters, and using an appropriate fitness measure to determine which to discard and which to use to create the next generation by genetic cross-over and random mutation. Repeating this process should result in increasingly fit populations.

For this study, the architecture of the networks and the learning algorithm were fixed to be standard Multi-Layer Perceptrons with one hidden layer, trained by gradient descent weight updating (back-propagation) with the Cross Entropy error measure (Bullinaria, 2003). The aim was to evolve various neural network topology and learning parameters to produce systems that suffer minimal catastrophic forgetting. Each network was initialized with different random weights from a different specific range, and then trained on the same set of initial patterns until it learned all those patterns (i.e. had all output activations within a particular tolerance of their target outputs), or until it had reached a maximum number of epochs of training. It was then trained with a new set of patterns in the same manner, and re-tested to determine how many of the original patterns it still remembered. The fittest individuals were those with the highest number remembered. The least fit half of the population was then removed, and each of the remaining individuals randomly selected a mating partner to produce one child, thus restoring the population size. The children each inherited innate characteristics (i.e. parameter values) from the range spanned by its two parents, and random Gaussian mutations were added to allow values outside that range (Bullinaria, 2003). The initial population had random innate parameters, and then for each new generation, a new global random set of training/remembering data was generated, and each individual had new random initial weights generated from their own innately specified ranges.

The following Section will make these ideas more concrete by specifying our simulations in more detail and presenting the results from a systematic sequence of experiments that explore the issued involved.

3. Simulation Results

For our main study we needed to fix a convenient training set of associated input and memory patterns, that was small enough for the simulations to run reasonably quickly, yet large enough to be representative. After some experimentation, we settled on random associations of 12 bit random binary patterns with 6 bits 'on'. New random data sets of this specification were generated for each generation. Each network was trained on 20 such patterns until the error on each output bit was less than 0.1, or the maximum of 1000 epochs was reached. It was then trained on a different set of 4 such patterns, and the number remembered correctly from the original 20 was measured using a tolerance of 0.2. Throughout, all our networks had 50 hidden units, which is plenty for this task, and we present population averages over 100 individuals.

It is appropriate to start by checking the baseline performances for standard

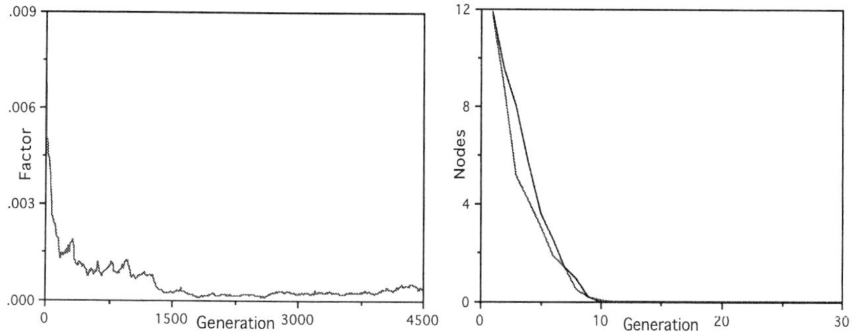

Figure 1: The parameters for hidden unit activation sharpening evolve to zero, indicating that this is not a good strategy for reducing catastrophic forgetting.

network training parameters. For 2500 individuals with different random training data sets, trained using back-propagation learning rates of 0.2 with random initial weights uniformly distributed in the range [–1, +1] the mean remembering rate was 69%. The associated standard deviation of 5% indicates the extent of the variance due to some data sets being 'easier' than others.

The first thing we wanted to explore was the suggestion of French (1991) that hidden unit activation sharpening could reduce the forgetting by developing semi-distributed representations in the hidden layer. The idea is that, at each epoch of training, the input to hidden weights are modified to bring the N_H highest activation hidden units closer to one, and the N_L lowest activations closer to zero, by a 'sharpening factor' of α times the difference. There are two variations to consider. First, when we force $N_H + N_L = N_{Hid}$ so all hidden activations get changed, the sharpening factor invariably evolves to zero, as seen on the left of Figure 1, leaving us with our standard network. If we let N_H and N_L evolved freely, they both evolve very quickly to zero, as seen on the right of Figure 1, again leaving us with our standard network. It seems that this kind of node sharpening does not really help with catastrophic forgetting for our class of training data. As a check, these parameters were left free to evolve in all our subsequent simulations, but in each case they chose to turn off node sharpening.

There are a number of traditional network parameters that one can evolve with the hope of improving performance. To get a feel for which are most effective, we shall consider each in turn before evolving them all at once. First we evolve the learning rates. It is now well established that allowing separate gradient descent step sizes η_L for each layer and bias set L is more efficient than a single parameter to control them all (Bullinaria, 2003). Figure 2 shows the evolution of these four parameters for our system, and the associated

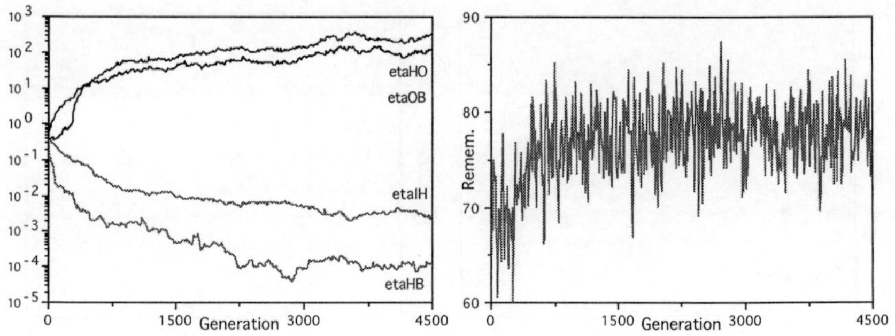

Figure 2: Evolution of the learning rates for the two weight layers and two sets of biases, and the associated improvement in remembering performance.

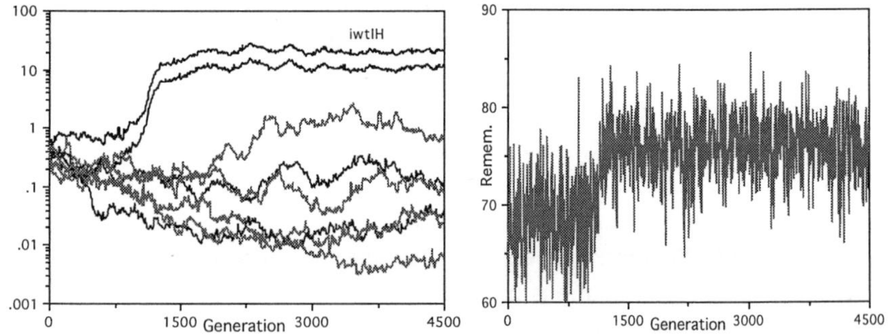

Figure 3: Evolution of the upper and lower limits of the four uniform random initial weight distributions, and the associated improvement in remembering performance.

improvement in remembering performance from the baseline 69% up to around 79%. Note the large differences in size between the four learning rates, and how far removed they are from the values traditionally used.

Associated with each learning rate is a random initial weight distribution. There are several options for specifying and evolving these, such as means and standard deviations of Gaussian distributions (μ_L, σ_L), or as lower and upper limits of uniform distributions ($-l_L$, u_L). Figure 3 shows how a set of uniform distributions evolve, and the associated improvement in remembering from the 69% baseline up to around 76%. It is clear that the sudden improvement in performance corresponds to a ten-fold widening of the input to hidden initial weight distributions.

One major advantage of the evolutionary approach is its ability to evolve simultaneously a number of parameters that interact in a complex manner. Allowing the learning rates and initial weight distributions to evolve together

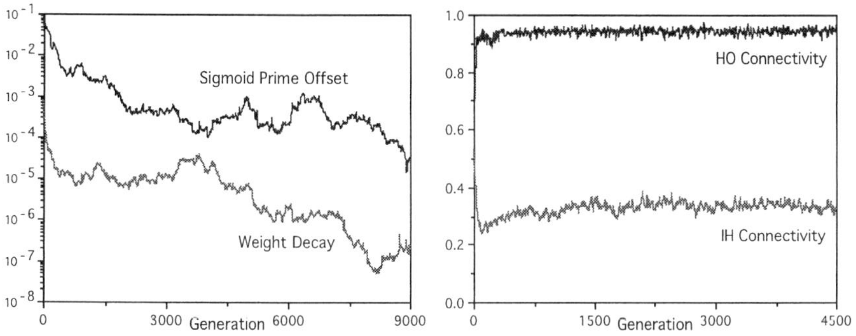

Figure 4: Evolution results in Sigmoid Prime Offset and Weight Decay parameters both taking on effectively zero values, and non-standard connectivity patterns.

leads to rather different patterns of results from when they are evolved separately. By coordinating their values, our system was able to improve its remembering performance up to around 88%. It is interesting to note that the improved remembering automatically brings with it faster learning, so there is no need to build that explicitly into the evolutionary fitness function.

Two more learning parameters that might conceivably affect our results are the *Sigmoid Prime Offset* which prevents saturation and poor learning at the hidden layer, and weight decay regularization which prevents over-fitting of the training data (Bullinaria, 2003). We see on the left of Figure 4 that, if we allow these to evolve, their parameters both take on values that are so low that they have no significant effect on the learning or remembering.

Another factor, that one might expect to reduce the interference that causes forgetting, is the connectivity between layers. We can evolve parameters that specify the proportion of possible connections that are used by the network, and do find that proportions significantly less than one emerge as seen on the right of Figure 4. There is almost full connectivity between the hidden and output layer, but only about one third of the input to hidden layer connections are used. However, there is virtually no improvement in the remembering performance. As a check, we tried evolving the connectivities with all other parameters held at the baseline values, but still there was little remembering improvement.

So far, our study has been based on how many of 20 original patterns were remembered after training on 4 new patterns. Now we need to explore the extent to which the number of new patterns affects the results. Figure 5 shows what happens with different numbers of new patterns. Not surprisingly, the baseline degree of forgetting as a percentage (left bar of each pair) increases with the number of new patterns. The important result is that for every case, evolving

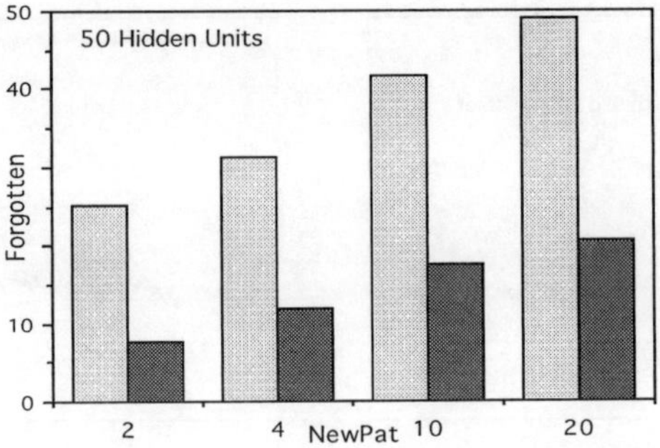

Figure 5: The more new patterns learned, the less of the original set are remembered, but in each case evolution results in improved performance over the baseline.

the network parameters, as described above, leads to a massive reduction in the amount of forgetting (right bar of each pair).

4. Conclusions

Through a systematic series of simulations, we have shown how, compared with traditionally formulated networks, an application of evolutionary techniques can significantly reduce the well known problem of catastrophic forgetting in neural network systems. We were surprised to find that evolving the learning rates and initial weight distributions alone could result in remembering performance increases that evolution of more esoteric factors cannot beat.

References

Bullinaria, J.A. (2003). Evolving Efficient Learning Algorithms for Binary Mappings. *Neural Networks*, 16, 793-800

French, R.M. (1999). Catastrophic Forgetting in Connectionist Networks. *Trends in Cognitive Sciences*, 4, 128-135

French, R.M. (1991). Using Semi-Distributed Representations to Overcome Catastrophic Forgetting in Connectionist Networks. In *Proceedings of the Thirteenth Annual Conference of the Cognitive Science Society,* 173-178. Hillsdale, NJ: LEA.

McCloskey, M., Cohen, N.J. (1989). Catastrophic Interference in Connectionist Networks: The Sequential Learning Problem. *The Psychology of Learning and Motivation*, 24, 109-165

Ratcliff, R. (1990). Connectionist Models of Recognition and Memory: Constraints Imposed by Learning and Forgetting Functions. *Psychological Review*, 97, 205-308

IMPROVING CELL ASSEMBLY CATEGORIES BY FATIGUE

CHRISTIAN HUYCK[A] HINA GHALIB[B]

[A] *School of Computer Science, Middlesex University, The Burroughs London, NW4 4BT, UK*

[A] *c.huyck@mdx.ac.uk,* [B] *h.ghalib@mdx.ac.uk*

Human concepts are represented by neural circuits called Cell Assemblies (CAs). We have a simulator based on neurons that are fatiguing leaky integrators, with Hebbian learning rules to modify the synaptic strength. A neuron is recruited into a CA by having its synaptic weight both to and from the CA increased. This causes the neuron to fire whenever the CA is ignited, and this supports the increased synaptic strength. A neuron may be incorrectly recruited into a CA via spontaneous neural activation or overlapping CAs. This paper aims to resolve the problem of incorrectly recruited neurons. We need to find a mechanism that allows incorrectly recruited neurons to eventually leave a CA. One way to do this is to reduce the amount of energy that is retained. Unfortunately this reduces the persistence of a CA. Another way is to modify the fatigue mechanism. Modifying the fatigue mechanism so the fatigue is reduced proportionally to the current fatigue level reduces the likelihood that an incorrectly recruited neuron fires. In the long run this reduces the synaptic strength between it and the CA. This enables the removal of the neuron from the CA.

1. Introduction

Cell Assemblies (CAs) are highly recurrent networks that can sustain activation even after the input has ceased. Hebb(1949) proposed that concepts in a human brain are represented by CAs, a set of neurons with a large number of excitatory synaptic interconnections.

CAs are categorizers and they exhibit persistence. They categorize because several different inputs ignite the same CA; each of the inputs of a particular category causes the net to move to a pseudo-stable state of neural activation. Thus the CA is an attractor state (Amit 1991) that attracts several initial inputs. Similarly, the state persists for a considerable time.

One of the key problems of research in CAs is their dual dynamics. The fast dynamic is categorization, which in CA terminology is known as CA ignition; this is well studied. Less studied is the slow dynamic of learning in CAs. As we show in this paper, the precise implementation of the fast dynamic has a strong influence on the slow learning dynamic of CA formation, and neural recruitment.

2. Categorization Using Cell Assemblies

An input presented to the network is categorized by igniting the CA that represents the category to which it belongs. CAs categorize a broad range of input patterns. However most of the categorization tasks face the problem of incorrectly recruited neurons.

2.1. *CANT*

CANT is a highly recurrent neural network model inspired by human neural functioning (Huyck 2000). It is based on neurons that are spiking fatiguing leaky integrators. We have implemented several categorization tasks including hierarchical categories, standard categorization benchmarks, and information retrieval tasks using this model.

The neurons in the CANT model are either excitatory or inhibitory; that is, all synapses leaving a neuron are either excitatory, or they are all inhibitory. In the simulations below, 20% of the neurons are inhibitory and 80% are excitatory. Neurons in a human brain are connected in a distance-biased way. If two neurons are closer together they are more likely to be connected. The CANT model adheres to this distance-biased connectivity. The neurons are connected in a toroidal topology. The inhibitory/excitatory nature of neurons, the ratio of inhibitory to excitatory, and the distance-biased topology makes the model biologically plausible (Churchland and Sejnowski 1992). Simulation in CANT goes in discrete time steps. Though discrete activation cycles are not biologically plausible this simplification makes CANT computationally effective.

2.2. *Learning*

We are committed to the idea of Hebbian learning. Hebbian learning is an unsupervised learning mechanism. The synaptic strength is based solely on the properties and firing behavior of the pre and post-synaptic neurons. In the simulations described in this paper, the strength of the synapse increases if both neurons fire, and decreases if the pre-synaptic neuron fires and the post-synaptic neuron does not. The decrease of strength is often called an Anti-Hebbian rule.

There is a long history of use of Hebbian rules. A simple rule is the correlatory rule, which causes the synapse to have a value that is approximately the likelihood that the post-synaptic neuron fires when the pre-synaptic neuron fires. Another variant of Hebbian learning is the compensatory learning rule. It modifies the correlation based on the total synaptic strength of the pre and post-synaptic neurons. The total strength of all the synapses leaving the pre-synaptic neuron is bound. If the total synaptic strength is below the limit, strengthening (learning) is increased and weakening (forgetting) is decreased. If the total synaptic strength is above the limit, strengthening (learning) is decreased and weakening (forgetting) is increased. This prevents particular neurons from having too much effect and helps all neurons to participate. This allows CAs for sparse patterns to be learned effectively.

2.3. *Bad Neuron Problem*

If a neuron never occurs in the pattern, it may still be incorrectly recruited into a CA via spontaneous neural activation or overlapping CAs. There is a need for a mechanism that forces bad neurons to eventually leave a CA. A bad neuron is one that we know should not be part of the CA.

This problem is particularly prevalent when there are overlapping CAs. That is when a neuron participates in more than one CA. Most simulations do not consider this case or they solely work with minor overlap. Overlap is crucial to the development of a good memory system based on CAs (Sakurai 1998).

3. Experiments

The experiments discussed in this paper are specifically designed to highlight the issue of the bad neuron problem. We present the first input pattern to the system for 800 cycles. These are presented in 50 cycle sets. In a set, an input pattern is presented for 20 cycles, and the net is allowed to run for 30 more. The fatigue and activation are erased, and a new set is started. The pattern consists of neurons 1 to 399. Any presentation of the first input pattern after 800 cycles leads to the ignition of its CA. This indicates that the system has learned the CA for this pattern. The second input pattern, which largely overlaps with the first, is presented to the system from 801 to 1999 cycles. It consists of neurons 1 to 439. If the first pattern is presented to the system after 1999 cycles it also activates some bad neurons, which are not part of the CA. These neurons have been incorrectly recruited into the CA of first input pattern as a result of repeated presentation of the second input pattern. Experiments with different decay and a different fatigue mechanism were conducted to investigate the impact on the bad neuron problem.

3.1. *Change Decay*

Biological neurons retain activation if they do not fire but the activation of non-firing neuron decays at each step. In CANT, each non-firing neuron looses activation at a constant rate called the decay rate d. This is described by equation 1, where A_t is the activation at time t.

$$A_t = A_{t-1}/d$$ **Equation 1**

Changing decay rate affects the number of bad neurons that exist in a network. A higher decay rate greatly reduces the amount of energy retained by the neurons. Thus, a neuron requires larger activation to fire. This has an impact on the firing frequency of neurons. The firing rate of neurons not stimulated externally decreases. Consequently, the synaptic strength between the neurons of the CA and the bad neuron is reduced. Though an increase in decay rate leads to a very low number of bad neurons it also affects the persistence. As the external input ceases, the neurons start to lose activation. They now require higher activation than provided by the reverberatory activity. As a result a CA is shut off soon after the external input is removed. Figure 1 shows that a decay rate of 1.2 or more reduces the number of bad neurons to 0. Measurements are taken once per set of 50. A Pearson value is a measure of persistence of a CA; we compare networks at cycles 15 and 35. We can see that a higher decay rate leads to very low Pearson value and thus a lack of persistence.

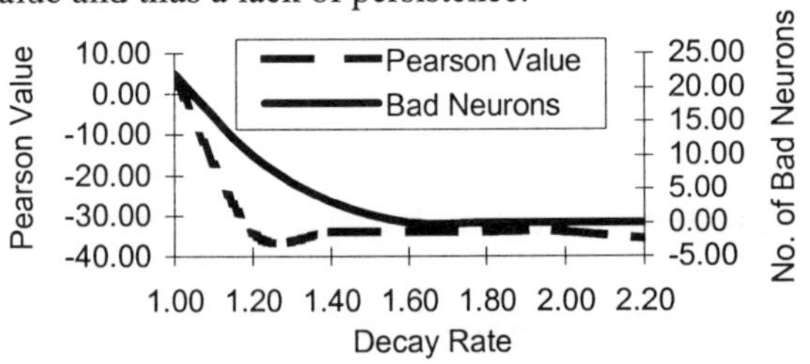

Figure 1: Effect of Decay on Persistence and Bad Neuron Problem

3.2. *Change Fatigue Mechanism*

In our simulation, fatigue is initially modeled by two constants and a dynamic fatigue value stored by each neuron. If a neuron fires its fatigue value is increased by a fatigue constant f_c. If it does not fire it is reduced by a second constant f_r to at most 0 described by equation 2.

$$f_t = f_{t-1} - f_r$$ **Equation 2**

The neuron only fires if the sum of its activation surpasses the firing threshold and f_t (fatigue at time step t). Thus, for a neuron that fired relatively recently, it becomes more difficult for it fire again. However, this difficulty in firing is gradually erased as the neuron rests.

Investigations revealed that the fatigue recovery mechanism could be modified to support well-learned CAs to recover quickly compared to weak CAs. In order to accomplish this, the fatigue recovery mechanism was changed such that fatigue recovery is made proportional to the fatigue of a neuron.

$$f_t = f_{t-1} * f_r$$ **Equation 3**

With this modified fatigue recovery mechanism, fatigue at time step t is equal to fatigue at the last time step multiplied by the fatigue recovery constant f_r. This mechanism encourages the highly fatigued neurons to recover quickly. It now takes longer for lightly fatigued neurons to recover. This encourages well-learned CAs to be strengthened and those that are weak will eventually be removed. The first input pattern is a well-learned CA. Therefore repeated presentation of the first input pattern after 2000 cycles leads to the removal of bad neurons. Figure 2 shows that after 2250 cycles the number of bad neurons approaches zero with the new fatigue recovery mechanism.

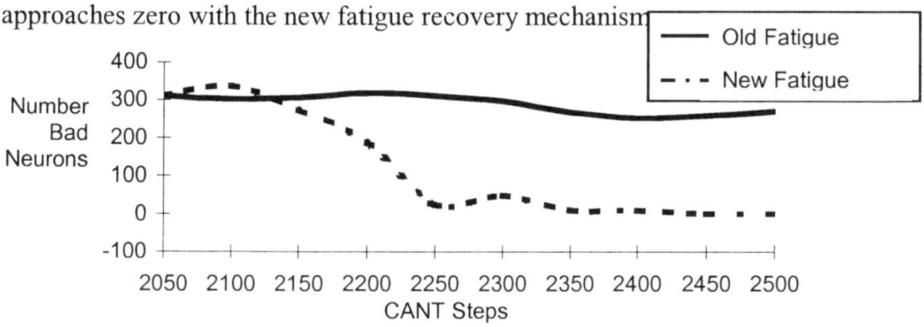

Figure 2: Effect of Fatigue Recovery Mechanism on Bad Neuron Problem

4. Conclusion

A network of simulated neurons is complex and has many interdependent parameters that affect the behavior of the network. CAs are subsets of neurons in such a network. A fatigue recovery mechanism based on the current fatigue level improves categorization by easing the removal of neurons from CAs while still allowing persistence. This modification of the short-term dynamic improves the long-term dynamic of learning. This should allow more CA attractor states to be formed. It should also improve the evolution of categories through CA recruitment of new neurons and CA fractionation into multiple CAs.

References

Amit D. (1989) Modelling Brain Function: The world of attractor dynamics Cambridge University Press.

Churchland, P. and T. Sejnowski (1992) The Computational Brain. MIT Press.

Hebb D. (1949). The Organization of Behavior. New York: John Wiley & Sons.

Huyck, C. (2000) Modelling Cell Assemblies. Proceedings of the International Conference on Artificial Intelligence ISBN: 1-892512-59-9 pp. 891-7

Palm G. (1990). Cell assemblies as a guideline for brain research. Concepts in Neuroscience, 1:1, 133-147

Sakurai, Y. (1998). The search for cell assemblies in the working brain. Behavioural Brain Research 91 pp. 1-13

COMPARING COMPUTATIONAL AND HUMAN MEASURES OF VISUAL SIMILARITY

T. M. GALE*, Y. SUN, R. ADAMS, AND N. DAVEY
Neural Net Group,
University of Hertfordshire,
Hatfield, AL10, 9AB, United Kingdom
** QEII Hospital,*
Welwyn Garden City, United Kingdom
E-mail: t.gale, comrys, r.g.adams, n.davey@herts.ac.uk

There have been many attempts to quantify visual similarity within different categories of objects, which a view to using such measures to predict impaired recognition performance. Although many studies have linked measures of visual similarity to behavioral outcomes associated with object recognition, there has been little research on whether these measures are associated with human ratings of perceived similarity. In this work, we compare similarity measures extracted from Principal Component Analysis, Isometric Feature Mapping and wavelets representations with ratings of human subjects. Our results show that features extracted by calculating the standard deviation of wavelet coefficients provides the closest fit to the human rating data of all the methods we applied here.

1. Introduction

There have been many attempts to quantify the visual similarity of different objects and categories, and to use such measures to predict impaired recognition performance, most notably for category-specific agnosias. Different approaches have included contour overlap and shared features (Humphreys *et al.*, 1988), shape overlap measures derived from non-standardised images (Tranel *et al.*, 1997), Euclidean distance measures derived from standardised pixellated images (Laws and Gale, 2002a) and measures derived from unsupervised neural network representations (Gale *et al.*, 2001; Gale *et al.*, 2003). Humphreys and Riddoch (2002) suggest that Euclidean distance-based representations of pixellated images are not useful analogues for human perceptual representation because the latter does not proceed on the basis of such detailed information. Nonetheless, Euclidean distance-based measures of pixellated images do predict naming accuracy (Laws and Gale,

2002a) as well as naming latency to degraded visual input (Laws *et al.*, 2002). Laws and Gale (2002b) have argued that although such measures may not mirror the kind of processing that occurs in the human visual system, they nonetheless capture information about shape overlap, spatial orientation and distribution of shading and, as such, can provide a useful metrics of visual similarity.

Although many studies have attempted to associate measures of visual similarity with performance on object recognition tasks (e.g. picture naming), there is little research on whether these measures correlate with human judgements of visual similarity. In a recent study (Gale *et al.*, 2004) we compared similarity measures extracted from self-organising map (SOM) representations of greyscale pictures with ratings of perceived visual similarity provided by human participants for the same set of pictures. There was a strong statistical association between the 2 measures, suggesting that the clustering of representations within the SOM may be a useful model for human perceptual representation of certain object categories. In the current study, we compare our previous findings with some other computational measures of visual similarity to investigate whether these approaches can also provide useful techniques for modelling aspects of human perceptual categorisation.

2. Method 1: Data from Human Participants

2.1. *Stimuli*

Seventy basic-level categories (e.g. apple, dog, guitar, vase, airplane, etc.) were selected to represent a broad range of objects (35 living, 35 nonliving). Three-hundred-and-seventy different exemplar pictures were collected to represent these categories and the number of exemplars representing each basic level category varied between 4 and 7 (means for living: nonliving = 5.31: 5.28). Pictures were collected from online galleries and CD-ROM encyclopaedias and all chosen images depicted the referent item in a consistent and typical orientation (for example, see Figure 1). The pictures were standardized as follows: first, extraneous background material was carefully removed and the depicted items were normalized for orientation within each basic level category (for example, all examples of 'dog' were viewed side-on and facing the same way); the size of the pictures was then manipulated such that each depicted item fitted within a grid of 64 pixels square (4096 pixels in total) whereby the maximal dimension of the object touched the borders of the square; finally the images were converted to 8-

bit greyscale. Figure 1 displays exemplars for one basic level category and the full picture set is available on request.

Figure 1. Example standardized images for the basic-level category 'beetles'

2.2. *Human Ratings of Visual Similarity*

A group of 24 human subjects rated the degree of visual overlap (VO) for each of the 70 basic level categories. The same standardized pictures that were used to test the computational models were presented in their basic level categories and participants were asked to rate the level of VO in each of the 70 categories. All ratings were collected on a 5-point scale ranging from 1 'very similar' to 5 'very dissimilar'.

3. Method 2: Data from Computational Experiments

3.1. *Representations*

In this work, we consider 3 different computational representations, namely, Principal Component Analysis (PCA), Isometric Feature Mapping (ISOMAP) (Tenenbaum *et al.*, 2000) and wavelets (Vidakovic, 1999).

PCA is considered here because of its simplicity and wide domain of application. The motivation for the ISOMAP is to find meaningful low-dimensional structures hidden in high-dimensional observed data. Rather than calculating the Euclidean distances as in multidimensional scaling, this approach preserves the *geodesic distances* between all pairs of data points in the manifold (Tenenbaum *et al.*, 2000).

Recently, psychologists have applyed wavelets for modeling aspects of the visual system (Granlund, 1978; Daugman, 1980; Watson, 1983). Considering an image as a signal, it can be decomposed into wavelets of different sizes. Here a *2-D haar wavelet* transformation is employed to the pixel image, decomposing it into different directions: *vertical* (VD), *horizontal* (HD) and *diagonal* detail (DD) images of different resolutions. The representations are derived from second order statistics: the standard deviation of the coefficients at each level and each direction.

3.2. *Computational Experiments*

Each picture size is 64 × 64, 4096 pixels in total. In the current study, we calculate similarity using $s(\mathbf{x}, \mathbf{y}) = \frac{1}{1+d(\mathbf{x},\mathbf{y})}$, where $\mathbf{x}$ and $\mathbf{y}$ are representations for different images; d denotes Euclidean distance. In addition, we compare results with previous experimental results using a self-organising map (SOM) (Gale *et al.*, 2004).

4. Experimental Results

Results are displayed in Table 1. As shown, it suggests that the wavelet representations give better fit to the human rating data.

Table 1. Comparing computational and human measures of visual similarity. (P-value is the significance value.)

representations	corr. coeff	P-value
Pixel_based	0.44	0.0001
PCA	0.39	0.0008
ISOMAP	0.36	0.002
SOM	0.35	0.003
Wavelets	**0.56**	0.0001

5. Discussion

Our results show that features extracted by calculating the standard deviation of wavelet coefficients provides the closest fit to the human rating data of all the methods we applied here. The correlations between the wavelet representation and human ratings is considerably higher than that of the raw data and human ratings. Although these correlations are far from perfect, they are of an order of magnitude greater than those found for other predictor variables in visual object naming experiments (for example, see (Laws *et al.*, 2002)). In the next stage, we shall consider using a different distance measurement in computing similarity, rather than a simple Euclidean distance. We will undertake further work to investigate why wavelets are better predictors for the human data.

References

Daugman, J. G. 1980. Two-dimensional spectral analysis of cortical receptive field profiles. *Vision Research* **20**, 847–856.

Gale, T. M., N. Davey, K. R. Laws, M. Looms, and R. J. Frank 2004. Self-organising map representations of greyscale images reflect human similarity judgements. In *Proceedings IEEE IS.*

Gale, T. M., D. J. Done, and R. J. Frank 2001. Visual crowding and category-specific deficits for pictorial stimuli: a neural network model. *Cognitive Neuropsychology* **18**, 509–550.

Gale, T. M., K. R. Laws, R. Frank, and V. C. Leeson 2003. Basic level visual similarity and category specificity. *Brain and Cognition* **53**, 229–231.

Granlund, G. H. 1978. In Search of a General Picture Processing Operator. *Computer Graphics and Image Processing* **8** (2), 155–173.

Humphreys, G. W., J. Riddoch, and P. T. Quinlan 1988. Cascade processes in picture identification. *Cognitive Neuropsychology* (5), 67–103.

Humphreys, G. W. and M. J. Riddoch 2002. Do pixel-level analyses describe psychological perceptual similarity? A comment on 'Category Specific Naming and the visual characteristics of line-drawn stimuli' by Laws and Gale. *Cortex* **38**, 3–5.

Laws, K. R. and T. M. Gale 2002a. Category-specific naming and the 'visual' characteristics of line drawn stimuli. *Cortex* **38**, 7–21.

Laws, K. R. and T. M. Gale 2002b. Why are our similarities so different? A reply to Humphreys and Riddoch. *Cortex* **38**, 643–650.

Laws, K. R., V. C. Leeson, and T. M. Gale 2002. The effect of 'masking' on picture naming latencies. *Cortex* **38**, 137–147.

Tenenbaum, J. B., V. d. Silva, and J. C. Langford 2000. A global geometric framework for nonlinear dimensionality reduction. *Science* **290**, 2319–2323.

Tranel, D., C. G. Logan, R. J. Frank, and A. R. Damasio 1997, Oct. Explaining category-related effects in the retrieval of conceptual and lexical knowledge for concrete entities: operationalization and analysis of factors. *Neuropsychologia.* **35** (10), 1329–1339.

Vidakovic, B. 1999. *Statistical Modeling by Wavelets.* John Wiley & Sons, Inc.

Watson, A. B. 1983. Detection and recognition of simple spatial forms. In *Physical & Biological Processing of Images*, pp. 100–114. Berlin: Springer-verlog.

PSYCHOLOGICAL ALIFE: BRIDGING THE GAP BETWEEN MIND AND BRAIN. ENACTIVE DISTRIBUTED ASSOCIATIONISM & TRANSIENT LOCALISM

ANTHONY F. MORSE

COGS, Department Of Informatics, University Of Sussex,
Falmer, Brighton, BN1 9RH, UK
anthonfm@sussex.ac.uk

Enactive Distributed Associationism (EDA) is a simple and general purpose approach to the modeling of developmental learning, requiring no explicit supervision, specialization, or any artificially imparted symbolic base. These biologically plausible models implement unsupervised self-organization and exhibit ongoing classical conditioning leading to successful environmental adaptation in complex real environments. Producing multiple psychological phenomena, this approach is highly appropriate for cumulative cognitive modeling, especially where complex ontogenetically developing traits are to be modeled (such as phobic responses). Unlike many Artificial Life models, EDA provides a high level of interpretability and an understanding of the structures developed in relation to their formal psychological properties. This cognitive penetrability combines Transient Localism with an analogy to the localist symbolic networks to form an approach that is free from any assumptions regarding what is to be modeled.

1. Introduction

In biological brains, functional organization results from massive amounts of neuronal re-structuring and re-organization throughout a lifetime of learning. By modeling these biological processes, it is suggested that models of psychological learning, able to account for multiple and diverse phenomena can be generated. Such an approach surely provides a potential means to the unification of psychological theories. Enactive Distributed Associationism (EDA) is one such model of learning, in which plasticity in the modeled biology is mirrored in the development of psychological, emotional and cognitive capabilities spanning much of psychology.

1.1. *Trading Spaces*

EDA utilises one of the simplest computational techniques, structured Hebbian plasticity, but applied to the activity space of a Liquid State Machine (Maass et al 2002a) rather than any designed symbolic base. The learning rule is interpreted as follows: Iff *a* and *b* commonly co-occur in the context *x(.)* then the presence of one will lead to the expectation of the other when in context *x(.)*. This is implemented using an adaptive resonance network to provide pattern recognition over the input neurones, and then applying the standard positive Hebbian learning rule between all inputs and those contexts / patterns found. The negative Hebbian learning rule is then applied directionally from the context / patterns back to the input layer. Morse 2003 provides further details of this algorithm and demonstrates its application in symbolic domains.

As indicated, the approach taken here follows the idea of trading spaces proposed by Clark & Thornton (1997) in which computation is traded against representation in order to solve harder problems.

1) We know that EDA can predict (given experience) ALL linear relationships, and some non-linear ones.
2) We know that EDA results in psychological functionality given an appropriate symbolic basis.
3) Liquid State Machine models provide powerful pattern recognition capabilities where the solution (to a stable symbolic interpretation) is linearly solvable.

Given all this, the implausible supervised learning of the Liquid State Machine is replaced with the EDA learning algorithm to provide an unsupervised biologically plausible model of cortical plasticity. By providing the cortical substrate rather than a symbolic basis, everything that is separable in the input stream plays its role without necessitating explicit representation.

2. Transient Localism

The theory of Transient Localism invites us to view the Liquid State Machine's activity vector as containing micro-features of the input. Although these features are not explicitly represented, they could be, and to do so would involve only a linear transformation, which being within the abilities of unsupervised associative learning provokes the question, why bother transforming this into a stable localist space at all? In fact the added subtlety of having it all at once provides a great deal more computational power than slim-line symbolic models can provide.

3. Enactive Distributed Associationism

The EDA network is embodied on a wheeled robot in a real environment (office) and equipped with Infra-Red (IR) and bumper sensors. No alterations in the EDA methodology are necessary to account for different forms of incoming data as the robot's sensors are randomly connected to a liquid state machine and provide a constant input perturbation. The resulting activity of the liquid state is then subject to distributed associationism (Morse 2003) which constructs spreading activation-like networks of contextualised micro-features. Through the addition of motor outputs and inhibitory bumper feedback, the network produces coherent adaptive behavior maximizing this rewarding feedback by minimizing bumper activity. This reward system provides a homeostatic response in that should the agent be punished, spreading inhibition will stop the current behavior and promote the most appropriate behavior to alleviate the punishment in this situation / context. Equally, an excited feedback node will prime and hence promote behaviors previously associated with that feedback.

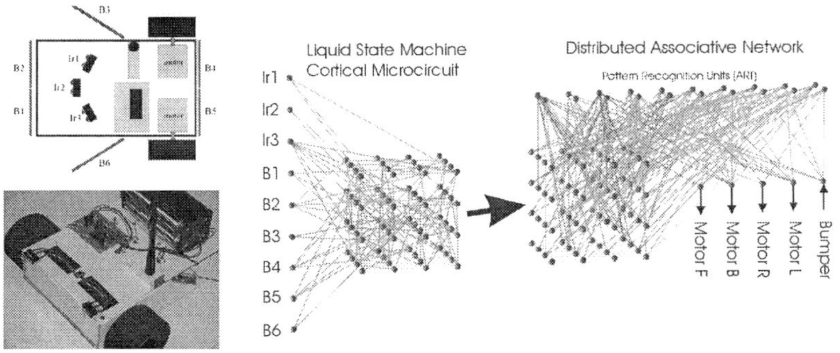

Figure 1. Showing the robot (left), and the Enactive Distributed Associative architecture (right)

4. Experimental Setup

Weak random noise is added to the activity of each motor neuron on every update cycle, providing the agent with initial experience of disorganised behaviour from which it can learn. As the agents' homeostatic / motivational drives vary in activity, they become associated with liquid state features consistent with their various states. This provides a basis for their future prediction by association to the liquid state machine. Once the agent has unwittingly produced behaviour adversely affecting its motivational drives (such as a collision), the future activity of that drive will, in relevant contexts, inhibit behaviour leading to predicted collisions. Assuming that the liquid state has made separable sufficient relevant features, the future behaviour of the agent will become organised and coherent with respect to the sensory-motor contingencies it has so far experienced. This set-up results in an embodied developmental agent displaying multiple psychological phenomena, and open to structured learning through human and environmental interaction.

4.1. *Classical Conditioning & Autonomous Behavior*

The EDA method is analogous to psychological descriptions of classical conditioning (Pavlov 1927). The unconditioned response (UR) of the agent is to change direction when it encounters the unconditioned stimulus (US) (bumper activity). Initially the agent fails to react to activity on the IR sensors, however as this is typically high during a collision it soon becomes a conditioned stimulus (CS). Thus the presence of IR activity (CS) will in future provoke direction changes before a collision occurs. Following classical conditioning, stimuli associated with punishment (either by experiment or implicitly by the physical nature of the environment) quickly become avoided. Equally, stimuli associated with rewarding behaviour become repeatedly exhibited. This is demonstrated in the robots successful acquisition of obstacle avoidance.

4.2. *Phobic Responses and Cognitive Behavioural Techniques*

The avoidance of stimuli consistently present during negative feedback leads to the possibility of coincidental or miss-placed associations provoking an avoidance reaction unnecessarily. This misplaced aversion is interpreted as a phobic response (i.e. avoidance due to miss-placed association, Wolpe (1958)). One possible treatment of phobic responses, from Cognitive Behavioral Therapy (CBT), gradually presents phobic stimuli to patients in a relaxed state. This encourages a disassociation of the conditioned stimulus from the fear response.
To model the CBT situation, the agent is provided with an additional input, normally active when no IR or bumper sensor is active. Following developmental learning this new input 'relaxes' the agent by reducing (via learnt inhibitory associations) the anticipation of disaster (activity of the bumper neuron). Having successfully learnt to avoid obstacles, the agent is found to reverse should an obstacle be in the range of any of its IR sensors. This conditioned response prevents it from travelling down narrow passages, where both of its side IR's would be simultaneously active. This is interpreted as a phobic response in that the agent faced with such a narrow passage incorrectly anticipates a bad outcome (collision) if it were to continue moving forward. Following the CBT treatment of phobic responses in human participants, subjects must first be relaxed, simulated by activating the new input unit. The resulting reduction in the levels of anxiety allows the agent to approach the passage more closely. Following several aborted attempts (providing valuable learning experience contradicting the misplaced association), the agent successfully enters and traverses the narrow passage and is subsequently able to

do this without necessitating further external activation of the additional 'relaxing' input. This demonstrates successful modeling of both phobic behavior and basic CBT treatment of that behavior, in that the removal of a phobia requires both a relaxed state (the additional input) and a gradual exposure to phobic stimuli in order to overcome the fear response.

5. Conclusions and Future Research

Enactive Distributed Association provides a simple model of learning based on plasticity and information available locally to neurons. Taking Maass's liquid state machine model of the cortex of real brains as the basis of simple structured Hebbian plasticity provides a general-purpose approach to ontogenetic developmental robotics and the modeling of psychological phenomena. Not only does this approach produce useful long-term adaptability, in combination with Transient Localism one can easily construct formal symbolic and psychological descriptions of the agents' behavior. This not only leads to the exhibition of all of the psychological phenomena previously displayed by spreading activation networks (priming, overt and covert recognition, schemata, various prosopagnosias etc.), but also envelops more complex developmental phenomena such as phobic reactions and their treatment. Future work will incorporate sensor and motor rich robots displaying many more diverse cognitive behaviors.

References

Clark, A., & Thornton, C. (1997). Trading spaces: Computation, representation, and the limits of uninformed learning. *Behavioral and Brain Sciences,* 20 (1), 57-92.

Maass, W., Natschläger, T., & Markram, H. (2002a). Real-time computing without stable states: A new framework for neural computation based on perturbations. *Neural Computation,* 14(11), 2531-2560.

Morse, A.F. (2003). Autonomous Generation Of Burton's IAC Cognitive Models. In Schmalhofer, Young, & Katz (Eds.), *Proceedings of EuroCogSci03 The European Cognitive Science Conference.* LEA Press

Palvov, I. (1927). Conditioned Reflexes: An Investigation of the Physiological Activity of the Cerebral Cortex

Wolpe, J. (1958). *Psychotherapy by Reciprocal Inhibition.* Stanford, CA: Stanford University Press.

Future Challenges: Thinking and Reasoning

MODELLING THINKING AND REASONING: THE CHALLENGE AHEAD

JONATHAN ST B T EVANS

Centre for Thinking and Language
School of Psychology
University of Plymouth
Plymouth PL4 8AA, UK
jonathan.evans@plymouth.ac.uk

Research on the psychology of reasoning has been dominated for the past 40 years by the deduction paradigm which assesses people's ability to reason logically. As a result, computational models developed to date implement syntactic or semantic forms of mental logic. More recently, however, theorists have given much greater emphasis to pragmatic influences on reasoning and developed the theory that human reasoning reflects the operation of two distinct cognitive systems. The challenges that this theory sets for computational modelling of human reasoning are identified and discussed.

1. The deduction paradigm

Over the past 40 years or so, research on the psychology of human reasoning has been dominated by the deduction paradigm (Evans, 2002). This method has its origins in *logicism*, the view once popular in psychology as well as philosophy that logic provides the basis for rational human thinking. Participants in psychological experiments are therefore given the premises of logical arguments and told to accept, for the purpose of the experiment, that these premises are true. On this basis, they are instructed to draw conclusions (or evaluate presented conclusions) only if they necessarily follow from the premises. The answers given can then be compared with standard textbook logic and marked as right or wrong.

Many hundreds of psychological experiments using the deduction paradigm have been reported in the psychological literature (for reviews, see Evans, 2002; Evans, Newstead, & Byrne, 1993; Manktelow, 1999). The major findings are as follows. First, people make many errors relative to a normative analysis by textbook logic. In particular, they draw more inferences than they are entitled to. People will often endorse conclusions that are plausible but not logically necessary, in conflict with the instructions they are given. Second, errors are not random but reflect systematic cognitive biases. Finally, participants are strongly influenced by prior belief and knowledge relevant to the context of the problems.

For example, they prefer to endorse conclusions that are believable and are reluctant to draw inferences from unbelievable premises. Such beliefs are irrelevant to the logical nature of the task as presented by the standard instructions of the deduction paradigm.

A good example is the belief bias effect in syllogistic reasoning. Participants are presented with two premises and a conclusion. The conclusion may or may not follow logically from the premises (valid or invalid) but also may or may not conform with prior belief (believable or unbelievable). In two cases, logic and belief are in agreement, but in two other cases they conflict. Example problems from the study of Evans, Barston and Pollard (1983) are shown in Table 1. When both logic and belief support the conclusion (valid-believable) most people (89%) say 'yes' – the conclusion follows. Similarly, when the conclusion is both invalid and unbelievable, few people (10%) accept the conclusion.

Table 1. Example problems and argument acceptance rates from the study of Evans, Barston and Pollard (1983).

<u>Valid argument, believable conclusion (NO CONFLICT – 89% yes)</u>
No police dogs are vicious
Some highly trained dogs are vicious
Therefore, some highly trained dogs are not police dogs

<u>Valid argument, unbelievable conclusion (CONFLICT – 56% yes)</u>
No nutritional things are inexpensive
Some vitamin tablets are inexpensive
Therefore, some vitamin tablets are not nutritional

<u>Invalid argument, believable conclusion (CONFLICT – 71% yes)</u>
No addictive things are inexpensive
Some cigarettes are inexpensive
Therefore, some addictive things are not cigarettes

<u>Invalid argument, unbelievable conclusion (NO CONFLICT – 10% yes)</u>
No millionaires are hard workers
Some rich people are hard workers
Therefore, some millionaires are not rich people

The interesting problems are those that put logic and belief into conflict. Acceptance of valid conclusions drops sharply to 56% when the conclusion is unbelievable. Conversely acceptance of invalid conclusions soars to 71% when the conclusion conforms with prior belief. These shifts illustrate the belief bias effect, which is strong for valid conclusions but statistically even stronger for invalid conclusions. I will return to discussion of the belief bias effect later.

In the past ten years or so, there has been increasing dissatisfaction with the deduction paradigm among leading psychological researchers (Evans & Over,

1996; Oaksford & Chater, 1998). This comes partly from an increasing belief that logic – or certainly elementary textbook logic – is far from being an appropriate normative model for real world reasoning which appears to be both probabilistic and defeasible in nature. The other major influence has been the findings with the paradigm itself. Despite the use of deductive reasoning instructions, massive influence of prior knowledge elicited by the context has been regularly observed. This has led many to the view that pragmatic and not deductive reasoning is the default system for human reasoning.

2. Existing computational models of human reasoning

Two major theories have, until recently, dominated accounts of deductive reasoning in humans. One is based on the idea that people possess 'natural logics' comprised of inference rule schemas together with effective algorithms for applying them (Braine & O'Brien, 1998; Rips, 1994). This approach is often known as 'mental logic'. Theorists assume that the content of any particular premises is stripped out and the underlying logical form recovered. Reasoning then proceeds as a mental proof with the toolbox of inference rules provided. Finally, any conclusions derived are translated back into the original context.

The best developed computational model of human reasoning is of this kind and known as the ANDS model. This has been programmed in PROLOG and compared with human reasoning protocols (Rips, 1994). The system employs both forward and backward search methods. The former are assumed to be straightforward and largely error free. The latter require suppositional reasoning strategies, such as *reductio ad absurdum*, and are assumed to be much more difficult. A limitation of the ANDS model is that it must know in advance the conclusion towards which people are reasoning. More generally, the mental logic approach has limited scope to account for the findings of reasoning experiments. It describes only correct reasoning and hence has no specific account of cognitive biases and content effects. The core mental logic has thus to be supplemented with a host of other mechanisms to account for pragmatic influences.

The most influential theory has been the mental model theory of reasoning (Johnson-Laird, 1983; Johnson-Laird & Byrne, 1991). This theory also accounts for deductive competence but without the use of inference rules. It is assumed that people construct and manipulate mental models that represent possibilities and that validity is decided by a semantic principle: a conclusion is valid if there are no counterexamples to it. Hence, people will endorse inferences unless they discover mental model of the premises that exclude the conclusion. Some authors see this as another form of mental logic. However, the mental models account is better equipped than inference rule theories in accounting for errors and biases. For example, problems requiring construction of multiple mental models will be more difficult due to working memory constraints. Some of the belief bias effects can be accounted for on the argument that people are less motivated to search for counterexamples to believable conclusions, and so on.

Mental model accounts have been proposed for syllogistic, relational and propositional reasoning problems (Johnson-Laird & Byrne, 1991). In each case, Johnson-Laird has programmed computational models in LISP to check the accuracy and consistency of the theory, an approach which he claims to have led to the discovery of new phenomena, such as 'illusory inferences' (Johnson-Laird & Savary, 1999). However, this theory also has also been subject to recent criticism, on the grounds that it is too closely tied to standard logic and provides an inadequate account of pragmatic influences in reasoning (Evans & Over, 2004).

3. Dual-process theories of reasoning

In the current literature, there is increasing interest in dual-process or dual-system theories of reasoning (Evans, 2003; Evans & Over, 1996; Sloman, 1996; Stanovich, 1999). The idea that there may be two distinct kinds of reasoning process has been proposed for many years. For example, Evans et al. (1983) argued that the belief bias data (Table 1) suggested a within-participant conflict between a logical and belief based reasoning process. When belief and logic are in conflict, sometimes one process 'wins' and sometimes the other, even within the same individual person.

The notion of dual processes has developed recently into the strong claim that there are two distinct cognitive systems responsible for these processes with differing evolutionary histories. System 1 - the heuristic system - is ancient and reflects our experiential learning about the world. System 2 – the analytic system – is recently evolved and unique to humans. It permits a different kind of thinking – abstract, decontextualised and hypothetical, that only humans engage in. Table 2 shows a list of characteristics that various theories have been attributed to the two systems.

Table 2. Characteristic attributed to heuristic and analytic reasoning systems in dual process theories

HEURISTIC/SYSTEM 1	ANALYTIC/SYSTEM 2
– Unconscious	– conscious
– Non-verbal	– verbal
– evolved early	– evolved late
– shared with animals	– uniquely human
– rapid, parallel	– slow, sequential
– high capacity	– low capacity
– pragmatic	– logical, abstract
– independent of working memory and IQ	– related to working memory capacity and IQ

A wide range of evidence has been produced to support dual-processing accounts of reasoning, with different methods bearing on different sets of the characteristics listed in Table 2. For example, the heuristic system being ancient is assumed to have been optimized by long evolution. The recent analytic system, however, is thought to be heritable and linked with measures of general intelligence. A recent extensive research program summarized by Stanovich (1999) has provided much support for this aspect of the theory. Participants who find normatively correct solutions to a wide range of reasoning and decision problems are generally significantly higher in general intelligence, as are those better able to inhibit pragmatic influences and avoid belief bias. Solution rates to problems that can be solved by reference to real world knowledge, on the other hand, are independent of general intelligence.

One key idea is that the heuristic system is massively parallel, preconscious and rapid in nature. Analytic reasoning, by contrast, is sequential and slow and requires access to a central working memory of limited capacity. There is evidence that people attend and apply analytic reasoning only to information preconsciously selected as relevant by the heuristic system (Ball, Lucas, Miles, & Gale, 2003; Evans, 1996). Recently, researchers have experimented with rapid response tasks, in which participants are required to give quick answers without time for slow analytic thinking to occur (Roberts & Newton, 2002; Schroyens, Schaeken, & Handley, 2003). These experiments find, consistent with the theory, that biases attributed to the heuristic system have a more prominent influence under these conditions.

If the two systems are indeed distinct and have differing evolutionary histories then they should be neurologically differentiated. Recent studies using neural imaging techniques have supported this hypothesis (Goel, Buchel, Rith, & Olan, 2000; Goel & Dolan, 2003; Houde et al., 2000; 2001). Particularly striking are the findings of Goel and Dolan (2003) who investigated belief bias with fMRI scanning. On belief-logic conflict problems (see Table 1), different brain areas were recruited when the conflict was resolved in favor or logic or in favor of belief.

A puzzling question arises from these findings. If there are, in effect, two minds in one brain competing for control of behavior, why do we have no conscious awareness of this conflict and why do we feel 'in control'? The answer appears to lie in the confabulatory nature of the analytic thinking system (Evans & Over, 1996; Stanovich, 2004). There is striking psychological evidence that when behavior can be demonstrated to be caused by unconscious processes, people will provide an account of their behavior which is, in fact, a rationalization. Maintaining an illusion of unity and control appears to be one major function of our analytic thinking system.

4. Can human reasoning be modelled?

I now address the challenges faced by those who might wish to model human reasoning processes. Programming computers to reason logically, following rules in a sequential algorithm, is relatively easy because that is the kind of thing they were designed to do. In terms of the dual-process theory, then, modelling analytic reasoning is not too difficult and corresponds with what has already been done (see section 2). Modelling the heuristic system, however, effectively involves the formidable 'frame problem'. Heuristic processes rapidly and effortlessly recruit the prior knowledge that is relevant to the problem in hand.

The problem really is analogous to that of natural language modeling. While this appeared to be tractable problem in the early days of artificial intelligence research, we now understand that for a computer to converse in natural language, it would need not just knowledge of the syntax and semantics of the language, but effectively to share the whole system of knowledge about the world that people carry inside their heads, together with a mechanisms for retrieving and applying it. In the same way, the problem of modeling human reasoning seemed tractable when reasoning was thought of as a logical system dealing in abstract representations, but now seems intractable once we accept that contextualized reasoning is what has to be addressed.

One way to look at the problems is in terms of cognitive architectures. In the early heady days of cognitive science (1970's and 80's) researchers were carried away with the idea that brains like computers were information processing systems and that cognitive theories could hence be implemented as working computer programs. What was overlooked at this time was the fact that while computers were designed with CPU chips engineered for serial processing of instructions, brains evolved with massively parallel architecture. This does not simply explain why the relatively slow biological systems in the brain have such vast computational power (for example in processing visual information) compared with their silicon counterparts. It also means that brains are very different in the way in which they process information.

Stanovich (2004) invokes this architectural difference in explaining why humans are poor at logical reasoning but excel in complex processing involving, say, pattern recognition or knowledge retrieval, while computers are the other way around. He argues that the brain evolved an architecture entirely suited to the old, heuristic system. It favors both massively parallel processing and a modular structure in which knowledge is encapsulated and highly contextualized. When the analytic system developed with its requirements for serial processing and abstract, decontextualised representations, it found itself with computational machinery highly unsuited to the task. This explains why few individuals excel in this kind of reasoning and why everyone is prone to 'biases' arising from the heuristic system as default contextualized reasoning intrudes on tasks where it is not appropriate.

Another problem concerns the interaction between the two systems. What appears to happen is the following. Heuristic systems responses are available rapidly and will control behavior *unless* the analytic system intervenes. However, the analytic system is inherently serial and limited capacity in nature: unlike the heuristic system, it can only do one thing at a time. It seems to be a kind of trouble-shooter that switches attention from one task to another as priorities require. So what system controls this interventional role? What decides where the analytic system is needed? Analytic thinking is volitional to a degree: we can make a conscious decision to attend to our driving, for example. However, there is good evidence that one function of the heuristic system is to direct the locus of attention of the analytic system. Also, when the analytic system is engaged, it does not necessarily determine the behavior. Sometimes a heuristic process 'wins' and the analytic processes rationalize and maintain an illusion of control. Put like this, the interaction between the systems sounds chaotic.

Despite the formidable problems discussed in this section, I do believe that computational modeling can play a part in the psychology of reasoning. We just need to limit our ambitions of what it can achieve. Dual systems accounts of the kind discussed in section 3 are what we might term 'grand theories'. They are quite hard to apply to particular tasks, in order to generate specific predictions. They also carry theoretical baggage that not all researchers who accept a dual-process framework will want to sign up to. For this reason, I am currently developing a more limited heuristic-analytic account of reasoning tasks which is designed to specify ways in which the two systems interact in controlling behavior on reasoning problems. If the principles can be sufficiently clearly specified, a theory at this level could certainly be specified in a working computer program. Enormously complex though the brain is, the way in which its cognitive systems are manifest in particular tasks and paradigms studied by psychologists is greatly simplified by the parameters of these tasks. Hence, I believe that the interaction of heuristic and analytic processes in particular cognitive tasks can and should be explored with the aid of computational models.

References

Ball, L. J., Lucas, E. J., Miles, J. N. V., & Gale, A. G. (2003). Inspection times and the selection task: What do eye-movements reveal about relevance effects. *Quarterly Journal of Experimental Psychology, 56A,* 1053-1077.

Braine, M. D. S. & O'Brien, D. P. (1998). *(Eds) Mental logic*. Mahwah, New Jersey: Lawrence Erlbaum Associates.

Evans, J. St. B. T. (1996). Deciding before you think: Relevance and reasoning in the selection task. *British Journal of Psychology, 87,* 223-240.

Evans, J. St. B. T. (2002). Logic and human reasoning: An assessment of the deduction paradigm. *Psychological Bulletin, 128,* 978-996.

Evans, J. St. B. T. (2003). In two minds: Dual process accounts of reasoning. *Trends in Cognitive Sciences, 7,* 454-459.

Evans, J. St. B. T., Barston, J. L., & Pollard, P. (1983). On the conflict between logic and belief in syllogistic reasoning. *Memory and Cognition, 11,* 295-306.

Evans, J. St. B. T., Newstead, S. E., & Byrne, R. M. J. (1993). *Human Reasoning: The Psychology of Deduction*. Hove & London: Erlbaum.

Evans, J. St. B. T. & Over, D. E. (1996). *Rationality and Reasoning*. Hove: Psychology Press.

Evans, J. St. B. T. & Over, D. E. (2004). *If.* Oxford: Oxford University Press (in press).

Goel, V., Buchel, C., Rith, C., & Olan, J. (2000). Dissociation of mechanisms underlying syllogistic reasoning. *NeuroImage, 12,* 504-514.

Goel, V. & Dolan, R. J. (2003). Explaining modulation of reasoning by belief. *Cognition, 87,* B11-B22.

Houde, O., Zago, L., Crivello, F., Moutier, S., Pineau, A., Mazoyer, B. et al. (2001). Access to deductive logic depends upon a right ventromedial prefrontal area devoted to emotion and feeling: Evidence from a training paradigm. *NeuroImage, 14,* 1486-1492.

Houde, O., Zago, L., Mellet, E., Moutier, S., Pineau, A. et al. (2000). Shifting from the perceptual brain to the logical brain: The neural impact of cognitive inhibition training. *Journal of Cognitive Neuroscience, 12,* 721-728.

Johnson-Laird, P. N. (1983). *Mental Models*. Cambridge: Cambridge University Press.

Johnson-Laird, P. N. & Byrne, R. M. J. (1991). *Deduction*. Hove & London: Erlbaum.

Johnson-Laird, P. N. & Savary, F. (1999). Illusory inferences: a novel class of erroneous deductions. *Cognition, 71,* 191-299.

Manktelow, K. I. (1999). *Reasoning and Thinking*. Hove, UK: Psychology Press.

Oaksford, M. & Chater, N. (1998). *Rationality in an Uncertain World*. Hove, UK: Psychology Press.

Rips, L. J. (1994). *The Psychology of Proof.* Cambridge, MA: MIT Press.

Roberts, M. J. & Newton, E. J. (2002). Inspection times, the change task, and the rapid-response selection task. *Quarterly Journal of Experimental Psychology, 54A,* 1031-1048.

Schroyens, W., Schaeken, W., & Handley, S. H. (2003). In search of counterexamples: Deductive rationality in human reasoning. *Quarterly Journal of Experimental Psychology, 56A,* 1129-1145.

Sloman, S. A. (1996). The empirical case for two systems of reasoning. *Psychological Bulletin, 119,* 3-22.

Stanovich, K. E. (1999). *Who is Rational? Studies of Individual Differences in Reasoning*. Mahway, NJ: Lawrence Elrbaum Associates.

Stanovich, K. E. (2004). *The robot's rebellion: Finding meaning the age of Darwin*. Chicago: Chicago University Press.

Author Index